Xin Cui
Chang-He Li

Magnetic Traction Nanolubricant Grinding

Force Model and Mechanisms

nova science publishers
www.novapublishers.com

DOI: https://doi.org/10.52305/PVGY4472

NOTICE TO THE READER

Library of Congress Cataloging-in-Publication Data

ISBN: 979-8-89530-217-0 (Hardcover)
ISBN: 979-8-89530-329-0 (eBook)

Published by Nova Science Publishers, Inc. † New York

Contents

Preface xi

Acknowledgments xv

About the Authors xvii

Abbreviations xix

Chapter 1 **The Biological Failure Mechanism and Preparation of Eco-Friendly Lubricant in Machining** 1

1.1. Introduction 2

1.2. Traditional Cutting Fluids 3

1.2.1. Preparation *3*

1.2.2. Classification *4*

1.2.3. Function *6*

1.2.4. The Problem of Cutting Fluids *7*

1.3. Eco-Friendly Cutting Fluids 11

1.3.1. Rust Inhibitors *12*

1.3.2. Bactericide *15*

1.3.3. Extreme Pressure Additives *16*

1.3.4. Base Oil *18*

1.4. Bio-Lubricant 20

1.4.1. Lubrication Mechanism *20*

1.4.2. Limitations *27*

1.4.3. Summary *32*

1.5. Nanolubricant 32

1.5.1. Nano-Enhanced Phases *33*

1.5.2. Preparation *37*

1.5.3. Stability *39*

References 44

Chapter 2 **Minimum Quantity Lubrication of Aeronautical Materials Using Carbon Group Nanolubricants: Mechanisms and Applications** 49
2.1. Introduction 50
2.2. Carbon Group Nanolubricant 51
2.2.1. Spheroidal Nanolubricant 52
2.2.2. Tubular Nanolubricant 53
2.2.3. Layered Nanolubricant 53
2.3. Cutting Mechanism of Minimum Quantity Lubrication 54
2.3.1. Antifriction and Antiwear Mechanism 55
2.3.2. Tool Failure Mechanism 56
2.3.3. Material Removal Mechanism 57
2.3.4. Surface Formation Mechanism 58
2.4. The Application of CGNs in MQL Turning 59
2.4.1. Turning Force 59
2.4.2. Coefficient of Friction (CoF) 61
2.4.3. Chip Morphology 61
2.4.4. Tool Life and Wear 62
2.4.5. Surface Integrity 64
2.4.6. The Turning Mechanism of NMQL Using Carbon Nano-Enhancers 65
2.5. The Application of CGNs in MQL Milling 68
2.5.1. Milling Force 68
2.5.2. Chip Morphology 69
2.5.3. Tool Life and Wear 70
2.5.4. Surface Integrity 71
2.5.5. The Milling Mechanism of NMQL Using Carbon Nano-Enhancers 74
2.6. The Application of CGNs in MQL Grinding 75
2.6.1. Grinding Force 77
2.6.2. Coefficient of Friction (CoF) 78
2.6.3. Chip Morphology 79
2.6.4. Tool Life and Wear 79
2.6.5. Surface Integrity 80
2.6.6. The Grinding Mechanism of NMQL Using Carbon Nano-Enhancers 81
References 86

Chapter 3 **How the Physicochemical Properties of Nanolubricants Affect the Grinding Performance of Difficult-to-Machine Materials** 91
3.1. Introduction 92

3.2. Nanolubricant Physical Properties and Grinding Performance 93
3.2.1. The Molecular Structure of Fatty Acids 95
3.2.2. Viscosity 96
3.2.3. Surface Tension 97
3.2.4. PH Value 97
3.2.5. Pour Point 98
3.2.6. Thermal Stability 98
3.3. Titanium Alloy Grinding 99
3.3.1. Force and Coefficient of Friction (CoF) 100
3.3.2. Wheel Wear 102
3.3.3. Grinding Temperature 103
3.3.4. Debris Morphology 104
3.3.5. Surface Integrity 105
3.3.6. An Analysis of Grinding Defects and Solutions 107
3.4. Nickel-Based Alloy Grinding 108
3.4.1. Force and Coefficient of Friction (CoF) 109
3.4.2. Wheel Wear 112
3.4.3. Grinding Temperature 114
3.4.4. Debris Morphology 115
3.4.5. Surface Integrity 116
3.4.6. Ann Analysis of Grinding Defects and Solutions 118
3.5. High Strength Steel Grinding 119
3.5.1. Force and Coefficient of Friction (CoF) 120
3.5.2. Wheel Wear 122
3.5.3. Grinding Temperature 124
3.5.4. Debris Morphology 125
3.5.5. Surface Integrity 127
3.5.6. An Analysis of Grinding Defects and Solutions 129
References 132

Chapter 4 **Principles and Equipment Designs of Magnetic Traction Nanolubricant (MTN) Grinding** 137
4.1. Introduction 137
4.2. Magnetic Traction Method Based - The Adsorption Effect 139
4.2.1. Principle 139

4.2.2. Device Design ... *140*
4.3. The Magnetic Traction Method - The Suction Effect ... 142
4.3.1. Principle ... *142*
4.2.2. Device Design ... *149*
4.4. Preparation and Physical Properties of Nanolubricant ... 164
4.4.1. Nanomaterials ... *164*
4.4.2. Preparation Method ... *166*
4.4.3. Viscosity and Density ... *167*
4.4.4. Saturation Magnetization and Relative Permeability ... *171*
References ... 172

Chapter 5 **The Distribution and Evolution of the Magnetic Field in Magnetic Traction Nanolubricant Grinding** ... 175
5.1. Introduction ... 175
5.2. Model of the Grinding Zone Magnetic Field ... 176
5.2.1. Theoretical Modelling of Magnetic Field Distribution ... *176*
5.2.2. Physical Modelling of Equipment ... *178*
5.3. The Effect of Permanent Magnet Properties on Magnetic Field Distribution ... 180
5.3.1. Magnetic Pole Arrangement ... *180*
5.3.2. Magnetic Strength ... *186*
5.4. The Effect of Device Parameters on Magnetic Field Distribution ... 188
5.4.1. Distance Between the Magnet and the Grinding Wheel (L) ... *188*
5.4.2. Magnet Height (H) ... *190*
5.4.3. Angle Between the Magnet and the Grinding Wheel (β) ... *192*
5.4.4. Grinding Wheel Thickness (B) ... *194*
5.5. Testing and Verififying the Magnetic Field ... 196
References ... 198

Chapter 6 **The Kinematics Mechanism of Microdroplets of Magnetic Nanolubricants in a Complex Airflow Field Around the Grinding Wheel** ... 201
6.1. Introduction ... 201

6.2. Dynamic Modeling of the Collision Adsorption of Magnetic Nanolubricants on the Grinding Wheel Surface 203
6.2.1. Governing Equations *203*
6.2.2. Physical Modeling *206*
6.3. The Physical Mechanism of Microdroplets: Magnetic Field Strength 208
6.3.1. Research Results *208*
6.3.2. Microdroplet Sliding Behavior *213*
6.3.3. Film Formation Behavior of Microdroplet Suction *215*
6.4. The Physical Mechanism of Microdroplets: Droplet Velocity 216
6.4.1. Research Results *216*
6.4.2. Film Formation Behavior of Microdroplet Spreading *220*
6.5. The Microdroplet Behavior Map Under the Combined Action of Magnetic Field Intensity and Droplet Velocity 221
References 225

Chapter 7 The Antifriction Mechanism and Tribological Test of Magnetic Nanolubricants 227
7.1. Introduction 227
7.2. A Novel Tribological Test: Simulating Grinding Conditions 228
7.2.1. An Innovative Experimental Method *228*
7.2.2. Experimental Equipment *232*
7.2.3. Experimental Scheme *232*
7.3. Results of Various Magnetic Nanolubricants 234
7.3.1. Coefficient of Friction (CoF) *234*
7.3.2. Wear Contour Curve and Wear Area *240*
7.3.3. The Surface Morphology of Wear Marks *242*
7.3.4. Antifriction and Antiwear Mechanisms *243*
7.4. Results of Different Mixing Ratios of Fe_3O_4/Graphene Nanolubricant 245
7.4.1. Coefficient of Friction (CoF) *245*
7.4.2. Wear Contour Curve and Wear Area *248*
7.4.3. The Surface Morphology of Wear Marks *249*
References 252

Chapter 8 **Modelling Infiltration Behavior and Grinding Force in Magnetic Traction Nanolubricant Grinding** ... 255
8.1. Introduction ... 256
8.2. Modelling Nanolubricant Infiltration Behavior in the Grinding Zone's Magnetic Field ... 257
8.2.1. Physical Mechanism ... 257
8.2.2. Nanolubricant Force Analysis and Velocity Distribution ... 259
8.2.3. The Numerical Solution of Infiltration Velocity and Flow ... 262
8.3. Geometric Modelling of the Grinding Wheel Based on a Truncated Hexahedron ... 265
8.3.1. Abrasive Morphology of the CBN Grinding Wheel ... 265
8.3.2. Abrasive Distribution and the Attitude Model ... 269
8.3.3. The Abrasive Matrix on the Grinding Wheel Surface ... 271
8.3.4. Dynamic Abrasive and Cutting Depths ... 272
8.4. Force Modelling a Single Abrasive in the Micro-Contact Region ... 274
8.4.1. Geometric Modelling of an Abrasive/Workpiece Interference ... 274
8.4.2. Mechanical Behavior and Stress Distribution of an Abrasive/Workpiece Interface ... 276
8.4.3. The Single Abrasive Interference Model ... 279
8.4.4. The Identification of Plastic Deformation and Material Removal ... 282
8.5. Modelling and Experimental Verification of the Grinding Force ... 283
8.5.1. The Grinding Force Model ... 283
8.5.2. Results for Different Grinding Depths ... 285
8.5.3. Results for Different Wheel Periphery Speeds ... 286
8.5.4. Results for Different Feed Speeds ... 287
8.5.5. Results for Different Magnetic Field Strengths ... 288
References ... 291

Chapter 9 **Enhanced Grindability and Mechanism in the Magnetic Traction Lubricant Grinding of Ti-6Al-4V** 293
9.1. Introduction 293
9.2. Experimental Setup of Ti-6Al-4V Grinding 294
9.2.1. Experimental Equipment *294*
9.2.2. Experimental Materials *296*
9.2.3. Experimental Scheme *296*
9.3. Results of General Grinding 298
9.3.1. Grinding Force *298*
9.3.2. Roughness of the Workpiece *302*
9.3.3. SEM and EDS Analysis *304*
9.4. Experimental Results of Large-Arc Length Grinding 307
9.4.1. Grinding Force *307*
9.4.2. Grinding Temperature *310*
9.4.3. Three-Dimensional Morphology and Surface Roughness *312*
References 316

Chapter 10 **The Influence of Magnetic Field Parameters on Grinding Performance: An Orthogonal Optimization Experiment** 319
Abstract 319
10.1. Introduction 319
10.2. Design of Orthogonal Optimization Experiment 320
10.2.1. Experimental Equipment *320*
10.2.2. Experimental Scheme *321*
10.2.3. Experimental Materials *322*
10.3. Results of Orthogonal Optimization Experiment 322
10.3.1. SNR Analysis *322*
10.3.2. Variance Analysis *327*
10.3.3. Optimization Results *330*
10.4. Experimental Verification and Discussion 330
10.4.1. Surface Profile Power Density Spectrum Analysis *330*
10.4.2. Three-Dimensional Morphology of Workpiece Surface *333*
References 335

Index 337

Preface

High-strength alloy materials have been applied to various key parts in the aerospace field for decades. For example, titanium alloy takes up almost 30% of the weight in an aerospace engine. With increasingly rigorous requirements in precision and surface quality, grinding is attracting more interest as a vital process in manufacturing. During the grinding process of high-strength alloy materials, the temperature of the grinding zone always remains at a high level for the high strength and low thermal conductivity of the material characteristics, resulting in the deterioration of the workpiece surface roughness, and even burns. To overcome this problem, the traditional pouring flooding method has introduced a large amount of grinding fluid, where the low effective flow rate, unfortunately, leads to a significant difference between the grinding zone and the basic cooling environment of the workpiece. Especially, the surface integrity may easily deteriorate with a large grinding arc under this condition. In recent research, nano-lubricant minimum quantity has been proven as an efficient and environment-friendly cooling lubrication method. The unique techniques in nano-lubricant minimum quantity, including atomizing jet and enhanced heat transfer mechanism, are becoming the most promising approach compared to flooding in the grinding process, and have been applied in the grinding of common materials preliminarily. Given this context, the insufficient infiltration problem of lubricant caused by the poor transport energy is still an unresolved challenge in the deep grinding and forming grinding of difficult-to-machine materials.

To handle the challenge mentioned above, this book proposes a novel magnetic traction nanolubricant (MTN) grinding method for the first time, which aims to improve the infiltration and cooling lubrication properties of nanolubricant in grinding zone with the utilization of magnetic field. To investigate the magnetic adsorption transport mechanism and the anti-friction mechanical behavior in the proposed method, the systematic theory is discussed, which is then verified by the experiments designed correspondingly. The book is divided into ten chapters and the main contents of the thesis are listed as follows:

Chapter 1 provides a comprehensive overview of preparation, function and environmental pollution problem traditional cutting fluids. Further, the sterilization, degradation and anti-friction mechanisms of eco-friendly cutting fluids are explained. The lubrication mechanism and limitations of bio-lubricant are reviewed in minimum quantity lubrication (MQL) machining. Finally, the nanolubricant is proposed for improving lubrication and cooling performance and the nano-enhanced phases, preparation and stability improvement methods of nanolubricants are concluded.

Chapter 2 provides a comprehensive quantitative assessment of tribological characteristics of carbon group nano-enhancers (CGNs) in minimum quantity lubrication machining of aerospace materials, by considering force, tool wear, chip, and surface integrity. Further, the influence of molecular structure and micromorphology of CGNs on film formation and lubrication behaviors in MQL turning, milling, and grinding are concluded. Finally, the optimized options are recommended by considering diverse processes and materials.

Chapter 3 analyses the antifriction and heat dissipation behaviors of biolubricant in a high temperature and pressure interface, by considering the physicochemical properties as the main factors. Further, a comparative assessment of grindability of titanium alloy, nickel-based alloy, and high-strength steel is carried out. Finally, the current challenges and potential methods are proposed to promote the application of biolubricant.

Chapter 4 proposes the novel magnetic traction nanolubricant (MTN) grinding method, in which the adsorption and suction effect are used to improve infiltration force of nanolubricant. Further, the magnetic traction device and magnetic nanolubricant are designed and prepared. Finally, the physical viscosity, saturation magnetization and relative permeability of magnetic nanolubricant is studied to characterize magnetic response performance.

Chapter 5 studies the magnetization mechanism of the grinding wheel and its influence on the distribution of the magnetic inductance line. Further, the influence of device parameters on changing trend of magnetic field intensity and gradient is explored. Finally, the optimized parameters of the magnetic traction device is obtained based on the evaluation of magnetic force and transport performance of the lubricant droplets.

Chapter 6 reveals the kinematic behavior of the magnetic nano lubricant in the jet zone and the dynamic behavior of collision and adsorption on the workpiece surface of the grinding wheel. Further, the effect of the magnetic field intensity and jet velocity on microdroplet sliding, film formation, and falling is studied. The collision-adsorption behavior map of microdroplets under the action of magnetic field intensity and jet velocity is obtained, which provides a theoretical basis for process optimization.

Chapter 7 researches the tribological characteristics of magnetic Fe_3O_4/graphene nanolubricant using the innovative tribological test method by simulating grinding conditions. Further, the ratio optimization of Fe_3O_4/graphene is studied by considering performance and economy. As result, minimum coefficient of friction (0.32), wear area (6146μm^2) and surface roughness (1.64μm) are obtained by optimized Fe_3O_4/ graphene (2:1).

Chapter 8 studies the infiltration mechanism of magnetic lubricant in grinding zone, and the model of lubricant infiltration velocity and flow rate with the influence of the magnetic field is established. Further, an improved grinding force model of magnetic traction nanolubricant is established for titanium alloy grinding and verified by experiments. The results show that the introduction of magnetic field can significantly increase the flow speed and flow rate of lubricant.

Chapter 9 studies the grindability and mechanism of MTN grinding. The general grinding results show that the Fe_3O_4/graphene nanolubricant obtained better cooling and lubrication performance than palm oil, Fe_3O_4 nanolubricant, and graphene nanolubricant. Further, the improvement in infiltration performance under magnetic field is verified by the large contact length grinding (LCLG) experiment.

Chapter 10 carried out the orthogonal optimization research of device parameters by considering influence of magnetic field on grinding performance. The influence weights of horizontal distance, vertical distance, and angle between the permanent magnet and grinding wheel on grinding force and surface roughness are studied. Finally, the optimized process parameter combination is obtained by using characterization parameters, including surface power spectral density function and surface morphology of titanium alloy workpiece.

In conclusion, we hope that this book will bring a new idea of sustainable grinding and enable readers to have a comprehensive knowledge and application of sustainable grinding of aerospace materials. We thank the editors of Nova Science Publishers for their hard work. We would like to express our gratitude to all of them. Finally, we are also grateful for the publication of this book with the support of the National Natural Science

Foundation of China (52105457, 52475469, 52375447), the Special Fund of Taishan Scholars Project (tsqn202211179).

Dr. Xin Cui,
Associate Professor
Qingdao University of Technology, China

Dr. Chang-He Li,
Professor
Qingdao University of Technology, China

Acknowledgments

The authors would like to acknowledge the support of the National Natural Science Foundation of China (52105457, 52475469, 52375447), the Special Fund of Taishan Scholars Project (tsqn202211179).

About the Authors

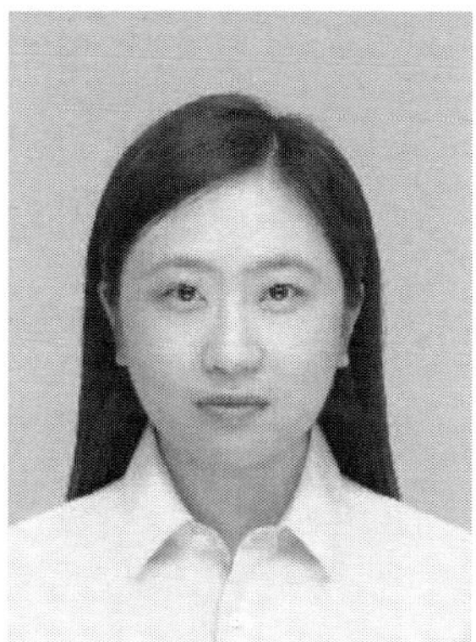

Xin Cui

Key Lab of Industrial Fluid Energy Conservation and Pollution Control, Ministry of Education, Qingdao University of Technology, Qingdao, China

Dr. Xin Cui is currently working as an Associate Professor at Key Lab of Industrial Fluid Energy Conservation and Pollution Control, Ministry of Education, Qingdao University of Technology. She mainly engaged in green and precision manufacturing processing work, is the core member of the create team of Intelligent and Sustainable Manufacturing. A total of 4 projects, including hosting a National Natural Science Foundation youth project, participating in 3 projects including the national key research and development program, Qingdao West Coast New Area source innovation project etc. She has published 17 SCI/EI accepted papers, 16 papers in Q1 and Q2. She applied for the authorization of 3 international invention patents. During the doctoral period, she has applied for 7 international patents including the US, Australia and South Africa. She has won 6 scientific research awards, including the Science and Technology Award of China Circular Economy Association, the Science and Technology Award of China Business Federation, and the China Patent Award, etc.

E-mail: Cxinxin5240@163.com

Chang-He Li
Key Lab of Industrial Fluid Energy Conservation and Pollution Control, Ministry of Education, Qingdao University of Technology, Qingdao, China

Professor Chang-He Li is a foreign academician of the Russian Academy of Engineering, a Highly Cited Chinese Researcher of Elsevier, a distinguished expert of Taishan Scholars in Shandong Province, member of the Standing Committee of the Abrasive Processing Professional Committee of the Chinese Mechanical Engineering Society, member of the Standing Committee of the Optical Finishing Professional Committee of the Chinese Mechanical Engineering Society, one of the World's Top 2% Scientists 2022. He is rated as Shandong Excellent Inventor, Shandong Excellent Scientific and Technological Worker, Shandong Famous Teacher of Teaching and Shandong Excellent Postgraduate Instructor. He has published 247 SCI/EI-accepted papers, including 77 in SCI/JCR(1-2), and 50 ESI highly cited papers, 20 hot papers, with a total of more than 12458 citations (Web of Science), and is ranked 25th in the list of academic influence of scholars in China in 2022 (only 70 people in China in the field of mechanical engineering). He is selected as the 24th academic influence of domestic scholars in 2022, and has a high academic status and popularity in the same field at home and abroad. He is granted 69 invention patents in the United States, South Korea, Australia, etc., 52 PCT international patents, 117 national invention patents, and 56 software copyrights. He is the Editorial Board member of Frontiers of Mechanical Engineering and Green Manufacturing Open, Guest Editor of International Journal of Advanced Manufacturing Technology, Chinese Journal of Mechanical Engineering, and Associate Editor of Chinese Journal of Mechanical Engineering.

E-mail: Sy_lichanghe@163.com

Abbreviations

CGNs	Carbon group nano-enhancers
CNMQL	Cryogenic minimum quantity lubrication
CoF	Coefficient of friction
LCLG	Large contact length grinding
MQL	Minimum quantity lubrication
MRF	Magnetorheological Finishing
MTN	Magnetic traction nanolubricant
NMQL	Nanolubricant minimum quantity lubrication

Chapter 1

The Biological Failure Mechanism and Preparation of Eco-Friendly Lubricant in Machining

Abstract

Traditional cutting fluids, which are prepared by non-renewable minerals and scarce water, are a necessity in the manufacturing industry and have been for hundreds of years. However, this is also at the core of machining-induced contamination due to environmental, health, and economic issues. Eco-friendly cutting fluids, bio-lubricant and nanolubricant are renewable and biodegradable lubricants act as a substitute for traditional cutting fluids for sustainable transformation and upgrading of the traditional manufacturing industry. In this chapter, the eco-friendly lubricant used in machining was reviewed in detail. Firstly, this chapter reviews the preparation, function and environmental pollution problem of traditional cutting fluids. Secondly, the sterilization, degradation and antifriction mechanisms of eco-friendly cutting fluids is revealed. Further, the lubrication mechanism and limitations of bio-lubricant minimum quantity lubrication is also addresssed. Finally, the nanolubricant is proposed for improving lubrication and cooling performance of minimum quantity lubrication machining. The nano-enhanced phases, preparation and stability improvement methods of nanolubricants are concluded. Through this chapter, the development, preparation process and mechanisms of Eco-friendly lubricant can provide reference for researchers.

Keywords: biolubricant, nanolubricant, minimum quantity lubrication, degradability, lubrication, cooling mechanism

1.1. Introduction

Manufacturing industry is the pillar industry of the national economy, and transformation and upgrading of the traditional manufacturing industry through energy saving and emission reduction in the processing process is the key development direction under the concept of sustainable development of the national economy. Achieving carbon peak and carbon neutrality is an important strategic goal for China in the next 40 years, and energy saving reform through manufacturing process is one of the development priorities of the manufacturing industry. Manufacturing powers have successively introduced carbon emission standards ISO 14067-2018 and other relevant laws, under the strict standards and legal requirements of the traditional manufacturing green transformation and upgrading has been imminent and imperative.

Mineral cutting fluid has been widely used in the industry for hundreds of years, and has solved the technical problems of cooling, lubrication and chip removal under the strong thermal coupling of the cutting interface. However, from the analysis of the mass use of mineral water-soluble cutting fluid, there are the following threats: (i) The grinding zone is a micro/nano interface with closed characteristics. In order to enhance the wettability of the grinding fluid, the amount of grinding fluid is greatly increased, most of the grinding fluid does not play a cooling and lubricating role and becomes a waste, and the use of traditional mineral cutting fluid consumes a lot of natural resources; (ii) Mineral cutting fluid due to the addition of toxic substances, caused a great threat to the health of workers; And the discharge into nature will cause serious environmental pollution; (iii) The use cost of cutting fluid is also very high, as shown in Figure 1, according to the statistics of automobile manufacturers, waste cutting fluid needs to be strictly harmless treatment before it can be discharged into the natural environment, and the use cost of cutting fluid is as high as 18~21% of the total cost, which is 3~5 times the cost of tools [1].

Scholars first considered not using cutting fluid and thus gave birth to dry cutting green processing technology, which can be traced back to Professor Klocke's keynote report at CIRP conference in 1997. Although dry cutting technology does not use cutting fluid, it is still based on the premise of ensuring part accuracy and tool life [2]. Therefore, it can be applied to light-load cutting processing, but for heavy-load machining, tool adhesion, wear and other conditions limit its use. In particular, grinding has the characteristics of high heat output, and the temperature of the grinding zone can reach more than 500°C, which inevitably causes serious workpiece surface burns [3].

Therefore, dry cutting does not meet the processability requirements of aerospace difficult-to-machine materials.

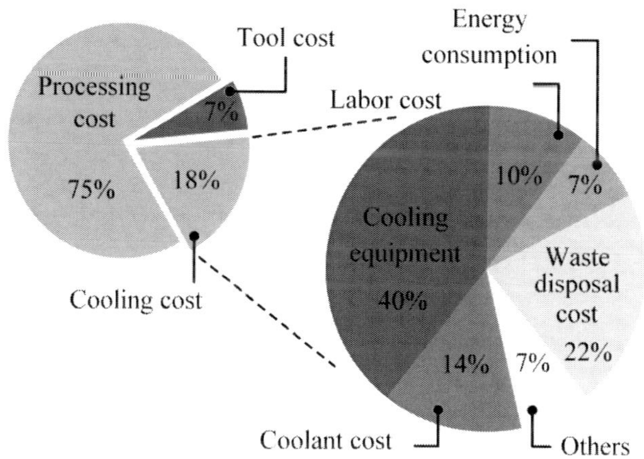

Figure 1. Distribution of costs in the automobile industry.

Due to the environmental and health hazards of the traditional flood and the bottleneck problems of poor cooling and heat transfer in dry cutting, there is urgent to find a clean grinding process for difficult-to-machine materials [4]. In previous studies, scholars have successively proposed solid lubrication [5], low-temperature cooling lubrication [6, 7], MQL [8, 9] and NMQL [10, 11], and conducted experiments and application tests. By reducing the amount of cutting fluid used, changing the type of lubricating medium, improving heat transfer capacity and other means, the problems and bottlenecks existing in traditional casting lubrication are changed, so as to achieve green and efficient processing.

In this chapter, the preparation process and parameter characterization of green cutting fluid, bio-lubricant, nano-lubricant and magnetic nano-lubricant will be introduced in detail.

1.2. Traditional Cutting Fluids

1.2.1. Preparation

The earliest cutting fluid was not the oil or oil water mixture as we know it, but water was used as the cooling fluid in metal processing. Later, simple water-based solutions such as soapy water lime water and animal and vegetable oils such as linseed oil were used. People slowly discovered that these simple composition cutting fluids could not meet the cooling and lubricating effect required in metal processing. After people extracted a large amount of lubricating oil from crude oil, oil based cutting fluids developed in a real sense. With the improvement of metal processing precision, the requirements for cooling and lubrication have become more stringent, and the cooling performance of oil-based cutting fluids can no longer fully meet the needs of high-speed cutting. People started to look at the excellent cooling performance of water again because the specific heat capacity of water is larger than that of oil, and the cooling performance is better. However, due to insufficient lubrication and antirust properties of water, people used various additives such as lubricants and rust inhibitors additive to meet the lubrication and antirust properties required in metal processing.

1.2.2. Classification

The current cutting fluid has a relatively mature application system in the field of metal processing and has become a multicategory product. Cutting fluids are mainly divided into oil-based cutting fluids and water-based cutting fluids. The main types are shown in Table 1.

Oil-based cutting fluid is made by mixing base oil with tribological additives and auxiliary additives, and does not need to be diluted with water when used. The base oil of oil based cutting fluid is divided into mineral oil and synthetic oil. Mineral oil is a mixture of liquid hydrocarbons obtained from the refining of fossil fuels such as petroleum. According to the source of its raw material oil, it can be divided into two types: paraffin based and naphthenic based. Synthetic oil mainly includes synthetic hydrocarbons and synthetic esters. Among these, mineral oil-based cutting fluid has the characteristics of excellent lubrication and antirust properties and is most widely used. Water-based cutting fluids are divided into emulsion type, semisynthetic type and synthetic type. Emulsified cutting fluid is an oil-in-water emulsion obtained by mixing base oil (mineral oil) with water containing emulsifiers. The content of mineral oil is greater than 60%, showing milky white. Synthetic cutting fluid takes inorganic salt and organic amine as the main body and adds various additives such as preservatives or defoamers to dissolve in water to form a transparent or slightly colored aqueous solution.

Table 1. Main types of cutting fluids

Types	Classifications	Ingredients	Advantages	Disadvantages	Applications
Oil based cutting fluids	Mineral oil based	Mineral oil, extreme pressure additives, etc.	Excellent lubrication performance and antirust performance	Poor cooling performance, easy to produce smoke, fire and other phenomena, difficult to handle waste liquid	Widely used in aerospace, nuclear industry and other fields, for low speed and low-pressure metal processing
	Synthetic oil based	Synthetic oil, extreme pressure additives, etc.			
Water based cutting fluids	Emulsified cutting fluid	Higher amount of mineral oil, emulsifier, rust inhibitor, etc.	Excellent lubrication performance, antirust performance and cooling performance, the waste liquid treatment is relatively simple	Excessive electrolyte content in water will lead to reduced emulsion stability, poor cleaning	For high speed and low-pressure metal processing
	Synthetic cutting fluid	Defoamers, surfactants, water soluble extreme pressure additives, rust inhibitors, etc.	Excellent cleaning performance, stability and antirust performance, not easy to rot	Poor machinability, insufficient lubricity, inferior antirust performance, difficult to handle waste fluid	Generally only used for grinding, rarely used for cutting
	Semisynthetic cutting fluid	A small amount of mineral oil, surfactants, extreme pressure agents, rust inhibitors, etc.	Excellent lubrication performance, antirust performance and cooling performance	Difficult to handle waste fluid, Biochemical oxygen demand (BOD) and chemical oxygen demand (COD) are not easy to decrease	Excellent comprehensive performance, widely used in metal processing

Semisynthetic cutting fluid is a combination of oil-in-water emulsion and synthetic fluid, which has the characteristics of emulsified cutting fluid and synthetic cutting fluid. Its service life is four to six times longer than that of emulsified cutting fluid. The mineral oil content is 5% to 40%, showing translucent.

1.2.3. Function

In metal processing, frictional heat will be generated between the tool, chip and workpiece, which is extremely detrimental to the service life of the tool and the surface quality of the workpiece. If the chips generated during cutting are not removed in time, machining quality will be affected. In addition, the antirust problem of the processed workpiece must be maintained during long-term placement. A cutting fluid with excellent performance needs to have lubrication, cooling, cleaning, antirust and other effects to solve these problems.

i. Lubrication effect. In metal processing, friction will be generated in the tool chip and tool workpiece. After adding cutting fluid, a lubricating film will be produced between the two friction pairs to reduce friction and wear. Owing to the high temperature and high-pressure characteristics of the cutting process, forming a fluid lubrication state is difficult, and the lubricating film formed is often a boundary lubricating film. The lubricating film can effectively reduce the cutting force and the wear of the tool, thereby increasing the life of the tool and improving the quality of the workpiece.
ii. Cooling effect. Owing to the shearing effect of the tool and the workpiece and the friction between the tool and the chips, high temperature phenomena are generated in the cutting area, and the cutting heat is the most unfavorable for cutting, easily causing damage to the surface of the workpiece. To avoid this phenomenon, the cooling effect of the cutting fluid is required. When the cutting fluid enters the cutting area, a part of the heat will be taken away by the cutting fluid through convection heat transfer, and the vaporization of the cutting fluid can take away part of the heat.
iii. Cleaning effect. Many small chips are generated during metal processing. If the chips are left in the cutting area for a long time, accelerating the tool wear and scratch in the workpiece is easy; hence,

it needs to be processed in time. After adding the cutting fluid, the cutting fluid can wash away the small chips and clean the cutting area when the cutting fluid flows through it. The cleaning effect and cooling effect of the cutting fluid are inseparable. The cutting fluid will reduce the temperature of the chips while playing the cleaning effect to prevent the flying chips from causing burns to the operators.

iv. Antirust effect. During the storage time after processing, the workpiece is likely to react with air media, resulting in rust or corrosion. To avoid this phenomenon, the cutting fluid is required to have antirust and anticorrosion effects. The cutting fluid forms a physical adsorption film or a chemical reaction film on the surface of the workpiece to delay the rust of the workpiece. In addition, the antirust effect of the cutting fluid is not only manifested on the workpiece but also has a maintenance effect on the machine tool and the cutting tool.

1.2.4. The Problem of Cutting Fluids

(i) Chemical composition. The current metal processing industry is developing in the direction of faster processing efficiency and more complicated processing conditions. The cooling and lubricating effect of general cutting fluids cannot meet the requirements of the metal processing industry. For this reason, researchers opt to add various additives in cutting fluids. For example, under high-speed, heavy-load conditions, extreme pressure additives can provide the cutting fluid the effect of reducing wear and pressure to meet these needs [12, 13]. In addition, the cutting fluid contains the necessary elements for the survival of microorganisms, which easily breed bacteria and microorganisms. A bactericide must be added to inhibit the growth of microorganisms in the cutting fluid.

The use of additives determines the performance of the cutting fluid and provides the cutting fluid with a sense of time. However, the use of several additives can cause serious water pollution problems. For example, the sodium phosphate in rust inhibitors can lead to eutrophication of water, which in turn triggers red tides; antioxidants have poor biodegradability and are prone to long-term pollution of water sources. The water pollution level and biodegradation rate of commonly used additives are shown in Table 2. In addition, several substances in additives can cause serious harm to workers. As shown in Table 3.

Table 2. Water pollution level and biodegradation rate of commonly used additives

Additives	Chemical materials	Biodegradation rates %	Water pollution levels	Testing methods
Rust inhibitor	Ashless Sulfonate	50	1	CECL-33-T82
	Tolytriazole	70	1	OECD302B
Extreme pressure additives	Calcium Dialkylbenzene Sulfonate	60	1	CECL-33-T82
Antioxidants	Alkyl Diphenylamin	9	1	OCED301D

In addition, the base oil of cutting fluid is divided into mineral oil, synthetic oil and vegetable oil. Synthetic hydrocarbons in synthetic oils have poor biodegradability, similar to mineral oils. When this type of cutting fluid is discharged into the natural environment, it will stay in the water and soil for a long time, causing adverse effects on the environment. The US Environmental Protection Agency pointed out that more than 10 ppm of oil in water will kill marine plants, and 300 ppm of oil will kill freshwater fish. Synthetic cutting fluids based on inorganic salts and organic amines are alkaline, which can cause symptoms such as dryness and degreasing of human skin. The biodegradability of common base oils is shown in Table 4. The use of toxic additives and base oils with poor biodegradability is the key to the pollution problem of cutting fluids.

(ii) Physical action. In addition to the pollution caused by the chemical composition of the cutting fluid, the way the cutting fluid is used will also pollute the human body and the environment. In many high-speed machining, oil-based cutting fluid easily forms oil mist, and water-based cutting fluid easily generates aerosol. The bacteria and microorganisms in the cutting fluid enter the workers respiratory system along with the oil mist or aerosol, causing people to suffer from diseases such as lung disease and eczema. In addition, these minute cutting fluid molecules emit pungent odors and stay in the air, causing pollution to the working environment. Picciotto et al. [14] found that suspended particles of oil-based cutting fluids are related to death from ischemic heart disease. Therefore, the defects of processing methods are sources of cutting fluid pollution.

Table 3. Toxic substances in additives and their harm mechanism

Additives	Chemical materials	Molecular formulas	Structural formulas	Hazard mechanisms
Rust inhibitor	Sodium Nitrite/Alcoholamine	$NaNO_2$ $C_5H_{13}NO_2$	Alcoholamine: 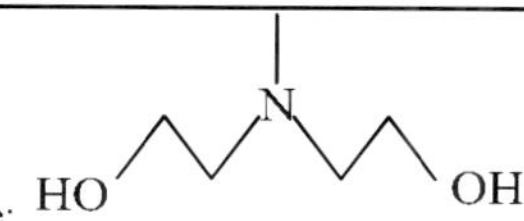	Sodium nitrite and amines can form nitrosamines. In acidic and high temperature environments, nitrosamines can form diazonium salts. This substance can produce carbocations and react with biological nucleophiles (such as DNA or enzymes) in cells, leading to cell cancer or cell apoptosis.
	P-tert-butyl benzoicacid	$(CH_3)_3CC_6H_4CO_2H$	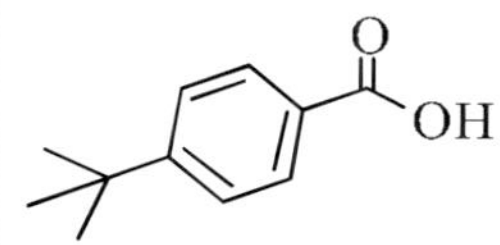	It will irritate the respiratory tract, skin, eyes, and be harmful to internal organs such as liver and kidney of animals.
Bactericide	Glutaraldehyde	$C_5H_8O_2$	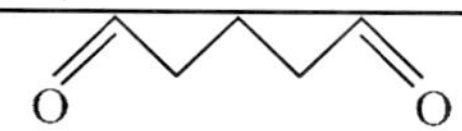	Glutaraldehyde is through its two aldehyde groups combined with organic molecules such as proteins in tissue cells, destroying the molecular structure of tissue cells, causing them to lose their biological activity and causing body damage.

Table 3. (Continued)

Additives	Chemical materials	Molecular formulas	Structural formulas	Hazard mechanisms
	4-chloro-3-methylphenol	C_7H_7ClO	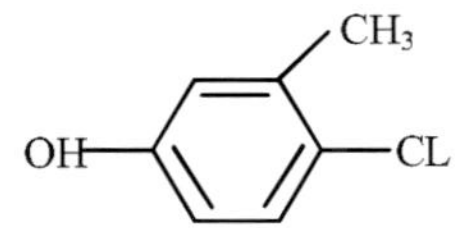	The biodegradability is poor, and it has great harm to aquatic organisms. It is decomposed by heat to liberate chlorine, which has a strong irritating effect on eyes, skin and mucous membranes.
	Ketohydroxyethylh exahydros-triazine	$C_9H_{21}N_3O_3$	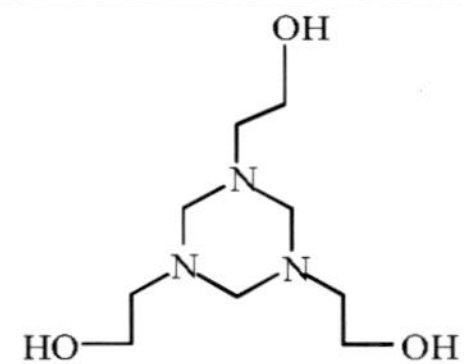	Slowly release harmful gas formaldehyde in cutting fluid
Extreme pressure additives	Short-chain chlorinated paraffin	$C_{10\text{-}13}$	--	Short-chain chlorinated paraffins can cause liver hypertrophy in rodents and peroxisome proliferation in liver cells. Studies have shown that short-chain chlorinated paraffin doses can increase the incidence of murine adenomas and cancers of the liver, thyroid and kidney.
	Dialkyldithio metal salt	Zinc dialkyl dithiophosp hates (ZDDP)	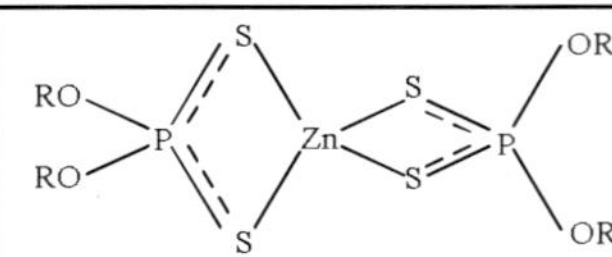	Contains sulfur which is harmful to organisms.

Table 4. Biodegradability of common base oils

Base oils	Biodegradation rates%
Polyester	80~100
Dibasic acid diester	60~100
Polyol ester	60~100
Phthalate diester	60~70
Polyolefin	≤20
Polyisobutylene	≤30
Polypropylene glycol	≤10
Alkylbenzene	≤10
Mineral oil	20~60

The amount of cutting fluid used in metal processing is substantial, and metal processing is inseparable from cutting fluid at present. To reduce the hazards of cutting fluids, developed countries have formulated many relevant legal standards, such as the regulations of the US labor safety and security national institutions: The limit is 3mg/m^3 for cutting fluids that do not contain polyaromatic hydrocarbons and 0.5 mg/m^3 for cutting fluids that contain polyaromatic hydrocarbons. In addition, the British Health and Safety Executive recommends that the droplet contents of oil-based cutting fluid and water-based cutting fluid are 3 and 1 mg/m^3, respectively. Although these established standards have alleviated the problem of cutting fluid pollution to a certain extent, manufacturers have increased production costs to comply with them, which is not conducive to the development of the metal processing industry. For this reason, how to solve the problem of cutting fluid pollution effectively has become a key issue in the field of metal processing. Researchers are looking for solutions to problems.

1.3. Eco-Friendly Cutting Fluids

On the one hand, the pollution of cutting fluid comes from its chemical composition. On the other hand, the pollution of cutting fluid comes from the processing methods that require cutting fluid. To comply with the concept of green development, combined with the causes of cutting fluid pollution, the development of eco-friendly cutting fluids and green processing methods has become a key issue in the field of metal processing [15]. This section reviews the current research progress of eco-friendly cutting fluids and green machining methods.

The pollution problems of traditional cutting fluids mainly come from base oils with poor biodegradability and additives that are harmful to organisms and the environment. According to the source of the pollution problem, researchers have developed some eco-friendly cutting fluids with good biodegradability and low harm to organisms and the environment. Bhaumik et al. [16] used nano minimum glycerin with good biodegradation and low biological hazard as a substitute for traditional cutting fluid. To outline the research progress of eco-friendly cutting fluids better, according to the pollution components of traditional cutting fluids, explanations will be made by summarizing several commonly used eco-friendly additives and base oils with good biodegradability.

1.3.1. Rust Inhibitors

In the field of metal processing, the corrosion of workpieces and machine tools often causes great economic losses to the enterprise. Adding rust inhibitors to the cutting fluid is an effective way to solve the problem. Many rust inhibitors are commonly used in cutting fluids, which are mainly divided into two categories: inorganic and organic. Inorganic rust inhibitors include inorganic salts such as sodium nitrite and molybdate, and their mechanism of action is mainly to form an insoluble passivation film on the metal surface to prevent the metal from rusting. The advantages of inorganic rust inhibitors are good rust prevention performance and excellent economic efficiency. However, most of their inorganic rust inhibitors cause great pollution to the environment and are easily toxic to organisms. Organic rust inhibitors are composed of hydrophilic polar groups and lipophilic nonpolar hydrocarbon groups, mainly including alcamines, carboxylic acids, amides, and boric acid ester. The mechanism of action is to make the polar end of the rust inhibitor adsorb on the metal surface through electrostatic adsorption and chemical adsorption. The nonpolar ends are randomly oriented and arranged on the side away from the metal surface, thereby forming a nonpolar hydrophobic film on the metal surface. Organic rust inhibitors can not only guarantee rust performance but have minimal harm to organisms and the environment. The research and development of organic rust inhibitors is an important part of the research on eco-friendly rust inhibitors.

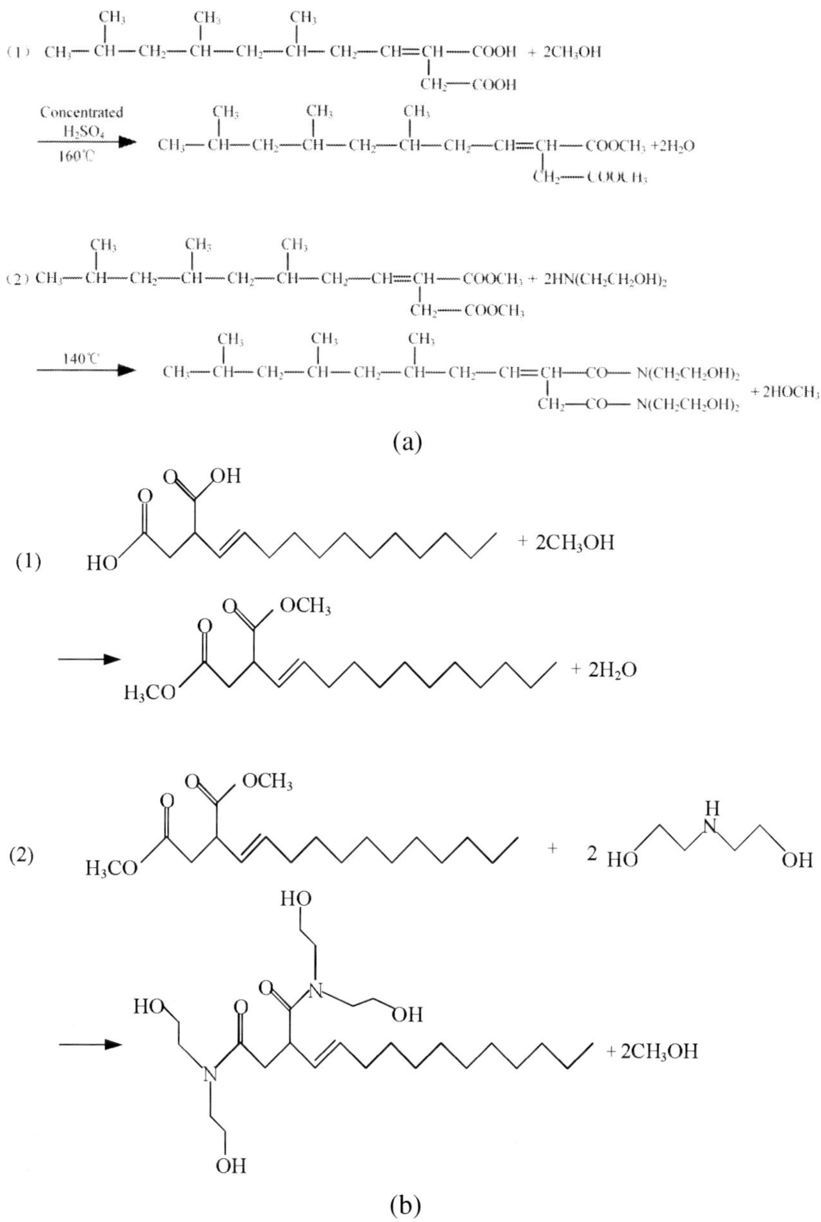

Figure 2. Synthesis process of rust inhibitor. (a) Dodecenyl succinic acid diethanolamide rust inhibitor, (b) Dodecenyl succinic acid diethanolamide.

Researchers have studied organic rust inhibitors. Amide organic rust inhibitors have the advantages of excellent environmental protection performance and good antirust performance, which has become the main direction of research on eco-friendly rust inhibitors. Zhou et al. [17] synthesized a dodecenyl succinic acid diethanolamide rust inhibitor with improved water solubility and antirust effect. The synthesis is shown in Figure 2(a). The polar group of the antirust inhibitor had a greater affinity for metal and could form a directional adsorption layer on the metal surface. Researchers carried out single-chip antirust experiments of cast iron with solution concentrations of 1%, 3%, 5%, and 7%, and the antirust performance met the GB/T6144-85 standard. Gan et al. [18] synthesized a water-soluble rust inhibitor dodecenyl succinic acid diethanolamide T746-X. The synthesis is shown in Figure 2(b), The results of the single-chip antirust experiment are shown in Figure 3. Five rust spots were observed on the chip of the semisynthetic cutting fluid that did not contain T746-X rust inhibitor. Two rust spots appeared on the chip containing the semisynthetic cutting fluid of T746-X with a mass concentration of 1.25 g/L. When the mass concentration of T746-X reached 3.75 g/L, the chip had no rust spots. In addition, the indicators of this antirust additive were in line with the industry standard of JB/T7453-2013.

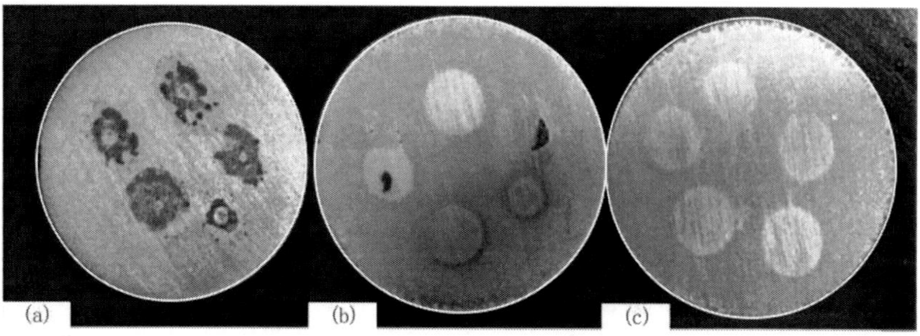

Figure 3. Results of single-chip antirust experiment with different contents of T746-X. (a) Did not contain T746-X, (b) Contains T746-X with a mass concentration of 1.25g/L, (c) Contains T746-X with a mass concentration of 3.75g/L.

In addition to amide rust inhibitors, boric acid ester rust inhibitors are environmentally friendly. Liu et al. [19] explored the compounding of boric acid ester rust inhibitors and other water-soluble rust inhibitors, and believed that selecting a suitable water-soluble rust inhibitor compound can make boric acid ester rust inhibitors have excellent antirust performance, microbial stability, and environmental protection performance. Wang et al. [20]

synthesized triethanolamine borate with excellent antirust properties with boric acid and triethanolamine. He used 20 steel and 45 steel to soak in an aqueous solution containing 2.5% triethanolamine borate for 30 days, and no rust was observed on the surface of the specimen.

1.3.2. Bactericide

The cutting fluid has the conditions for the growth of microorganisms and bacteria, and easily spoils and deteriorates. At present, the most popular and effective way to control the growth of microorganisms in cutting fluids is to add bactericides. Common bactericides include phenolic compounds, bromine and chlorine compounds, and formaldehydere leasing compounds. However, these traditional bactericides have poor stability and bactericidal performance. For this reason, how to enhance the bactericidal performance and stability of bactericides is one of the key research directions in the current research on eco-friendly bactericides.

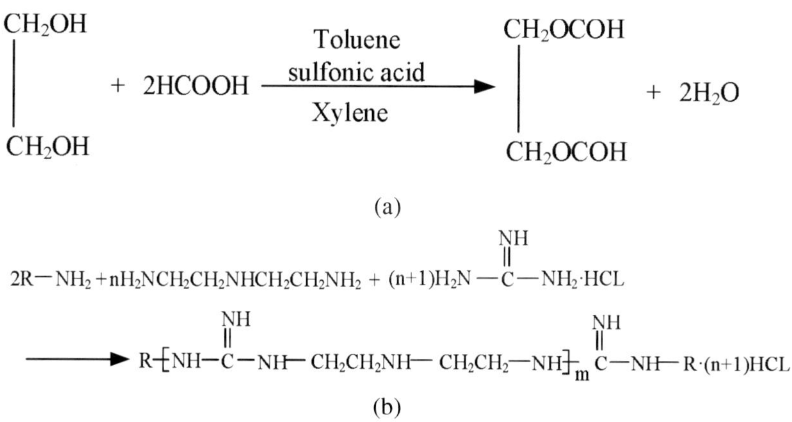

Figure 4. Synthesis of bactericides. (a) Ethane- 1,2-diyl diformate, (b) New organic guanidine bactericide through condensation reaction, n=1-20, R=CH3(CH2)m, and m=3-15.

Researchers have developed efficient, stable eco-friendly bactericides based on the existing problems of bactericides. Ezzat et al. [21] used the glycols(s) and formic acid as raw materials, paratoluene sulfonic acid as catalyst, and xylene as an azeotropic solvent to produce ethane-1,2-diyl diformate, after esterification, the best product with antibacterial effects. The synthesis is shown in Figure 4(a). Wang [22] used guanidine hydrochloride, diethylenetriamine, long-chain organic amines and other amine-containing monomers as raw materials to synthesize a new organic guanidine bactericide through condensation reaction. The synthesis is shown in Figure 4(b). When the dosage of the organic guanidine bactericide GA was 30 mg·L−1, the killing rate of blank heterotrophic bacteria, reducing bacteria and iron bacteria were all greater than 99%.

1.3.3. Extreme Pressure Additives

Extreme pressure additives are an important additive in cutting fluid. The extreme pressure additives form an extreme pressure lubricating film on the surface of the workpiece so that the cutting fluid can be used under extreme harsh conditions such as high temperature and high pressure, which broadens the scope of application of the cutting fluid. This behavior has an irreplaceable effect on the cooling and lubrication of processing methods such as cutting and forming of difficult-to-machine materials. According to the film forming mechanism, extreme pressure agents can be divided into two types: One is capable of direct chemical reaction with the metal surface to form a reactive film, such as sulfur or chlorine-containing compounds, metal sulfides, and metal chlorides. The other forms films by thermal degradation during processing, such as phosphate disulfide and phosphorus containing organic compounds. However, these extreme pressure additives have poor biodegradability, and several extreme pressure agents are carcinogens, which do not meet the requirements of green development. Finding extreme pressure additives with good biodegradability and low biological hazards is a key issue in the field of metal processing.

With the research and development of green extreme pressure additives, researchers have found that boric acid ester extreme pressure additives are expected to become ecofriendly additives. Yuan et al. [23] used dodecanoic acid, diethanolamine and boric acid as raw materials to prepare dodecanoic acid diethanolamide borate by esterification and condensation reactions. The synthesis is shown in Figure 5(a). The researchers added 11.3%dodecanoic acid diethanolamide borate into the semisynthetic cutting fluid. The prepared cutting fluid had good extreme pressure and wear resistance, and its friction factor can be as low as 0.056. In addition, the semisynthetic metal cutting fluid was formulated with dodecanoic acid diethanolamide boric acid ester, and all its indicators met the GB/T6144-2010 standard. Meng et al. [24] used ricinoleic acid, boron, and diethanolamine as raw materials, and prepared an alcohol amine borate antiwear additive using the azeotropic principle. The synthesis is shown in Figure 5(b). He used a four-ball friction tester to investigate the tribological performance of alcohol amine boric acid ester antiwear additives. The results showed that when the additive dosage was 200 $\mu g \cdot g^{-1}$, the antiwear effect of the base oil was better. The minimum wear scar diameter was 419 μm.

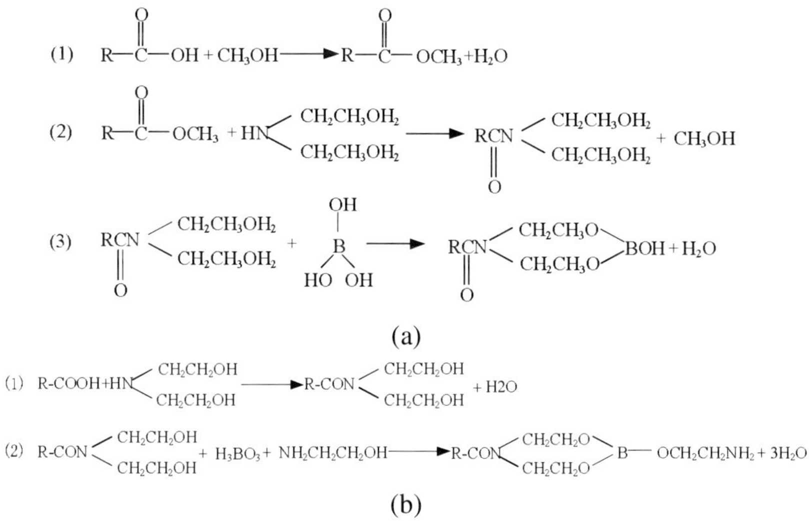

Figure 5. Synthesis of extreme pressure additive. (a) Dodecanoic acid diethanolamide borate, (b) Alcohol amine borate antiwear additive.

Foreign patents have also invented several eco-friendly extreme pressure additives. Among them, polyesters, rosin acid and its derivatives have become important eco-friendly extreme pressure additives that replace chlorinated paraffin.

The research on eco-friendly additives is not limited to the above three additives. For example, to enhance the cooling and cleaning performance of the water-based synthetic cutting fluid, a surfactant that can reduce the surface tension of the cutting fluid needs to be added to the cutting fluid. To solve the foam problem of cutting fluid, defoamers must be added. These additives are mostly inorganic additives, which require green research and development. To follow the concept of sustainable development, researchers must continue to develop new eco-friendly additives to improve the performance of the cutting fluid continuously, making the cutting fluid more environmentally while maintaining excellent performance.

1.3.4. Base Oil

Vegetable oil has gradually become the choice of green cutting fluid base oil due to its natural advantages such as good biodegradability, high yield, nontoxicity, and sufficient resources. Using vegetable oil as the base oil of the cutting fluid can make the processing technology more eco-friendly while enhancing the antiwear and lubricating performance of the processing area. The main component of vegetable oil is triglyceride formed by the combination of fatty acid and glycerin. Different vegetable oils have different types and contents of fatty acids, which will affect their physical and chemical performance. The fatty acid can react with the metal surface to form a monolayer or multilayer fatty acid soap adsorption film, which has antiwear and antiwear effects; when the vegetable oil molecules are arranged vertically, tightly and evenly on the surface of the metal, a single-molecule boundary layer lubricating film with excellent lubricating performance is formed on the surface of the metal. Fatty acids are divided into saturated fatty acids and unsaturated fatty acids. The fatty acid content of common vegetable oils and the molecular structure of fatty acids are shown in Table 5. Oleic acid, linoleic acid, and linolenic acid are unsaturated fatty acids, whereas stearic acid and palmitic acid are saturated fatty acids. For fatty acids, the lubricity of saturated fatty acids is better than that of unsaturated fatty acids.

Based on the physical and chemical properties of vegetable oils, researchers have developed vegetable oil-based ecofriendly fluids with excellent lubricating properties. Among them, soybean oil has the fattiest acid content and has become the research focus of eco-friendly fluid base oil. Lodhi and Kumar [25] used soybean oil as the base oil of the cutting fluid and tested it with a four-ball tester. The results showed that under a load above 1600 N, vegetable oil-based cutting fluids can still maintain a lower friction coefficient than traditional cutting fluids. Guo et al. [26] studied the lubrication mechanism of castor oil mixed with other vegetable oils (soybean oil, corn oil, and palm oil) as a base oil for MQL grinding, and they found that mixed base oil had a better lubricating effect than castor oil. Moreover, the soybean oil/castor oil mixed base oil had the best lubrication effect, and the tangential force and normal force were reduced by 27.03% and 23.15%, respectively.

Table 5. Fatty acid content of several common vegetable oils and molecular structure of fatty acids

Categories	Molecular formulas	Structural formulas	Fatty acid content of vegetable oil (%)				
			Soybean oil	Rapeseed oil	Castor oil	Corn oil	Peanut oil
Oleic acid	$C_{17}H_{33}COOH$	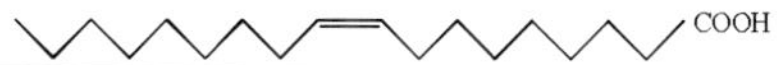	22-31	40-60	3-9	26-40	53-71
Linoleic acid	$C_{17}H_{31}COOH$	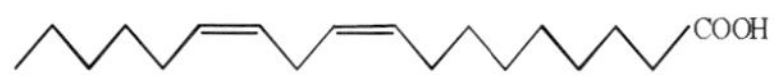	49-55	11-23	2.0-3.5	40-55	13-27
Linolenic acid	$C_{17}H_{29}COOH$		6-11	5-13	-	1-2	-
Palmitic acid	$C_{17}H_{31}COOH$		7-10	1.5-6	-	9-19	6-9
Stearic acid	$C_{17}H_{35}COOH$		3-5	0.5-3	-	1-3	3-6

In addition, most of the friction interface was based on boundary lubrication in part processing. However, under boundary lubrication conditions, adding nanoparticles to vegetable oils can substantially improve their lubricating properties. Li et al. [27] added graphene nanoparticles to vegetable oil-based cutting fluids to enhance their cooling and lubricating properties. Under the conditions of graphene nanoparticle mass fraction of 0.1%, cutting fluid flow rate of 60 mL/h, and gas pressure of 0.6 MPa, the optimal milling surface roughness was 0.406 μm.

Organic additives and biodegradable vegetable base oils are the main components of eco-friendly cutting fluids. First, the polar groups of amide and boric acid organic rust inhibitors have a greater affinity for metals and can be firmly adsorbed on the metal surface to form a directional adsorption layer, and the antirust performance of organic additives is further improved by the chemical modification of diethanolamine and dodecenyl succinic acid. Then, the esterification reaction of ethylene glycol and formic acid, and the condensation reaction of guanidine hydrochloride, diethylenetriamine and other amine group-containing monomers can obtain additives with excellent bactericidal properties. Second, the dodecanoic acid diethanolamide borate obtained by the esterification condensation reaction of fatty acid and methanol, and the azeotropic principle synthetic alcohol amine borate all have good extreme pressure performance. Finally, due to the excellent physical and chemical properties of saturated fatty acids, vegetable oils have become the primary choice for cutting fluid base oils. Among them, the high fatty acid content of soybean oil has become the focus of research. In addition, nanoparticles can further improve the lubricating properties of vegetable oils.

1.4. Bio-Lubricant

1.4.1. Lubrication Mechanism

Different vegetable oils have different physical and chemical properties; therefore, they have different properties during the processing process [28]. Therefore, understanding the lubricating mechanism of vegetable oil from the viscosity, surface tension, and molecular structure of vegetable oil is the key to achieving better lubricating performance of vegetable oil.

1.4.1.1. Viscosity

Viscosity is one of the main factors affecting the cooling and lubricating properties of vegetable oil-based cutting fluids. It is a variable exchange and adhesion caused by irregular intermolecular movement and is a method to measure the flow resistance of vegetable oil. The viscosity of vegetable oil is high at room temperature, and its size is mainly affected by the composition of particles and the mutual attraction between particles. The viscosity of vegetable oil is related to the temperature, pressure, shear rate, fatty acid composition, etc. The viscosity of vegetable oil is positively correlated with the amount of monounsaturated fatty acids and negatively correlated with the content of polyunsaturated fatty acids [29]. Figure 6 shows the viscosities of several common vegetable oils at 40°C and 100°C.

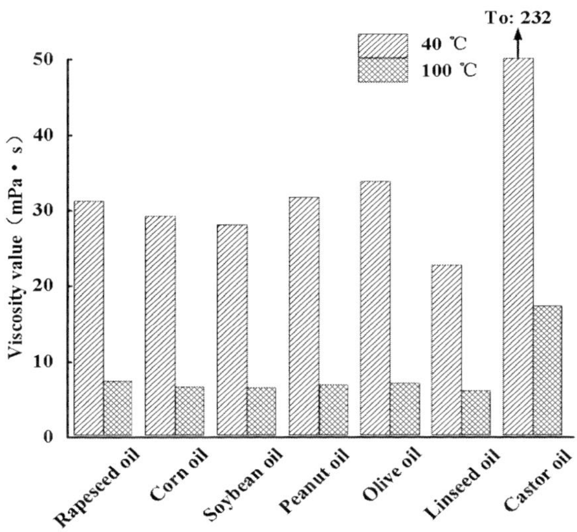

Figure 6. Viscosity of several common vegetable oils at 40°C and 100°C.

Different cutting fluids have different lubricating effects, owing to their different viscosities. Viscosity mainly affects the heat exchange and lubrication effect of the cutting fluid at the interface between the tool and chip.

In terms of heat exchange, cutting heat is an important physical phenomenon during the cutting process. In addition to 1%–2% of the energy consumed by cutting to form new surfaces and potential energy in the form of lattice distortion, 98%–99% of the energy is converted into heat energy. The influence of cutting fluid viscosity on the heat exchange between the tool and workpiece interface is also called the cooling effect or heat exchange performance of the cutting fluid [30]. In a specific processing process, the cutting fluid enters the processing area of the tool and workpiece interface at a certain angle, forms a layer of dense lubricating oil film at the tool and workpiece interface, and flows at a certain speed along the tool feed direction. Owing to the relative motion between the tool and workpiece, the fluid layer in the lubricating oil film also exhibits relative motion. Therefore, the lubricating oil film formed during machining conforms to the theory of the convective heat transfer of the flowing fluid, as shown in Figure 7. The Reynolds number Re in MQL grinding is typically much greater than 2300. Therefore, the lubricant film formed at the interface between the tool and workpiece carries out convective heat transfer in a turbulent manner, and the size of the heat transfer coefficient is closely related to the viscosity. In a lubricating oil film with convective heat transfer in the form of turbulent flow, there is a thin layer with a significant temperature change. Prandtl is known as the thermal boundary layer. The thermal boundary layer of turbulent flow can be divided into a viscous bottom layer and a turbulent branch layer. According to heat transfer theory, the temperature gradient of the thermal boundary layer is the largest in the lower viscous layer and changes gently in the turbulent branch layer. The heat-transfer capability of the cutting fluid depends on the viscosity of the lower layer. Therefore, the greater the viscosity of the cutting fluid, the thicker the viscous bottom layer, that is, the lower the temperature of the cutting fluid per unit time, that is, the lower the heat transfer coefficient.

In lubrication, the viscosity of vegetable oil affects the permeability of the lubricating oil, and the permeability of the cutting fluid further affects its heat exchange performance during machining. As a result of inertia, vegetable oil droplets enter the interface between the workpiece and tool through the nozzle and can continue to advance for a certain distance owing to inertia. Because of the different adhesions, the running distance of vegetable oil droplets with different viscosities is different. Vegetable oil with high viscosity has a short mobility and moving distance owing to its high adhesion, making it difficult to penetrate the gap between the workpiece and tool. Thus, low-viscosity vegetable oils can effectively penetrate these gaps. Therefore, there is a negative correlation between the viscosity and permeability of the cutting fluid.

Singh et al. [31] studied the grinding effect of three vegetable oils (rapeseed, soybean, and olive oils) as lubricants when grinding Ti-6Al-4 V-ELI. Among the three oils, rapeseed oil obtained the lowest grinding force (F_t=4.68 N, F_n=15.25 N), friction coefficient (0.307), and specific grinding energy (17.16 J/mm^3). The viscosity of rapeseed oil reached 38 mPa·s at 40°C, while that of soybean and olive oil was 26 mPa·s and 32 mPa·s, respectively. The results show that viscosity plays a key role in improving tribological properties.

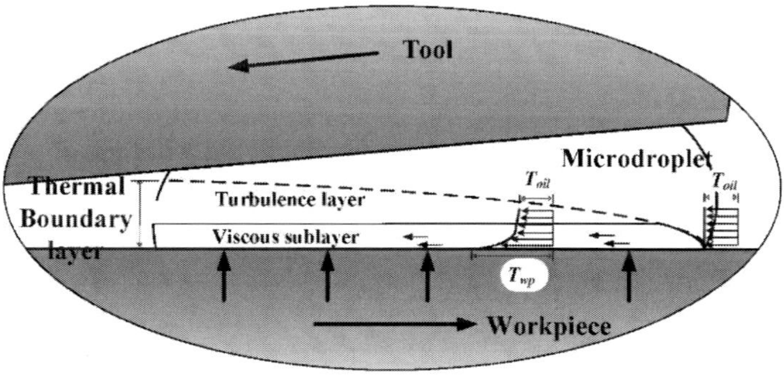

Figure 7. Convective heat transfer of fluid at the tool workpiece interface.

In addition, the viscosity of vegetable oil is not constant; it decreases with an increase in the cutting temperature [32]. The viscosity reduction followed the power-law model. Table 6 below shows the variation trend of viscosity (mPa·s) of several vegetable oils between 35–180°C with temperature. The viscosity value is obtained from the slope of fitting the experimental shear stress–shear rate data with Newton's viscosity law equation (the specific data are obtained from the fitting slope of the author's experimental shear stress–shear rate data and Newton's law viscosity equation).

Table 6. Variation trend of viscosity (mPa·s) of several vegetable oils between 35–180°C with temperature

Temperature (°C)	Viscosity (mPa·s)				
	Rapeseed oil	Corn oil	Soybean oil	Peanut oil	Olive oil
35	42.49	37.92	38.63	45.59	46.29
50	25.79	23.26	23.58	27.45	27.18
65	17.21	15.61	15.73	17.93	18.07
80	12.14	10.98	11.53	12.66	12.57
95	9.01	8.56	8.68	9.40	9.45
110	7.77	6.83	7.17	7.47	7.43
140	5.01	4.95	4.58	5.14	5.29
180	4.65	3.33	3.31	3.26	3.44

1.4.1.2. Surface Tension

In minimum quality lubrication processing, the surface tension is the key factor affecting the size of the droplets, and the size of the surface tension is positively related to the size of the droplets. A cutting fluid with low surface tension has a better lubrication effect.

On the one hand, low surface tension means that the droplets formed are small in size, large in number and large in specific surface area. A small droplet size implies that more droplets enter the gap between the tool and workpiece, and a higher specific surface area also represents a larger heat exchange area. On the other hand, according to the theory of convective heat transfer, droplets can be divided into thermal boundary layers and main flow areas during the process of thermal convection. The thickness of the thermal boundary layer remains unchanged. However, before absorbing sufficient heat, the grinding fluid in the mainstream area is rapidly discharged from the grinding area. In other words, the grinding fluid in the mainstream area will not provide a satisfactory heat exchange effect. When the contact angle decreases, the thermal boundary layer expands, and the proportion of abrasive fluid in the mainstream area decreases. This result explains why MQL droplets with small contact angles exhibit high cooling efficiencies.

Table 7. Variation trend of surface tension (mN/m) of several vegetable oils with temperatures between 20°C and 200°C

Temperature (°C)	Surface tension (mN/m)				
	Rapeseed oil	Corn oil	Palm oil	Sunflower oil	Soybean oil
20	33.8	33.8	–	34.0	33.9
40	32.0	32.2	31.5	32.3	32.2
60	30.5	30.6	30.0	30.7	30.6
80	29.0	29.1	28.5	29.2	29.1
100	27.5	27.6	27.1	27.7	27.6
120	26.0	26.1	25.7	26.3	26.1
140	24.5	24.6	24.4	24.8	24.7
160	23.1	23.2	23.0	23.0	23.2
180	21.7	21.8	21.6	21.6	21.8
200	20.3	20.4	–	20.2	20.4

Different types of vegetable oils have significant effects on the viscosity of vegetable oils but do not affect the surface tension. The surface tension increases with increasing vegetable oil density and decreases linearly with increasing temperature. Table 7 shows the variation trend of the surface tension of several vegetable oils at temperatures between 20°C and 200°C (The specific data are multiplied by the surface tension value obtained by the author based on the drop weight method by a certain correction factor ψ.) [33].

1.4.1.3. Molecular Structure

In the process of vegetable oil used for processing lubrication, vegetable oil can form a layer of lubricating oil film on the surface of the workpiece, which can be divided into physical film formation and chemical film formation through different mechanisms of lubricating oil film formation. Physical film formation: During metal processing, owing to the strong extrusion, the vegetable oil on the workpiece surface changes from the original liquid lubrication state to the boundary lubrication state. Polar atoms contained in vegetable oils, such as S, O, N, and P, and polar groups, such as ester bonds (-COOR), carboxyl groups (-COOH), and OH, have strong metal affinity. They can be physically adsorbed on the material surface of a metal workpiece through van der Waals forces forming a layer of molecular film, which has antiwear and antifriction effects, thereby improving the lubrication of the cutting area [34]. Due to the role of polar groups in vegetable oil, it is easier to form a high-strength lubricating oil film, which has stronger antiwear, antifriction, cooling, and lubrication effects in the processing process. Chemical film forming: Under the action of high temperature (100°C) during processing, polar molecules in vegetable oil easily undergo a "metal saponification reaction" with the metal surface. During the reaction, the H atom of the carboxyl group (-COOH) in the fatty acid reacts with the metal surface atom to form a single layer of a semi-chemically bonded oily lubricating film. An oily lubricating film can be composed of a single molecular layer or a multimolecular layer. It is adsorbed on the surface of the workpiece material as a boundary film, thus prolonging tool life [35].

Owing to the different types of fatty acids, the saturation degree and carbon chain length of various fatty acids are different. Fatty acids with different carbon chain lengths and fatty acid saturations have different oil film formation intensities.

Regarding the relationship between the carbon chain length and lubricating performance, the cohesion between molecules is proportional to the number of carbon atoms, and the number of carbon atoms in the carbon chain has an important influence on the molecular cohesion. Therefore, the strength and lubrication performance of lubricating oil films formed using fatty acids with longer carbon chains are stronger than those formed using fatty acids with shorter carbon chains. However, for saturated fatty acids, when the carbon number is greater than 16, the friction resistance and wear resistance of the formed lubricating oil film reach their peak and remain unchanged. In this case, the lubrication effect does not change with an increase in the number of carbon atoms in the saturated fatty acids. For unsaturated fatty acids, owing to the existence of polar unsaturated bonds, the density of the adsorption film decreases under the adsorption of the olefin bond, thereby reducing the strength and lubrication performance of the lubricating oil film.

Regarding the relationship between different saturations and lubricating performance, the influence of different fatty acid saturations on lubricating performance can be attributed to two aspects. First, the presence of C=C makes unsaturated fatty acids easily oxidize, which leads to the degradation of vegetable oil, thereby reducing the oxidation resistance and thermal stability of vegetable oil and finally leading to the failure of the formed physically adsorbed oil film. Moreover, the higher the C=C ratio is, the worse the oxidation resistance and thermal stability, thus reducing the lubricating performance of vegetable oil. On the other hand, the attraction between fatty acid molecules is affected by the shape of the fatty acid molecules. Saturated fatty acid molecules do not have C=C, and the molecular chains are mostly straight; therefore, intermolecular binding is high. However, the position where C=C exists in unsaturated fatty acids leads to bending of carbon chains, reducing the degree of intermolecular binding. In addition, because of the existence of olefin bonds in fatty acid molecules, the olefin bonds exhibit extremely strong molecular adsorption, leading to a reduction in the density of the physical adsorption film, thus reducing the strength of the lubricating oil film. Figure 8 shows the three-dimensional model of saturated and unsaturated fatty acid molecules.

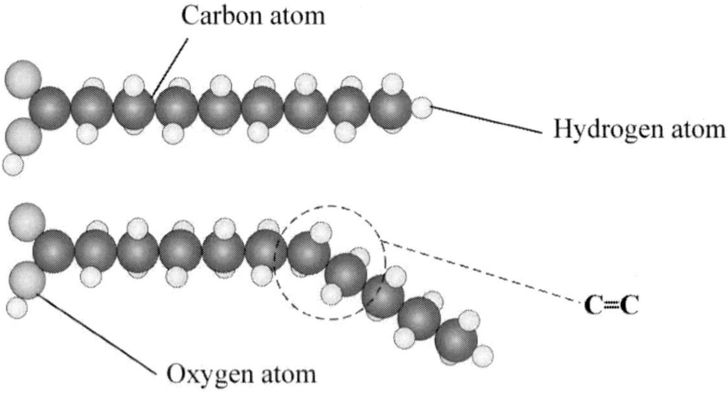

Figure 8. Three-dimensional model of saturated and unsaturated fatty acids.

Therefore, for the same carbon chain length, the lubricating oil film formed by saturated fatty acids during processing is stronger and has better lubrication performance than that formed by unsaturated fatty acids during processing. Simultaneously, the workpiece surface roughness after processing is lower. At the same time, under the same carbon chain length, the lubricating oil film formed by monounsaturated fatty acids has better strength and lubrication performance than that formed by polyunsaturated fatty acids and can obtain a lower surface roughness.

1.4.2. Limitations

Vegetable oils exhibit satisfactory physicochemical properties as lubricants. However, the lack of antioxidant properties remains an unavoidable defect in vegetable oils.

1.4.2.1. Autoxidation

Although the specific triglyceride structure of vegetable oil determines that it has better lubricating performance than mineral oil, the structure of triglycerides also causes vegetable oil to have poorer antioxidant performance than mineral oil. Poor oxidation and thermal stability are inevitable defects in vegetable oil-based lubricating oils [36]. The higher the C=C in fatty acids, the higher the degree of unsaturation, and the more easily vegetable oil is oxidized [37]. For example, castor oil contains a large amount of ricinic acid (monounsaturated fatty acid); therefore, it has better oxidation stability than vegetable oil with a high content of polyunsaturated fatty acids. Similarly, rapeseed oil (RO) contains approximately 64.42% monounsaturated fatty acids, so it has relatively high oxidation stability. A study on the tribological properties of oxidized rapeseed oil showed that oxidation adversely affects its antifriction performance [38]. As shown in Figure 9.

Oxidation is the main reaction of vegetable oil when it is used as a lubricant base oil. Oxidation causes the content of unsaturated fatty acids in vegetable oil to decrease, and the contents of foam, color, viscosity, density, specific heat, free fatty acids, polar substances, and polymer compounds in vegetable oil increase with time [39]. The trend of this change is shown in Figure 10.

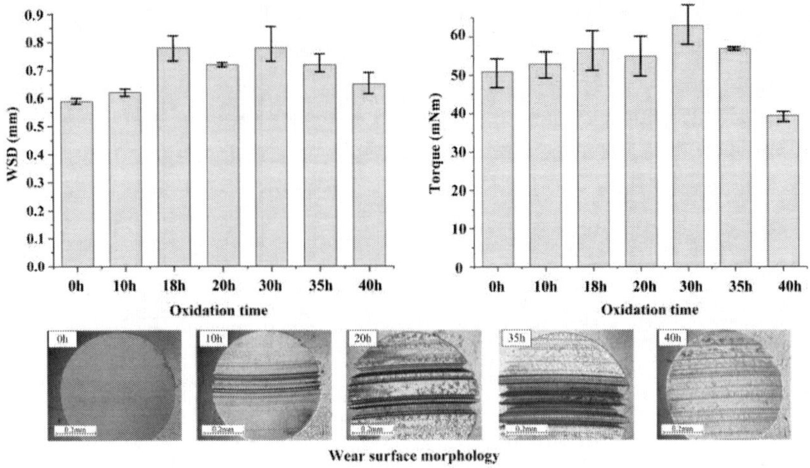

Figure 9. Wear and average torque measurement of nonoxidized and oxidized rapeseed oil.

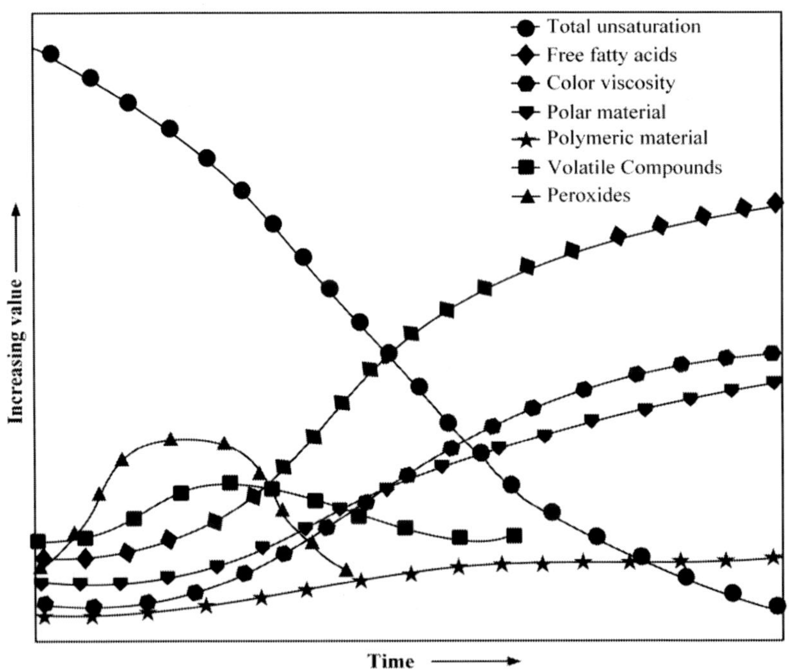

Figure 10. Change trend of the content of various substances under 1.38 kg in the oxidation process of vegetable oil.

Because a large amount of C=C in unsaturated fatty acids can be used as the active site of the oxidation reaction, vegetable oil has very low stability. The higher the degree of unsaturation, that is, the more C=C, the more sensitive the oil is to oxidation. Low oxidation stability means that, if untreated, the oil will quickly oxidize, thicken, and polymerize during use. Adverse effects on the lubrication performance. For eighteen carbon triglycerides (C-18), the relative oxidation instability of saturated, mono-, di-, and tri-unsaturated triglycerides is approximately 1:10:100:200.

There are four types of vegetable oil oxidation: automatic oxidation of vegetable oil, enzymatic oxidation, photooxidation, and metal oxidation. Among them, the automatic oxidation of vegetable oil is an oxidation reaction without any external influence and is the main method of vegetable oil oxidation.

The oxidation of vegetable oils involves three stages: initiation, propagation, and termination, which are complex and involve both continuous and discontinuous reactions. In vegetable oil molecules, the hydrogen atoms in methylene are more likely to detach due to the lower C-H bond strength in methylene (-CH2-) next to C=C. The initial oxidation phase of vegetable oils involves the detachment of hydrogen atoms from the methylene next to C=C, leading to the formation of alkyl radicals (R·). These alkyl radicals react with oxygen to form alkyl peroxide radicals (ROO·). Primary antioxidants can neutralize these newly formed oxidation products, but if they are not used or cannot completely neutralize the free radicals, further oxidation stages occur. During the propagation phase of oxidation, alkane peroxide radicals attack the C-H bond in the remaining fatty acid molecules to obtain hydrogen atoms and new free radicals, leading to the formation of primary oxidation compound hydroperoxide (ROOH). This process generates a series of new oxidation reactions. Hydrogen peroxide is formed in later stages, which is the rate-limiting step in the autooxidation of vegetable oils. The strength of the broken carbon–hydrogen bonds determine the rate constant of the rate-limiting step. The decomposition of hydroperoxide leads to the formation of more free radicals. The newly formed alkoxy radicals (RO) can repeatedly attack the C-H bond of fatty acids, generate more hydroperoxide, and accelerate the oxidation process. The type and content of free radicals generated during the oxidation of vegetable oils depend on the composition of fatty acids, and the growth rates of free radicals vary significantly. However, not all free radicals accelerate the oxidation process. Some free radicals react with each other to terminate the oxidation process. Hydroperoxide accumulates continuously during the oxidation process and decomposes into countless volatile and nonvolatile secondary oxides when superimposed to a certain extent. Volatile organic compounds are mostly chain-broken hydrocarbons and alcohols and have little effect on the wear resistance of vegetable oils because of their low carbon chain length and low polarity. Among nonvolatile organic compounds, epoxy compounds exist longer than other nonvolatile organic compounds due to their higher oxidation stability and lower tendency to precipitate. Epoxidation has been proven to be an effective chemical modification method that has beneficial effects on the oxidation stability and viscosity of vegetable oils.

Certain secondary oxides can also participate in the polymerization reaction to produce macromolecular weight compounds, which eventually form precipitates on the lubricating surface of the workpiece and significantly improve the viscosity of vegetable oils. This represents the final stage of the vegetable oil oxidation process. The self-oxidation mechanisms of vegetable oils are summarized in Table 8.

In the early stages of vegetable oil oxidation, hydroperoxide is the main oxidation compound of vegetable oil. The presence of hydroperoxide inhibits the antiwear property of the vegetable oil cutting fluid. With the oxidation process, the viscosity and surface tension of oxidized vegetable oil also increase with an increase in hydrogen peroxide content. Excessive viscosity and surface tension lead to poor flow performance of vegetable oil, which also reduces its lubricating performance.

Table 8. Autoxidation mechanism of vegetable oil

Initiation	$RH \rightarrow R^{\cdot} + H^{\cdot}$
Propagation	$R^{\cdot} + O_2 \rightarrow ROO^{\cdot}$ $ROO^{\cdot} + RH \rightarrow ROOH + R^{\cdot}$
Branching	$ROOH \rightarrow RO^{\cdot} + {}^{\cdot}OH$ $RO^{\cdot} + RH + O_2 \rightarrow ROH + ROO^{\cdot}$ ${}^{\cdot}OH + RH + O_2 \rightarrow H_2O + ROO^{\cdot}$
Termination	$ROO^{\cdot} + ROO^{\cdot} \rightarrow ROOH + O_2$ $ROO^{\cdot} + R^{\cdot} \rightarrow ROOH$ $R^{\cdot} + R^{\cdot} \rightarrow R\text{-}R$
Peroxide decomposition	$ROOH \rightarrow$ various lower molecular weight compounds
Polymerization	$ROOH \rightarrow$ various higher molecular weight compounds

1.4.2.2. High-Temperature Oxidation

The thermal stability of lubricating oil is an important parameter for its application at high temperatures, and it also affects the tribological properties of vegetable oils. The flash point is a parameter that characterizes thermal stability. When the flash point is lower than the grinding temperature, the vegetable oil is invalid. Additionally, the flash point is also an important parameter for evaluating the fire risk of flammable liquids. Combustible liquids (vegetable oils) may encounter hot surfaces (workpiece–tool interfaces), which may cause fire and explosion accidents. The flash points and autoignition temperatures of several common vegetable oils are listed in Table 9.

Table 9. Flash point and autoignition temperature of several common vegetable oils

Vegetable oil	Flash point	Ignition point	Autoignition temperature
Soybean oil	316°C	352°C	402°C
Tung oil	330°C	355°C	410°C
Sunflower oil	320°C	345°C	401°C
Rice oil	322°C	346°C	398°C
Peanut oil	327°C	351°C	406°C

The high temperature has sufficient energy to destroy the C-C or C-H covalent bonds in the acyl skeleton and form various free radicals. For example, alkoxy radicals are mainly found at 90°C, and alkyl radicals are mainly found at 120°C. A free radical oxidation chain reaction occurs, which increases the oxidation rate of vegetable oils.

During high-temperature oxidation, the content of ROOH, the primary oxidation product of vegetable oil, increases with increasing reaction temperature. The peroxide value of vegetable oil at 60°C is approximately twice that at room temperature. However, when the temperature is higher than 150°C, the primary oxidation product ROOH decomposes to form volatile small molecules, the decomposition rate is mostly faster than the generation rate, and the content of ROOH gradually decreases. In addition, high temperatures lead to the oxidation and polymerization of vegetable oils. The thermal decomposition of vegetable oil can be divided into three steps, which are related to the decomposition of polyunsaturated, monounsaturated, and saturated fatty acids. In the first stage of the thermal decomposition of vegetable oil, the polyunsaturated fatty acids in vegetable oil begin to decompose. In the heating process, polyunsaturated fatty acids (such as linoleic acid) in vegetable oil react with heat to generate volatile compounds such as dimers, trimers, and polymers, which are continuously removed by the steam generated during the heating process. The second step of the thermal decomposition of vegetable oil corresponds to the decomposition of monounsaturated fatty acids such as oleic acid. During the reaction, the double bond of unsaturated fatty acids in vegetable oil breaks, leading to the saturation of triglyceride molecules. The third step of thermal decomposition corresponds to the thermal decomposition of saturated fatty acids such as palmitic acid.

The decomposition of fatty acid carbon chains caused by high temperatures is an important reason for the decline in the properties of vegetable oil. Kim et al. [40] studied the stability and polymerization of biodiesel through high-temperature accelerated oxidation. During the accelerated oxidation at 95°C, the unsaturated fatty acid methyl esters in biodiesel decreased to varying degrees over time, which indicated that the oxidation led to the structural changes in unsaturated fatty acid methyl esters. In the process of high-temperature accelerated oxidation at 180°C, oxidized unsaturated fatty acid methyl ester and unoxidized unsaturated fatty acid methyl ester are polymerized through the Diels–Alder reaction, and unsaturated fatty acid methyl ester forms a dimer and oxidized dimer and other polymerization products through the Diels–Alder reaction.

1.4.2.3. Extreme Pressure Performance

Pure vegetable oil is superior to mineral oil in terms of antiwear property, friction property, bonding load capacity, and fatigue resistance under low load, but the effectiveness of vegetable oil-based cutting fluid is significantly reduced under extreme loads.

1.4.3. Summary

Scholars from around the world have proposed different modification schemes to improve the performance of vegetable oils. Chemical modification and additive modification are two methods used to enhance the antioxidant, extreme pressure, and antiwear properties of vegetable oil. The author has evaluated the effectiveness of these methods in improving vegetable oil performance.

1.5. Nanolubricant

The performance of nanolubricants during machining is influenced by the physical properties of the nano-enhanced phase, the type of base fluid, the preparation method, and the dispersion stabilization [41]. This section briefly reviews the commonly used base fluids and nano-enhanced phases for MQL, reveals the mechanism of action of the molecular structure cooling and lubricating properties of vegetable oils and nano-enhanced phases, and examines the commonly used nanolubricants preparation methods. The main challenge of nanolubricants lies in their poor stability. This section concludes by describing the kinetic behavior and dispersion-stabilizing mechanism of nano-enhanced phases in nanolubricants.

1.5.1. Nano-Enhanced Phases

Biolubricants have exceptional lubricating film-forming properties, but the grinding process for difficult-to-machine materials can easily lead to lubricating film failure, and the performance of biolubricants can be substantially limited [42]. Thus, the nano-enhanced phase is an effective way to solve the above difficulties, and the incorporation of this phase considerably improves the extreme-pressure performance of pure vegetable oils. Therefore, the synergistic effect of the two can be effectively utilized. The surface effect of nano-enhanced phases is shown in Figure 11.

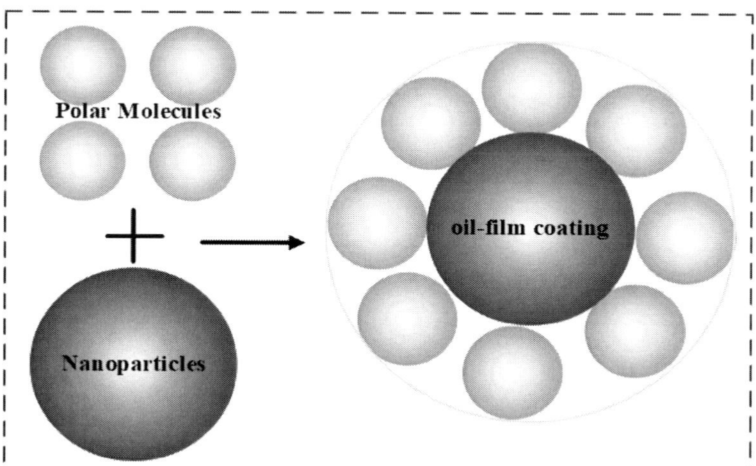

Figure 11. Schematic of nano-enhanced phases surface effect.

In addition, nano-enhanced phases can reduce the degree of friction in the cutting zone, uniformly withstand compressive forces, improve heat dissipation, and reduce stress concentrations [43]. Yildirim et al. [44] increased tool life by 105.9% compared with dry cutting after machining nickel-based high-temperature alloys using hexagonal boron nitride (HBN) nanolubricant with a mass fraction of 0.5 wt.%. The effect of nanolubricants is influenced by the type, size, shape, and concentration of the nano-enhanced phases. Scholars have conducted numerous experimental studies on the above variables.

1.5.1.1. Shapes

Minor differences in the roles of different shapes of nano-enhanced phases are observed. Table 10 shows the commonly used nano-enhanced phases. Stratiform structure nano-enhanced phases produce laminar exfoliation during friction, extrusion, and shear. For stratiform nano-enhanced phases, Tevet et al. [45] conducted tribological characterization of nano-enhanced phases with different structures and found their rolling, sliding, and exfoliation behaviors under boundary lubrication conditions. This characterization can improve the machined surface quality and reduce cutting forces and temperatures. HBN has weak van der Waals forces between the molecular layers due to its structure, which reduces friction by sliding through these molecular layers against each other during machining. In addition, GNPs have the advantages of high thermal conductivity and strength, making them the most promising material for cooling and lubrication. However, GNP is restricted by its high price. For spherical nano-enhanced phases, Musavi et al. [46] found that spherical copper oxide nanolubricant in A286 high-temperature alloy turning machining has superior lubrication performance, which resulted in improved surface quality because the spherical nano-enhanced phases remained in a steady state during the process. Therefore, the friction and wear behavior of the tool–workpiece–chip interface could be improved by increasing the sphericity of the nano-enhanced phases. In addition, nano-enhanced phases provide rolling and surface repair and increase lubricant viscosity. Jamil et al. [47] investigated TiO_2 nanoparticle-reinforced coolant and found that the near-spherical nano-enhanced phase reduced rolling resistance. Therefore, spherical nano-enhanced phases are suitable for tribological and lubrication applications and are highly desirable for mechanical manufacturing processes. Threadiness nano-enhanced phases are also widely used in nanolubricants due to their excellent thermal conductivity, which can markedly reduce the temperature during processing. However, the unavoidable agglomeration of nanolubricants during their use reduces their tribological properties in the machining region.

Table 10. Commonly used nano-enhanced phases for machining (Notes. CBN, cubic boron nitride; ND, diamond)

Categories	Nano-enhanced phases
Spherical nano-enhanced phases	Al_2O_3, SiO_2, ZrO_2, SiC, ND
Threadiness nano-enhanced phases	CuO_2, MnO_2, TiO_2, ZnO, CNTs
Stratiform nano-enhanced phases	CBN, GNP, MoO_3, HBN, MoS_2

1.5.1.2. Size

Nanolubricants are colloids of particles that are at least less than 100 nm in dimension with a base solvent, and the nano-enhanced phases offer the following advantages over conventional oil emulsion suspensions used for heat dissipation. i) High specific surface area yields effective heat transfer performance. ii) Brownian motion in the nanolubricant allows for a more stable dispersion of the nano-enhanced phases with less clogging compared with conventional fluid particles. iii) Wettability, concentration, and thermal conductivity can be adjusted by varying the concentrations of nanolubricants. Small nano-enhanced phases have been previously proposed to provide a smooth surface under nanolubricant trace lubrication conditions. However, some scholars have discovered experimental phenomena that contradict these conclusions. Dubey et al. [48] used two different particle sizes in the formulation of nanolubricants for turning and machining AISI 304 steel. Their experimental results showed that the experimental values of surface roughness obtained for the 40 nm diameter were lower for the 40- and 30-nm diameter Al_2O_3 nanolubricant conditions.

Therefore, the mechanism of action of different sizes of nano-enhanced phases at the friction interface is also different, and the scale of the roughness peak at the friction interface should be considered. In the case of a large cutting dosage, the large size of the nano-enhanced phases should be first considered when reducing the machining process parameters, such as the cutting force. Meanwhile, in the case of finishing, the use of small nano-enhanced phases is recommended when ensuring the surface integrity of the parts and tool life. Simultaneously, combining different sizes of nano-enhanced phases with base oils of corresponding viscosities can improve the machining quality.

1.5.1.3. Concentration

Nano-enhanced phases lead to superior results in vegetable oil biolubricants; however, the gain effect is also affected by the concentration. Talib et al. [49] found that HBN particles provided a lubricating film in Figure 12, which reduced friction and prevented wear. However, the remaining particles in the rough valleys restricted the movement of nearby particles as the HBN concentration increased, resulting in large forces that led to frictional wear. Additional damage areas are formed when 0.1 and 0.5 wt.% HBN particles slide along the contact surface, which leads to abrasive wear. Thus, 0.05 wt.% HBN particles are the optimal concentration. Agglomeration occurs when the concentration of nanolubricant exceeds certain conditions, subsequently affecting the stability of the nanolubricant and leading to deterioration of the processing quality.

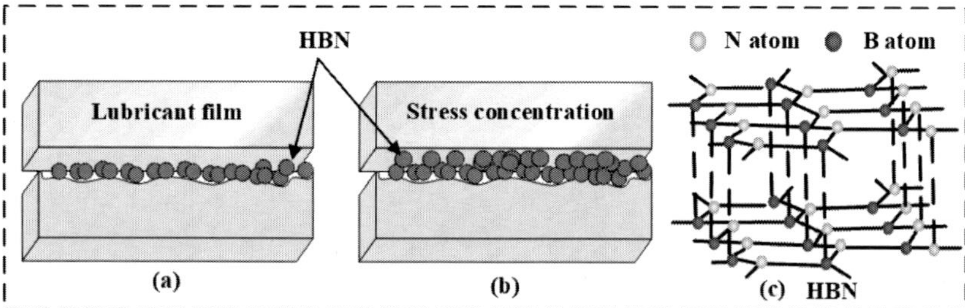

Figure 12. (a) Lubricating film, (b) Frictional wear, (c) molecular structure of the hexagonal boron nitride (HBN).

Pal et al. [50] prepared nanolubricant with different concentrations (0.5, 1.0, and 1.5 wt.%) by mixing Al_2O_3 nano-enhanced phases with sunflower oil. The results showed that compared with the flood condition, the optimal concentration of 1.5 wt.% reduced thrust, torque, surface roughness, and drill tip temperature by 44%, 67%, 56%, and 26%, respectively, at the 30th hole. In addition, different concentrations lead to viscosity variations of the nanolubricant. The thin lubrication film formed by low viscosity cannot withstand the stress from the extra loading of particles, which increases friction and damages the sliding surfaces. Sen et al. [51] found that the maximum viscosity was reached at 3% Al_2O_3 concentration with superior lubricating properties. However, the nano-enhanced phases can agglomerate and affect the thermal conductivity of the Al_2O_3 nano-enhanced phases in palm oil. The concentration of the nano-enhanced phases has a direct relationship with the contact angle, which decreases sharply with increasing nanolubricant concentration when the wetting area increases. However, the contact angle also increases further when a certain concentration is exceeded because the excess nano-enhanced phases aggregate and then deposit after losing dynamic stability.

The results of numerous studies have shown that optimal concentration values exist and produce optimal effects with the application of different nano-enhanced phase processes. Modulation of the nanolubricant performance can be achieved by optimizing the concentration of the nano-enhanced phases. Overall, the shape, size, and concentration of the nano-enhanced phases will affect the machining performance. The good or poor machining performance is the result of their joint action, and no complete parameter is available for the nano-enhanced phases. Therefore, future exploration should further optimize the parameters of the nano-enhanced phases to provide a complete parameter for the processing of different workpiece materials.

1.5.2. Preparation

The preparation of the nanolubricant is the most important step in the experimental study, that is, "combining" the base fluid with the nano-enhanced phases. The preparation stage is crucial for enhancing the thermophysical properties of nanolubricants. Therefore, fine preparation is required to avoid chemical changes, agglomeration, and the inhomogeneous suspension of nano-enhanced phases. The two main methods of synthesizing nanolubricants are as follows: one-step and two-step.

Physical (e.g., vapor deposition, laser ablation, and ultrasound-assisted submerged arc) and chemical (via reducing agents) methods are commonly used in a one-step process. Simultaneous preparation and dispersion of nano-enhanced phases in the base fluid prevent oxidation of the nano-enhanced phases, disperse them homogeneously, enhance stability, and avoid storage, drying, and nanolubricant transportation in the process, which reduces agglomeration and thus improves stability.

The two-step preparation involves the preparation and dispersion of nano-enhanced phases into the base fluid. First, nano-enhanced phases are prepared via vapor deposition, chemical reduction, or mechanical milling and then dispersed into the base liquid via the mechanical dispersion method, ultrasonication, and addition of dispersants. The mechanical dispersion method distributes nano-enhanced phases in the liquid medium using external mechanical energy such as shear or collision force, ball milling, mechanical stirring, and other methods. The ultrasonic dispersion method places the nanolubricant directly in the ultrasonic field and uses ultrasonic waves of suitable frequency to facilitate the movement of the nano-enhanced phases. This approach overcomes the mutual attraction between the nano-enhanced phases and destroys the equilibrium between the original particles as well as between the particles and molecules of the substrate, dispersing the nano-enhanced phases into the substrate liquid. Mechanical dispersion and sonication disrupt the original kinetic behavior of the nano-enhanced phases (van der Waals and Brownian forces) and reduce the interactions between the nano-enhanced phases. The addition of the dispersant increases the repulsive force between the nano-enhanced phases; a greater repulsive force between the particles than the attractive force prevents the agglomeration between the nano-enhanced phases to improve stability. Figure 13 shows the two-step method used to prepare nanolubricants.

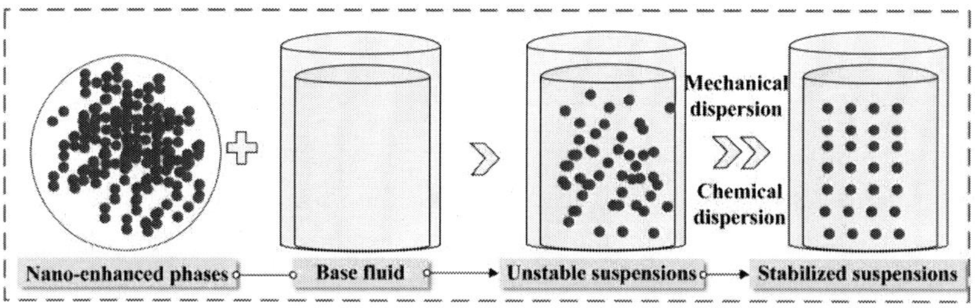

Figure 13. Two-step preparation.

Overall, the one- and two-step methods have advantages and disadvantages. The one-step method has the advantage of good dispersion and high suspension stability but suffers from the limitations of high cost and difficulty in controlling the size of nano-enhanced phases. Thus, this method is only applicable to low-pressure base fluids. The two-step method is the most cost-effective method for obtaining superior particle size control and is suitable for mass production. However, these steps are highly complex and require additional time and experimental equipment. In specific applications, choosing the appropriate preparation method according to different experimental requirements and the use of nanolubricant conditions is necessary. Thus, a two-step method should be used in the field of machining to prepare nanolubricants.

1.5.3. Stability

Nanolubricants are complex mixtures of nano-enhanced phases and biolubricants. They tend to aggregate due to their high surface activity, and the lack of suspension stability leads to increased cluster formation and deposition rates. Agglomeration causes the nano-enhanced phases to settle and clog the nozzles, thus reducing the thermophysical properties of the nanolubricant. The stability of nanolubricant depends on the properties of the nano-enhanced phases, the type of base fluid, the preparation method, and the subsequent treatment. Therefore, the stability of nanolubricants should be carefully considered.

1.5.3.1. Destabilizing Factors

Microscopically, nanolubricants belong to the liquid–solid two-phase system. Thus, a distinct phase interface exists between the liquid molecules and the nano-enhanced phases, and a buoyant force of the base liquid is observed on the particles. Brownian motion of nano-enhanced phases refers to the irregular way in which microscopic particles move in liquids or gases due to the thermal motion of molecules. The nano-enhanced phases are small and light; thus, their Brownian motion is fast, enabling their distribution over a large area in a short time. The nano-enhanced phases in the base fluid are in irregular motion and are thus subjected to Brownian forces. Van der Waals force potential exists for mutual attraction. Numerous atoms between neighboring particles are attracted to each other due to their large percentages on the surface of nano-enhanced phases; therefore, van der Waals forces emerge between particles. This force varies according to the distance between two particles; it usually increases as the distance decreases and is directly related to the geometry, size, and surface properties of the particles. Van der Waals gravity and electric double-layer repulsion depend on the size of the nano-enhanced phases. Numerous atoms are present on the particle surface when the particle size is small, which increases the density of active sites. Agglomeration occurs when van der Waals attraction exceeds electric double-layer repulsion. Several researchers have reported that nanolubricants are less stable at high particle concentrations. The increase in particle concentration results in large particle cluster sizes, which is due to low interparticle distances and increased van der Waals gravity, directly affecting the settling velocity. In addition, the nano-enhanced phases have gravity. Under the joint action of these forces, the metal particles are prone to collision, agglomeration, and settlement. Therefore, nanolubricants should reduce the frequency and efficiency of collision to minimize the impact of agglomerates and maintain good stability.

1.5.3.2. Stability Strategies

Nanolubricants are usually treated using physical or chemical methods to achieve stability. Physical methods aim to disperse the nano-enhanced phases in the liquid medium using external mechanical energy, such as shear or collision force, which destroys the original kinetic behavior of the nano-enhanced phases and reduces the interaction between nano-enhanced phases. Chemical methods change the surface properties of suspended nano-enhanced phases and inhibit the formation of particle clusters to obtain a stable suspension.

Commonly used physical stabilization methods include mechanical mixing and ultrasonic treatment. The device is shown in Figure 14. Mechanical stirrers are important tools for promoting nanolubricant stability. Ultrasonication is one of the methods used to improve the stability of nanolubricants and shows better stability than magnetic stirring. Ultrasonication places the nanolubricant directly into the ultrasonic field and uses ultrasonic waves of suitable frequency to facilitate the movement of nano-enhanced phases. Therefore, this method overcomes the mutual attraction between the nano-enhanced phases and destroys the balance between the original particles as well as between the particles and molecules of the base liquid, dispersing the nano-enhanced phases into the base liquid. Elsheikh et al. [52] added CuO nano-enhanced phases to rice bran vegetable oil at a volume concentration of 1vol.% and stirred the mixture under a magnetic stirrer for 4 min. The prepared solution was then sonicated for 2h using an ultrasonic generator to ensure uniform dispersion. Therefore, the results reveal the optimal ultrasonic treatment time for each nanolubricant. However, no accurate range of treatment times exists for enhanced coolants of different nano-enhanced phases to obtain nanolubricants with the desired optimum properties. Multiple physical methods can be used to synergize the treatment to disperse the nanolubricant and obtain improved processing properties.

The influence of the nature of the nano-enhanced phases (hydrophilic or hydrophobic) should be considered during the selection of the base fluid, which is either polar or nonpolar. Hydrophilic nano-enhanced phases such as metal oxide particles (CuO, SiO_2, and Al_2O_3) are readily dispersed in polar solvents such as water, whereas hydrophobic nano-enhanced phases (CNT, GNP) are readily dispersed in nonpolar base fluids. In these cases, the use of additional stabilizers is no longer necessary because the stability can be sufficiently improved using physical methods. However, different chemical stabilization techniques are required to form stable nanolubricants by mixing hydrophilic nano-enhanced phases in nonpolar solvents or hydrophobic nano-enhanced phases in polar solvents. Figure 14 shows the chemical stabilization methods, which include electrostatic stabilization (surfactants and covalent surface functional groups) and spatial stabilization (ionic polymers or ionic liquids).

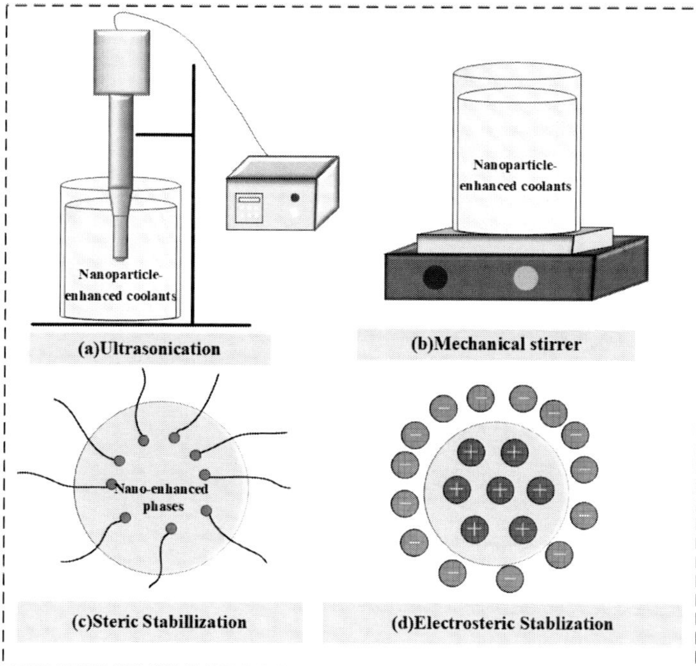

Figure 14. Physical and chemical stabilization of nanoparticle-enhanced coolant device. (a) ultrasonication, (b) mechanical stirrer, (c) steric stabilization, and (d) electrosteric stabilization.

The electrostatic stabilization technique is mainly based on providing a surface charge to the particles. Ion adsorption creates a double electric layer around the nano-enhanced phases, and electrostatic stabilization is achieved when electrostatic repulsive forces play a dominant role and balance the van der Waals attraction between two similarly charged particles. Surfactants with different charges are added to the nanolubricant suspension to improve electrostatic stability. Table 11 shows common surfactants, where the amphoteric surfactant includes cationic and anionic hydrophilic groups that can form cations and anions based on the pH of the medium.

Table 11. Surfactants commonly used in machining

Types	Surfactants
Nonionic	Alkylphenol ethoxylates, octyl phenol ethoxylate, polyvinylpyrrolidone
Cationic	Sodium dodecyl sulfate (SDS), sodium dodecylbenzene sulfonate (SDBS), sodium oleate
Anionic	Gum arabica (GA), dodecyl trimethyl ammonium bromide, Cetyltrimethylammonium bromide (CTAB), cetyltrimethylammonium chloride
Hermaphroditic	Lecithin, hydroxy sulfobetaine

However, the exact type of surfactant to be added has not been specifically demonstrated, while most surfactants (SDBS, SDS, GA, CTAB, and polyvinylpyrrolidone) are organic in nature and have some biodegradability issues. These surfactants may deteriorate over time due to biological activity and operating conditions. Therefore, the stability of surfactants may be a major challenge when considering long-term applications. Spatial stabilization of nanolubricants can be achieved using nonionic surfactants and polymers. This stabilization technique involves noncovalent functionalization. Spatial stabilization can stabilize nanolubricants with high particle concentrations, but electrostatic stabilization cannot. Furthermore, this stabilizing effect cannot be predicted using the ζ-potential method because the polymer does not affect the charge of the nano-enhanced phases. Currently existing limitations include the sensitivity of spatial stability to temperature changes, stabilizer concentrations, and absorption spectra.

Overall, the plant-based oil, molecular movement, and external factors of nanolubricants affect their stability. The selection and preparation methods for nanolubricant should consider their application properties and stability. A broad study of hybrid stabilization methods, that is, a combination of different physical and chemical methods, is needed to improve the stability of nanolubricants under different operating conditions. The above methods of stability enhancement can help improve the availability of nanolubricants in additional applications.

Conclusion

Organic additives and biodegradable vegetable base oils are the main components of eco-friendly cutting fluids. First, the polar groups of amide and boric acid organic rust inhibitors have a greater affinity for metals, and can be firmly adsorbed on the metal surface to form a directional adsorption layer. Through chemical modification of diethanolamine and dodecenyl succinic acid, the antirust performance of organic additives is further improved. Then, the esterification reaction of ethylene glycol and formic acid, and the condensation reaction of guanidine hydrochloride, diethylenetriamine and other amine group-containing monomers can obtain additives with excellent bactericidal properties. Second, the dodecanoic acid diethanolamide borate obtained by the esterification condensation reaction of fatty acid and methanol, and the azeotropic principle synthetic alcohol amine borate all have good extreme pressure performance. Finally, due to the excellent physical and chemical properties of saturated fatty acids, vegetable oils have become the primary choice for cutting fluid base oils. Among them, the high fatty acid content of soybean oil has become the focus of research. In addition, nanoparticles can further improve the lubricating properties of vegetable oils.

In the field of mechanical processing, the replacement of mineral oil-based lubricants with vegetable oils is a growing trend. However, the physical and chemical properties of different vegetable oils can have varying effects. Here are some key points to consider: (1) High viscosity vegetable oils have stronger heat exchange performance, reducing friction at the contact interface and lowering temperatures to improve lubrication. However, high viscosity can also mean higher adhesion, which reduces fluidity and permeability. (2) Vegetable oils with long carbon chains and low unsaturated fatty acid structures can create strong lubricating oil films that minimize wear and improve surface finish. (3) The presence of C=C is a major factor leading to the oxidation of vegetable oils. This increases the content of free fatty acids, polar substances, and polymer compounds, which weakens the lubricating oil film and antiwear performance.

Vegetable oils with high viscosity and high percentages of monounsaturated fatty acids, such as rapeseed oil and olive oil, are popular choices for vegetable oil-based lubricants. However, all vegetable oils lack antioxidant performance, which is an unavoidable flaw.

References

[1] Pusavec F, Kramar D, Krajnik P, et al. (2010). Transitioning to sustainable production - part II: evaluation of sustainable machining technologies. *J. Clean Prod.*, *18*(12), 1211-1221. 10.1016/j.jclepro.2010.01.015.

[2] Wang XM, Song YX, Li CH, et al. (2023). Nanofluids application in machining: a comprehensive review. *Int. J. Adv. Manuf. Technol.*, 52. 10.1007/s00170-022-10767-2.

[3] Jia DZ, Li CH, Zhang YB, et al. (2017). Specific energy and surface roughness of minimum quantity lubrication grinding Ni-based alloy with mixed vegetable oil-based nanofluids. *Precis. Eng.-J. Int. Soc. Precis. Eng. Nanotechnol.*, *50*, 248-262. 10.1016/j.precisioneng.2017.05.012.

[4] Xu WH, Li CH, Zhang YB, et al. (2022). Electrostatic atomization minimum quantity lubrication machining: from mechanism to application. *Int. J. Extreme Manuf.*, *4*(4), 43. 10.1088/2631-7990/ac9652.

[5] Agarwal S, Venkateswara Rao P. (2007). Performance improvement of sic grinding using solid lubricants. *Machining Science and Technology*, *11*(1), 61-79.

[6] Liu MZ, Li CH, Cao HJ, et al. (2022). Research Progresses and Applications of CMQL Machining Technology. *China mechanical engineering*, *33*(5), 529-550.

[7] Liu MZ, Li CH, Yang M, et al. (2023). Mechanism and enhanced grindability of cryogenic air combined with biolubricant grinding titanium alloy. *Tribol. Int.*, *187*, 1-22. 10.1016/j.triboint.2023.108704.

[8] Tang LZ, Zhang YB, Li CH, et al. (2022). Biological Stability of Water-Based Cutting Fluids: Progress and Application. *Chin. J. Mech. Eng.*, *35*(1), 24. 10.1186/s10033-021-00667-z.

[9] Wu XF, Xu WH, Ma H, et al. (2023). Mechanism of electrostatic atomization and surface quality evaluation of 7075 aluminum alloy under electrostatic minimum quantity lubrication milling. *Surface Technology*, *52*(6), 337-350.

[10] Zhang YB, Li HN, Li CH, et al. (2022). Nano-enhanced biolubricant in sustainable manufacturing: From processability to mechanisms. *Friction*, *10*(6), 803-841. 10.1007/s40544-021-0536-y.

[11] Cui X, Li CH, Zhang YB, et al. (2023). Comparative assessment of force, temperature, and wheel wear in sustainable grinding aerospace alloy using biolubricant. *Front. Mech. Eng.*, *18*(1), 33. 10.1007/s11465-022-0719-x.

[12] Saini V, Bijwe J, Seth S, et al. (2021). Interfacial interaction of PTFE sub-micron particles in oil with steel surfaces as excellent extreme-pressure additive. *J. Mol. Liq.*, *325*, 14. 10.1016/j.molliq.2020.115238.

[13] Elinski MB, LaMascus P, Zheng L, et al. (2020). Cooperativity Between Zirconium Dioxide Nanoparticles and Extreme Pressure Additives in Forming Protective Tribofilms: Toward Enabling Low Viscosity Lubricants. *Tribol. Lett.*, *68*(4), 17. 10.1007/s11249-020-01346-1.

[14] Picciotto S, Peters A, Eisen EA. (2015). Hypothetical Exposure Limits for Oil-Based Metalworking Fluids and Cardiovascular Mortality in a Cohort of Autoworkers: Structural Accelerated Failure Time Models in a Public Health Framework. *Am. J. Epidemiol.*, *181*(8), 563-570. 10.1093/aje/kwu484.

[15] Wang XM, Li CH, Zhang YB, et al. (2022). Research progress on key technology of enabled atomization and supply system of minimum quantity lubrication. *Surface Technology*, *51*(9), 1-14.

[16] Bhaumik S, Paleu V, Sharma S, et al. (2020). Nano and micro additivated glycerol as a promising alternative to existing non-biodegradable and skin unfriendly synthetic cutting fluids. *J. Clean Prod.*, *263*, 11. 10.1016/j.jclepro.2020.121383.

[17] Zhou H, Chen Y, Chen W, et al. (2008). Synthesis and antirust performance study on dodecencylsuccinic acid diethanolamide. *37*(02), 12-14.

[18] Gan WL, Chen WM, Li X, et al. (2019). Application of modified dodecenyl succinic acid in semi-synthetic cutting fluid. *Lubrication Engineering*, *44*(6), 132-137.

[19] Liu J, Liu W, Zhou HY. (2014). Preparation and application of environmental friendly water soluble corrosion inhibitor. *Lubrication Engineering*, *39*(1), 119-124.

[20] Wang YJ, Chu ZJ, Wang HZ. (2013). Synthesis of triethanolamine borate and evaluation of Its antirust performance. *Materials Protection*, *46*(11), 29-31.

[21] Ezzat H, Nagy S, Shawky ES, et al., Synthesis of antibacterial additive for metal working fluids application, 2018, pp. 05011-.

[22] Wang XF. (2017). Synthesis and Performance Evaluation of Water-based Cutting Fluid Bactericide. *Technol Dev Chem Ind*, *46*(06), 26-28.

[23] Yuan B, Yi S, Du TJLE. (2015). Development of Twelve Acid Diethanolamide Boric Acid Ester and Its Application in Semisynthetic Cutting Fluid. *40*(12), 99-102.

[24] Meng XT, Huang W, Cong YF, et al. (2015). Preparation of alcohol amine borate antiwear additive. *Contemporary Chemical Industry*, *44*(9), 2113-2115.

[25] Lodhi APS, Kumar D. (2021). Natural ingredients based environmental friendly metalworking fluid with superior lubricity. *Colloid Surf. A-Physicochem. Eng. Asp.*, *613*, 9. 10.1016/j.colsurfa.2020.126071.

[26] Guo SM, Li CH, Zhang YB, et al. (2017). Experimental evaluation of the lubrication performance of mixtures of castor oil with other vegetable oils in MQL grinding of nickel-based alloy. *J. Clean Prod.*, *140*, 1060-1076. 10.1016/j.jclepro.2016.10.073.

[27] Li M, Yu TB, Zhang RC, et al. (2020). Study on the milling surface quality of GH4169 alloy based on graphene-enhanced MQL method. *Journal of Northeastern University (Natural Science)*, *41*(3), 387-392. 10.12068/j.issn.1005-3026.2020.03.015.

[28] Gao T, Zhang YB, Li CH, et al. (2021). Grindability of carbon fiber reinforced polymer using CNT biological lubricant. *Sci Rep*, *11*(1), 1-14. 10.1038/s41598-021-02071-y.

[29] Kaur A, Singh B, Kaur A, et al. (2019). Chemical, thermal, rheological and FTIR studies of vegetable oils and their effect on eggless muffin characteristics. *J. Food Process Preserv.*, *43*(7), 11. 10.1111/jfpp.13978.

[30] Semyonova A, Khomutov N, Misyura S, et al. (2022). Dynamic and kinematic characteristics of unsteady motion of a water-in-oil emulsion droplet in collision with a solid heated wall under conditions of convective heat transfer. *Int. Commun. Heat Mass Transf.*, *137*, 17. 10.1016/j.icheatmasstransfer.2022.106277.

[31] Singh H, Sharma VS, Dogra M. (2020). Exploration of graphene assisted vegetables oil based minimum quantity lubrication for surface grinding of TI-6AL-4V-ELI. *Tribol. Int.*, *144*, 13. 10.1016/j.triboint.2019.106113.

[32] Ike E. (2019). The study of viscosity-temperature dependence and activation energy for palm oil and soybean oil. *Global Journal of Pure Applied Sciences*, *25*(2), 209.

[33] Esteban B, Riba JR, Baquero G, et al. (2012). Characterization of the surface tension of vegetable oils to be used as fuel in diesel engines. *Fuel*, *102*, 231-238. 10.1016/j.fuel.2012.07.042.

[34] Golodnizky D, Rosen-Kligvasser J, Davidovich-Pinhas M. (2021). The role of the polar head group and aliphatic tail in the self-assembly of low molecular weight molecules in oil. *Food Struct.-Neth.*, *30*, 11. 10.1016/j.foostr.2021.100240.

[35] Pranav P, Sneha E, Rani SJIL, et al. (2021). Vegetable oil-based cutting fluids and its behavioral characteristics in machining processes: a review.

[36] Owuna FJ. (2020). Stability of vegetable based oils used in the formulation of ecofriendly lubricants – a review. *Egyptian Journal of Petroleum*, *29*(3), 251-256.

[37] Kaur A, Singh B, Kaur A, et al. (2021). Impact of intermittent frying on chemical properties, fatty acid composition, and oxidative stability of 10 different vegetable oil blends. *J. Food Process Preserv.*, *45*(12), 16. 10.1111/jfpp.16015.

[38] Kreivaitis R, Padgurskas J, Gumbyte M, et al. (2011). The influence of oxidation on tribological properties of rapeseed oil. *Transport*, *26*(2), 121-127. 10.3846/16484142.2011.586109.

[39] Choe E, D. BM. (2007). Chemistry of Deep-Fat Frying Oils. *Journal of Food Science*.

[40] Kim JK, Jeon CH, Lee HW, et al. (2018). Effect of Accelerated High Temperature on Oxidation and Polymerization of Biodiesel from Vegetable Oils. *Energies*, *11*(12), 11. 10.3390/en11123514.

[41] Cui X, Li CH, Ding WF, et al. (2022). Minimum quantity lubrication machining of aeronautical materials using carbon group nanolubricant: From mechanisms to application. *Chin. J. Aeronaut.*, *35*(11), 85-112. 10.1016/j.cja.2021.08.011.

[42] Chu AX, Li CH, Zhou ZM, et al. (2023). Nanofluids Minimal Quantity Lubrication Machining: From Mechanisms to Application. *Lubricants*, *11*(10), 52. 10.3390/lubricants11100422.

[43] Gong P, Zhang YB, Wang CJ, et al. (2024). Residual stress generation in grinding: Mechanism and modeling. *J. Mater. Process. Technol.*, *324*, 39. 10.1016/j.jmatprotec.2023.118262.

[44] Yildirim CV, Sarikaya M, Kivak T, et al. (2019). The effect of addition of hBN nanoparticles to nanofluid-MQL on tool wear patterns, tool life, roughness and temperature in turning of Ni-based Inconel 625. *Tribol. Int.*, *134*, 443-456. 10.1016/j.triboint.2019.02.027.

[45] Tevet O, Von-Huth P, Popovitz-Biro R, et al. (2011). Friction mechanism of individual multilayered nanoparticles. *Proc. Natl. Acad. Sci. U. S. A.*, *108*(50), 19901-19906. 10.1073/pnas.1106553108.

[46] Musavi SH, Davoodi B, Niknam SA. (2019). Effects of reinforced nanoparticles with surfactant on surface quality and chip formation morphology in MQL-turning of superalloys. *J. Manuf. Process.*, *40*, 128-139. 10.1016/j.jmapro.2019.03.014.

[47] Jamil M, Khan AM, Hegab H, et al. (2020). Milling of Ti-6Al-4V under hybrid Al_2O_3-MWCNT nanofluids considering energy consumption, surface quality, and tool wear: a sustainable machining. *Int. J. Adv. Manuf. Technol.*, *107*(9-10), 4141-4157. 10.1007/s00170-020-05296-9.

[48] Dubey V, Sharma AK, Pimenov DY. (2022). Prediction of Surface Roughness Using Machine Learning Approach in MQL Turning of AISI 304 Steel by Varying Nanoparticle Size in the Cutting Fluid. *Lubricants*, *10*(5), 16. 10.3390/lubricants 10050081.

[49] Talib N, Sasahara H, Rahim EA. (2017). Evaluation of modified jatropha-based oil with hexagonal boron nitride particle as a biolubricant in orthogonal cutting process. *Int. J. Adv. Manuf. Technol.*, *92*(1-4), 371-391. 10.1007/s00170-017-0148-x.

[50] Pal A, Chatha SS, Sidhu HS. (2022). Performance Evaluation of Various Vegetable Oils and Distilled Water as Base Fluids Using Eco-friendly MQL Technique in Drilling of AISI 321 Stainless Steel. *Int. J. Precis. Eng. anuf.-Gr Tech.*, *9*(3), 745-764. 10.1007/s40684-021-00355-2.

[51] Sen B, Mia M, Gupta MK, et al. (2019). Influence of Al_2O_3 and palm oil-mixed nano-fluid on machining performances of Inconel-690: IF-THEN rules-based FIS model in eco-benign milling. *Int. J. Adv. Manuf. Technol.*, *103*(9-12), 3389-3403. 10.1007/s00170-019-03814-y.

[52] Elsheikh AH, Abd Elaziz M, Das SR, et al. (2021). A new optimized predictive model based on political optimizer for eco-friendly MQL-turning of AISI 4340 alloy with nano-lubricants. *J. Manuf. Process.*, *67*, 562-578. 10.1016/j.jmapro. 2021.05.014.

Chapter 2

Minimum Quantity Lubrication of Aeronautical Materials Using Carbon Group Nanolubricants: Mechanisms and Applications

Abstract

The usage of nano-lubricants, especially non-toxic carbon group nano-enhancers (CGNs), in minimum quantity lubrication could solve the technical bottleneck of MQL facing aeronautical difficult-to-machine materials. However, the machining mechanisms under lubrication of CGNs are unclear at the complex interface between tool and workpiece, which characterized by high temperature, pressure, and speed, has limited application in factories and therefore necessitates in-depth understanding. To fill this gap, this chapter concentrates on the comprehensive quantitative assessment of tribological characteristics based on force, tool wear, chip, and surface integrity in titanium alloy and nickel alloy machining and attempts to answer mechanisms systematically. Firstly, to establish evaluation standard, the cutting mechanisms and performance improvement behavior covering antifriction, antiwear, tool failure, material removal, and surface formation of MQL are revealed. Secondly, the unique film formation and lubrication behaviors of CGNs in MQL turning, milling, and grinding are concluded. The influence law of molecular structure and micromorphology of CGNs is also answered and optimized options recommended by considering diverse boundary conditions. Finally, in view of CGNs limitations in MQL, the future development direction is proposed, which needs to be improved in thermal stability of lubricant, activity of CGNs, controllable atomization and transportation methods, and intelligent formation of processing technology solutions.

Keywords: aerospace materials, carbon nanoparticles, grinding, milling, turning, lubrication mechanism

2.1. Introduction

Nanolubricant minimum quantity lubrication (NMQL) cutting/grinding, as a new high-efficiency and green technology, initially solved the problem of parts forming for aerospace difficult to machine materials [1, 2]. Due to the addition of nano-enhancers, the tribological characteristics of tool/workpiece interface at high temperature, high pressure and high speed changed substantially, and the lubrication performance parameters such as cutting force, coefficient of friction (CoF), tool wear and surface quality changed greatly. Therefore, it is very important to understand the influence mechanism of physicochemical properties and micro morphology of nano-enhancers on interfacial tribological properties, and to select environmentally friendly, high-performance, low-cost nano-enhancers for further application. At present, some scholars have carried out experimental research on Molybdenum disulfide (MoS_2), aluminium oxide (Al_2O_3), carbon nanotubes (CNTs), diamond (ND), fullerene and other nano-enhancers [3, 4].

With the continuous development of material science, carbon group nano-enhancers (CGNs) provide new opportunities for NMQL. According to the 'green additive standard of lubricants' drawn up by German Blue Angel, CGNs are almost the best option among the various nano-enhancers mentioned. Furthermore, CGNs with different dimensions have unique molecular structure, antifriction, and antiwear performance, as well as higher chemical stability and thermal stability. Based on the preceding analysis, CGNs have the potential to be excellent lubricants for NMQL because of their excellent environmental protection, chemical inertness, antifriction, antiwear, and heat transfer characteristics. Early in 2008, scholars tried to reveal the lubrication mechanism of CGNs-based nanolubricant by tribological tests. Joly-Pottuz et al. [5] analyzed the tribological performance of the diamond and speculated on the formation of a diamond-like tribofilm.

As shown in Figure 1, review starts with the physicochemical properties of common CGNs (section 2.2), followed by the antifriction and antiwear mechanisms of NMQL and its influence on processability, which establishes a standard for machining evaluation (section 2.3). The application in turning, milling and grinding, and lubrication property of CGNs are evaluated, and selection standard of CGNs for various conditions are summarized (section 2.4–2.6).

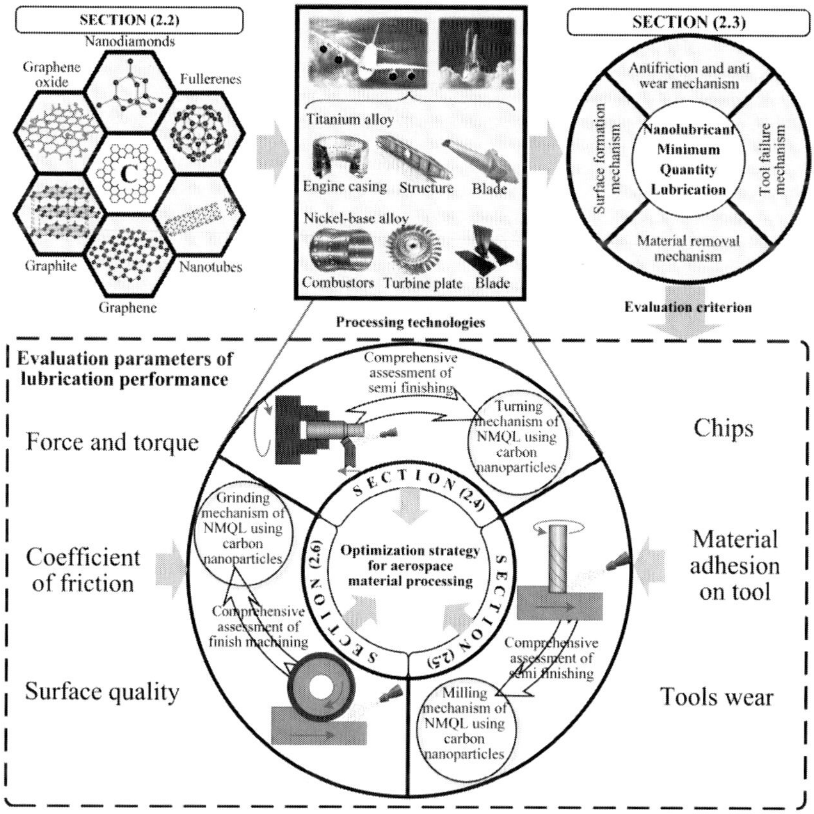

Figure 1. Structure of this section.

2.2. Carbon Group Nanolubricant

Some studies involving CGNs application in NMQL, antifriction, antiwear, tool failure, material removal mechanism, and surface formation mechanisms remain blank. Particularly for the typical turning, milling, and grinding conditions with different boundary conditions, the application of CGNs also shows various levels of processing performance.

As common types of CGNs, fullerene, ND, CNTs, and graphene (GR) have become research hot-spots due to their unique structure and excellent properties. Fullerene and ND are spherical, CNTs are tubular, and GR is layered. According to previous studies, CGNs have specific self-lubricating properties and the potential to become ‘molecular ball (shaft)’ lubricating additives, which are widely considered in the field of fluid and solid lubrication. Physicochemical properties, micromorphology, and molecular structure of CGNs are presented in Table 1 [6-9].

Table 1. Physicochemical properties, micromorphology and molecular structure of different CGNs

CGNs	Thermal conductivity (W/mK)	Young's modulus	Mohs hardness	Micro-morphology	Molecular structure
Fullerene (C60, zero dimensional)	0.4	62-107 GPa	Higher than ND	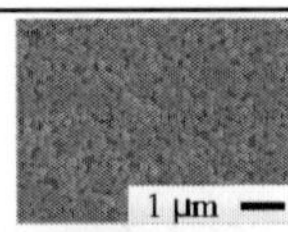	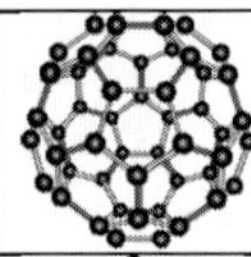
Nano-diamond (ND, zero dimensional)	2300	1000–1100 GPa	10		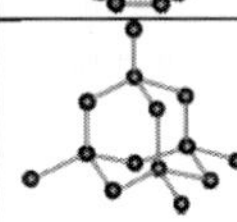
Carbon nanotubes (CNTs, one dimensional)	3000	704.5 GPa	Close to ND	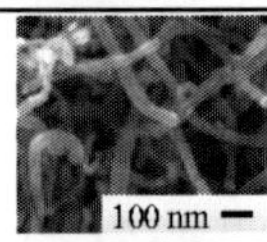	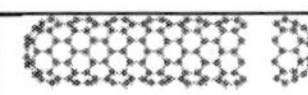
Graphene (GR, two dimensional)	5300 (monolayer)	1.0 TPa	Higher than ND	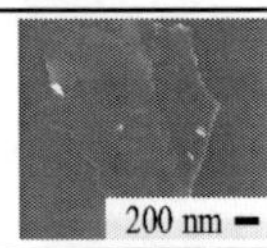	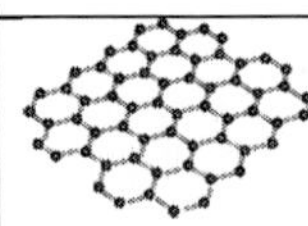
Graphite (GNP, two dimensional)	150–1500 (inversely proportional to thickness)	--	1–2	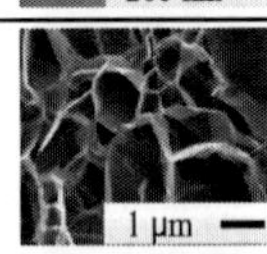	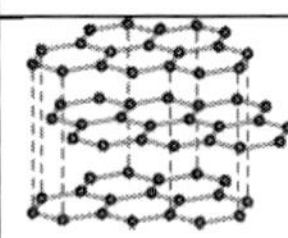
Graphene oxide (GO, two dimensional)	Lower than monolayer graphene	--	--		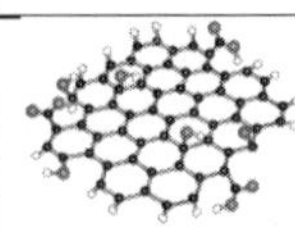

2.2.1. Spheroidal Nanolubricant

Zero-dimensional nano-enhancers refer to the three dimensions of nano-enhancers below 100nm, such as spherical fullerene, spherical diamond and so on. Tiwari et al. [10] reviewed the types and applications of zero-dimensional CGNs in detail and described experimental techniques involved in the synthesis and fabrication processes. There is no surprise that if spherical nano-enhancers can be introduced into minimum quantity lubrication (MQL) they can act like 'ball rollers in the bearings' at nanoscale, significantly reducing friction, so as to improve tool life, temperature, and force etc.

ND has a unique spherical shape and abundant surface functional groups. It is a supramolecular structure formed by a single crystal diamond "core" and many functional groups around its outer surface. Diamond is composed of carbon atoms, and each carbon atom forms covalent bonds with the surrounding four carbon atoms. ND particles have unique physical and chemical properties such as maximum hardness, high thermal conductivity, high wear resistance and good chemical stability. The diamond is not easily corroded at room temperature, with good stability and corrosion resistance.

2.2.2. Tubular Nanolubricant

One-dimensional nano-enhancers refer to nano-enhancers with two scales below 100nm, such as tubular carbon nanotubes. After the discovery of C60, with the increase of n value, Cn crystal gradually develops into the ellipsoid. When n is large enough, the shape of Cn becomes tubular in the middle and hemispherical in both ends. The structure of CNTs is a hollow cylinder surrounded by graphite surface, which has excellent self-lubricating performance. As an additive, Kharlamova [11] reported that CNTs can reduce the CoF of the material and effectively improve the friction and wear resistance. Compared with C60, CNTs have low cost and excellent mechanical and physical properties. The theoretical Young's modulus is 1.8TPA, the bending strength is 14.2 GPA, the tensile strength is 100 times that of steel, and the density is only 1/6 of that of steel [12].

2.2.3. Layered Nanolubricant

Two-dimensional nano-enhancers refer to nano-enhancers with a scale of less than 100nm, such as layered graphene, nano-graphite, etc. GR is a new carbonaceous material formed by the close arrangement of single-layer carbon atoms, with a single-layer two-dimensional honeycomb lattice structure. It is the thinnest two-dimensional material (thickness is only 0.335nm). Due to its special microstructure, graphene has excellent mechanical, thermal conductivity and electrical properties, and is expected to have significant application prospects in microelectronics, information, energy, materials and biomedical fields [13, 14]. Since the discovery of graphene in 2004, the tribological properties of graphene are not detailed enough. Only graphene films, as composite additives and lubricant additives have been studied.

Graphene oxide (GO), as an oxide of graphene, still maintains the layered structure of graphite, and introduces a large number of oxygen-containing functional groups (hydroxyl, carboxyl, epoxy, etc.) in the oxidation process. The existence of oxygen-containing functional groups makes graphene oxide as an additive having good dispersion in lubricating fluid. There are abundant oxygen-containing functional groups in the molecular structure, which interact with the metal surface through electrostatic interaction, hydrogen bonding and dispersion force, which makes graphene oxide acting as a lubricating additive has a strong adsorption on the metal surface [15, 16].

As mineral rich in the earth, graphite (GNP) has the characteristics of high temperature resistance, corrosion resistance and self-lubricating. It has been widely studied and applied in solid lubricating materials and various lubricants. The carbon atoms in the middle layer and between the layers in the crystal structure are combined by weak Л electronic bond. This weak binding force is easy to shear between layers. This is a common saying that the low friction performance of graphite is caused by sliding within particles. Therefore, as an excellent solid lubricant, the research and application of graphite in liquid media has become a research hotspot in the field of lubrication.

2.3. Cutting Mechanism of Minimum Quantity Lubrication

With the addition of nano-enhancers, the interface antifriction and antiwear characteristics change, and the material removal constitutive relation will also change under the action of thermal softening effect, which will affect the force, CoF, tool life, tool wear, workpiece surface quality and other cutting performance evaluation parameters [17]. Although scholars have observed the changes of the above experimental phenomena through experiments, there is no relevant mechanism to explain the change rules, and no unified index system has been established to comprehensively measure the processing performance of NMQL. Therefore, this part of the content for the above status quo analysis (see Figure 2a).

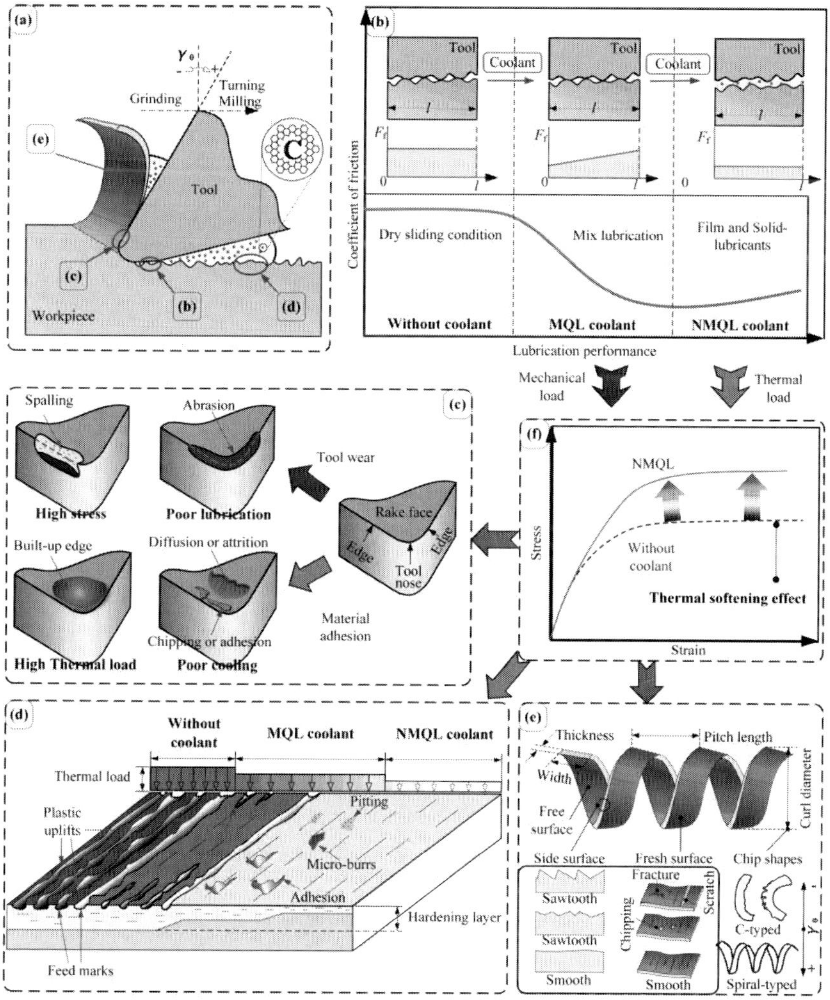

Figure 2. Cutting mechanism of NMQL. (a) Processing forms, (b) State of lubrication under different lubrication conditions, (c) Tool failure mode, (d) Workpiece surface defect form, (e) Chip morphology, (f) Stress-strain curves under different lubrication conditions.

2.3.1. Antifriction and Antiwear Mechanism

In the process of material removal, compared with MQL, the lubrication state of tool/workpiece interface changes essentially after adding nano-enhancers [18]. The lubrication state of tool/workpiece interface changes from boundary lubrication under MQL condition to oil film lubrication state under nanofluids. When nano-enhancers enter the tool/workpiece interface, the space between the tool and the workpiece is increased, and the direct contact between the tool and the workpiece is avoided during boundary lubrication, and the friction between the tool and the workpiece is reduced. On the other hand, the nano-enhancers support two contact pairs at the tool/workpiece interface, resulting in a larger and continuous space. Therefore, it is easier for the nanofluid to infiltrate the tool/workpiece interface, which improves the wettability of the nanofluid, thus forming a continuous lubricating oil film more easily (see Figure 2b and 2f).

2.3.2. Tool Failure Mechanism

Tool wear and material adhesion are two main forms of tool failure. For aerospace material precision machining, tool premature failure not only increases the cost, but also reduces the machining accuracy and workpiece surface quality [19, 20]. In the process of material removal, the tool failure mainly occurs at the tool nose and the main cutting edge, and the friction between the tool and the workpiece increases after blunt failure. After adding nano-enhancers, a dynamic or static protective oil film is formed in this area, thus delaying the tool failure and increasing the sharpness of the tool nose, which is of great significance (see Figure 2c).

2.3.2.1. Abrasion

In the process of cutting aerospace difficult-to-machine materials in the aerospace industry, the tool bears higher load, which leads to stress concentration and crack under high pressure interface, and then the material peels off after the crack expands on the tool surface. After adding nano-enhancers, it can fill the surface cracks of the tool at the initial stage of crack formation, restraining the further propagation of the cracks and ensuring the integrity of the tool. In addition, the nano-enhancers form dynamic and static oil film on the tool as a protective layer, which effectively reduces the CoF and delays the tool wear.

2.3.2.2. Material Adhesion

Due to the extremely low thermal conductivity and high strength of difficult-to-machine materials in the aerospace industry, the tool nose/workpiece contact area not only generates high heat during cutting, but also is difficult to disperse through the workpiece material, resulting in rapid accumulation of cutting heat in the first deformation zone, and the increase of material plasticity. After adding nano-enhancers, a lubricating oil film is formed, which reduces the friction force in the processing area and thus reduces the temperature. Thus, the thermoplastic flow of workpiece material and the accumulation of tool and workpiece surface are reduced, and the formation of built-up edge is avoided.

2.3.3. Material Removal Mechanism

With the addition of nano-enhancers, the cutting force and interfacial CoF change significantly, which leads to the fundamental change of material removal mechanism (chip formation). Chip morphology is the main basis for analyzing material removal [21], which can be carried out from four aspects: thickness, sawtooth, chip curl diameter and fresh surface (see Figure 2e).

2.3.3.1. Thickness

The chip thickness decreases with the addition of nano-enhancers. This is due to the significant improvement of the cooling and lubricating properties of nano-enhancers, and the improvement of lubrication properties reduces the production of friction heat. Therefore, when the temperature of the cutting zone is significantly reduced, the plasticity of the workpiece in the material removal process decreases, and the generation of built-up edge on the tool is weakened. The tool maintains a sharp edge, which reduces the chip thickness.

2.3.3.2. Sawtooth

When the lubrication condition is poor, the chip will break under the cutting action of the main cutting edge, which is easy to produce sawtooth. This will further aggravate the wear of the main cutting edge and reduce the surface quality of the workpiece. The sawtooth mainly appears on the side and free surface of chip. The side sawtooth is mainly caused by the tearing of the workpiece material, while the free surface sawtooth is formed by the extrusion of the rake face, which is related to the force direction of material removal. When the material is removed, the higher the deformation is, the higher the sawtooth height (or even fracture) will be formed. With the addition of nano-enhancers, the lubrication state between tool and workpiece is improved, and the wear rate of the main cutting edge decreases, and nano-enhancers will accumulate in the wear pits of the main cutting edge. At the same time, the interface CoF and friction angle decrease, which makes the deformation degree of the material removal process weaken. As a result, the material breaks near the sharper main cutting edge, resulting in sawtooth free chips.

2.3.3.3. Chip Curl Diameter

Because of the decrease of CoF, the friction angle of the second deformation zone and the deformation degree of the first deformation zone, the chip becomes thinner and the curl diameter decreases. In this case, the chip extrusion force, friction area and friction heat of the rake face in the second deformation zone decrease, and the heat required for material removal in the first deformation zone also decreases. This is very meaningful for difficult-to-machine materials.

2.3.3.4. Fresh Surface

The roughness of fresh chip surface represents the tribological characteristics between the rake face and the chip. In the process of material removal and chip front outflow, the larger friction force and shear force that form at the tool/chip interface, lead to adhesion between the rake face and the chip. At the same time, the hard particles on the rake face will scratch the workpiece material, leaving a certain degree of scratch on the front of the chip. In addition, a large amount of heat generated by the tool/chip interface will increase the number of scratches. In addition, the scratch indicates that the tool has been worn. The addition of nano-enhancers protects the tool and reduces scratches.

2.3.4. Surface Formation Mechanism

The surface integrity of workpiece is mainly affected by the factors such as force, temperature and tool wear degree, and the most important method to improve the surface integrity is to restrain the surface defects formed by machining. For titanium alloy and nickel-base alloy, the common surface defects are adhesion, burn, plastic accumulation and scratch. The first three are due to excessive temperatures in the cutting zone (see Figure 2d). The adhesions formed on the machined surface come from two aspects: on the one hand, the material at the tool nose is rolled and welded to the machined surface after being subjected to high cutting heat. On the other hand, in the process of material removal, friction heat is generated between the flank surface and the machined surface of the workpiece. And the workpiece material peels off under the thermal softening effect, while the workpiece material is rolled and welded as adhesion, and the surface pit is formed at the material spalling place. In addition, the plastic accumulation of the workpiece material is serious under the thermal softening, which reduces the material removal rate and the surface quality of the workpiece. Even, the cutting/grinding heat generated on the surface of the workpiece accumulates in the surface layer of the workpiece and cannot be diffused, resulting in burns. The scratch is caused by the decrease of the straightness of the tool's main cutting edge. At the same time, with the continuous processing of the tool, the scratch will continue to expand and deepen, and even the material will appear plastic and uplift. The reason for the above phenomenon is that the cooling and lubrication performance of the tool/workpiece contact area is insufficient, and the addition of nano-enhancers will bring great changes. This is due to the formation of a nano lubricating film between the flank surface and the workpiece surface, which reduces the friction and heat generation and protects the newly formed surface of the workpiece. This has been confirmed by the EDS analysis of the workpiece surface.

2.4. The Application of CGNs in MQL Turning

Turning is one of the main machining forms in hard machining materials, and it is a semi-finishing method, mainly used in the machining of rotary parts and cylindrical parts in the hard machining materials of aerospace [22]. It has been confirmed that C-group nano-enhancers play a significant role on tribological properties by previous single factor experiments. However, previous studies missed the law of multiple factors on tribological properties. Aiming to guide the production practice, this section puts forward a comprehensive review for experimental results and concludes turning mechanism.

2.4.1. Turning Force

Compared with MQL, the large force reduction of 45.65% direction was observed when using cryogenic minimum quantity lubrication (CNMQL) for Ti-6Al-4V using graphene oxide [23]. Achieving 22.22% force reduction for Ti-6Al-4V [19, 24] with employment of graphite and graphene. Marques et al. [25] used graphite and MoS_2 nanofluids in turning Inconel 718. Compared with MQL, the maximum cutting force decreased by 10.64% and 8% respectively (see Figure 3). With the increase of cutting speed, the cutting force fluctuates greatly under the condition of MoS_2 nanofluid, while the cutting force changes smoothly under the condition of graphite nanofluid.

The results showed that the turning force of titanium alloy and nickel-base alloy decreased obviously after adding layered nano-enhancers. The decreasing order of turning force was MoS_2 disulfide < graphite = graphene < graphene oxide. On the one hand, C-group layered nano-enhancers show better lubrication performance than MoS_2 disulfide. This is because their intermolecular bond energy is smaller, so they are more conducive to antifriction and antiwear, which will be analyzed in detail later. On the other hand, among the three C-group nano-enhancers, graphene oxide has the largest reduction in cutting force because it contains more functional groups and is easier to adsorb on the surface of workpiece and tool, thus forming a stable oil film and reducing the friction force in the cutting zone.

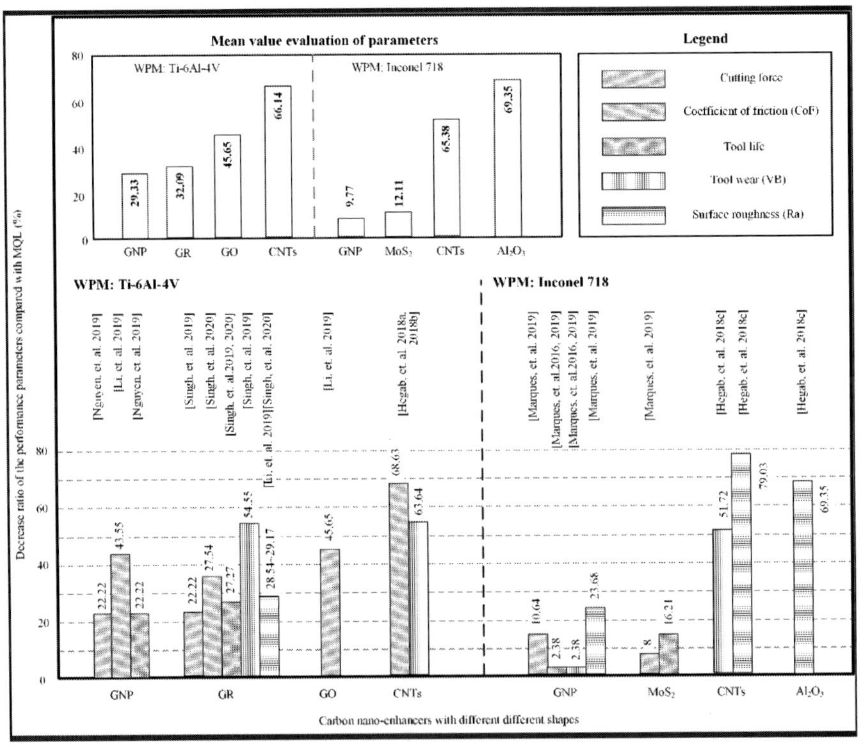

Figure 3. Analysis of experimental data in turning.

2.4.2. Coefficient of Friction (CoF)

CoF is the evaluator of lubrication at the tool-workpiece interface and can be calculated based on measured normal and tangential forces. The reduction of CoFs were observed including 27.54%、43.55% and 68.63% in V-MQL turning of Ti-6Al-4V with employment of graphene, graphite and CNTs, respectively (see Figure 3) [23, 26, 27].

The results showed that the decreasing degree of CoF is graphene<graphite<CNTs. Compared with layered nano-enhancers, tubular nano-enhancers exhibit better friction reduction performance in the turning area, which is due to the different film-forming and lubrication mechanisms of different nano-enhancers in the cutting area.

2.4.3. Chip Morphology

Compared with pure oil, when multi-walled carbon nanotubes (MWCNTs) were added, the chip thickness became smaller, the chip morphology was more uniform than that of pure oil, and the sawtooth height was reduced [26]. Moreover, the chip thickness of 4wt% MWCNTs lubricant was smaller than that of 2wt% MWCNTs lubricant, which eliminated the non-homogeneous serrated chips. Li et al. [23] also confirmed that the addition of graphene oxide can effectively reduce the area/degree of built-up edge in the rake face. Singh et al. [27] analyzed sawtooth chip morphologies under different lubrication conditions. When graphene with 1.0wt% was added, the chip curl diameter was the smallest when compared with MQL (see Figure 4a).

According to the analysis of the experimental results of the above chip morphology, it can be seen that the addition of graphene and CNTs improves the interfacial lubrication characteristics and reduces the CoF and material removal friction angle, which shows the advantages of adding nano-enhancers. In addition, when the concentration of nano-enhancers increases, the lubrication performance will also change.

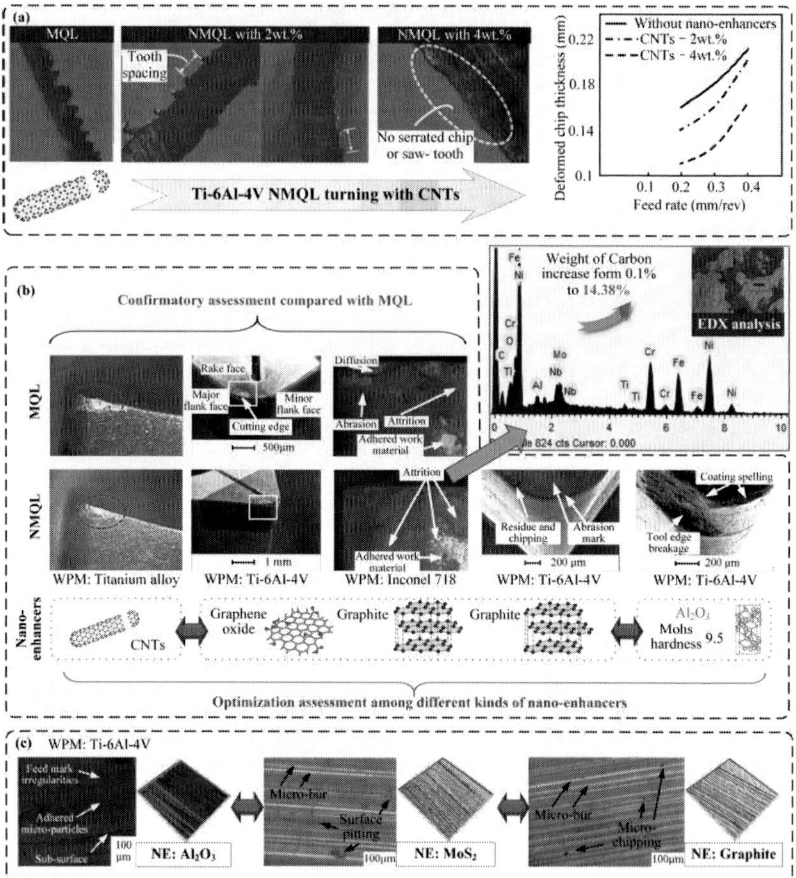

Figure 4. Chip, tool and workpiece analysis of NMQL turning using C-group nano-enhancers. (a) Micromorphology of chip, (b) Micromorphology and EDS analysis of tools, (c) Micromorphology of workpiece surface.

2.4.4. Tool Life and Wear

2.4.4.1. Tool Life

In the experiment of turning Ti-6Al-4V with graphite particles as additive, the tool life was increased by 22.22% compared with MQL [24]. When the concentration of graphite particles was further changed, the tool life first increased and then decreased. And the tool life reached the maximum when the mass fraction was 0.5%. When the concentration of graphite particles increased to 2%wt, the lifetime decreased greatly, which is only 23.96% of

0.5wt%. When graphene was used as additives in turning Ti-6Al-4V, the tool life was increased by 27.27% compared with MQL [19, 27]. Marques et al. [25] respectively used graphite and MoS_2 disulfide nanofluids to carry out turning Inconel 718 experiments. The results showed that the tool life under graphite nanofluid lubrication was 2.38% and 16.21% longer than MQL and MoS_2 nanofluid respectively (see Figure 3).

According to tool life analysis, the inhibition effect of different kinds of nano-enhancers on tool wear in the turning process is: MoS_2 <graphite<graphene. This is the same as the conclusion obtained in Section 4.1, which confirms the advantages of C-group layered nano-enhancers. At the same time, graphene also has better lubrication performance than graphite, which is due to the better film-forming performance of graphene with fewer molecular layers in turning.

2.4.4.2. Tool Wear (VB)

For turning Ti-6Al-4V, compared with MQL, obviously improved tool wear at the flank surface was observed where VB values were decreased by 54.55% with employment of graphene [19]. When CNTs were used as additives, VB decreased by 63.64% [28]. When CNTs were used as additives, VB decreased by 51.72% (see Figure 3) [29]. According to VB analysis, CNTs showed better friction reduction performance than graphene, and both of them are much better than graphite. This is also consistent with the results of CoFs, which show that the tubular nano-enhancers can obtain lower interfacial CoF, thus greatly reducing tool wear.

2.4.4.3. Tool Wear and Material Adhesion

When turning Inconel 718, the phenomenon of abrasion, diffusion and attrition was found. The phenomenon improved after adding graphite. The same law was obtained by Gutnichenko et al. [30]. In the process of turning titanium alloy, the addition of MWCNTs significantly reduced abrasion [28]. In another group of experiments, grapheme was used to avoid the phenomenon of chipping, attrition. Marques et al. [25] found that there was a lubricating oil film on the tool surface through EDS analysis, and the C content increased to 14.38% (the content of C element in titanium alloy itself was 0.1%). This layer of lubricating film may come from the workpiece material or graphite. Gupta et al. [31] and Singh et al. [27] prevented slight adhesion under MQL conditions by adding graphite and graphene, respectively. Li et al. [23] found that there was built-up edge in MQL, which disappeared after adding graphene oxide. Revuru et al. [32] found that under MQL condition, the tool even failed due to plastic deformation, which was effectively avoided by adding graphite particles. In addition, due to the different micro morphology and physical properties of nano-enhancers, the tool wear form will also be changed. Different from the layered nano-enhancers, Gupta et al. [31] found that the tool wear was attrition when spherical Al_2O_3 was used (see Figure 4b).

Through the observation of the tool surface morphology, it is found that the tool wear and surface adhesion are improved after adding nano-enhancers. In addition, a large amount of C element is found on the tool surface by EDS analysis, which also confirmed that nano-enhancers can form lubricating film on the tool surface, but the current research do not analyze the film forming mechanism of different nano-enhancers.

2.4.5. Surface Integrity

2.4.5.1. Surface Roughness

Machined surface roughness *Ra* of Ti-6Al-4V was reduced by 29.17% and 28.54% in graphene nanofluid tests of Singh et al. [27] and Li et al. [23], respectively. For turning Inconel 718, Hegab et al. [29] carried out the experiment of CNTs and Al_2O_3 nanofluid lubrication turning. Compared with MQL, the surface roughness *Ra* of NMQL was reduced by 79.03% and 69.35%, respectively. Marques et al. [25] used graphite particles for turning Inconel 718 and found that *Ra* decreased by 23.68% (see Figure 3). The results showed that the order of the influence of different shape nano-enhancers on surface roughness is graphite< Al_2O_3< CNTs, that is, tubular is better than spherical, better than layered.

2.4.5.2. Morphology of Machined Surface

Gupta et al. used Al_2O_3, MoS_2 and graphite to carry out titanium alloy turning experiments. After adding graphite, the surface of the workpiece produced finer and more regular and clearer cutting marks than MQL and the other two kinds of nano-enhancers, which showed that graphite has more stable friction reducing performance, and the vibration caused by cutting force changes will not occur during the cutting process. Further zoom in on the furrow [31]. Compared with Al_2O_3 and MoS_2, the feed mark on the surface of the workpiece under graphite was more regular and obvious. Adhered microparticles of cutting material and various surface and subsurface layers disappeared, small micro burrs and surface pitting were significantly reduced, and the surface was very smooth and clear (see Figure 4c). The reason is that graphite has higher heat transfer coefficient than the other two particles, which makes the heat transfer out of the cutting zone and reduces the temperature of the cutting zone. Therefore, the microstructure changes little and the cutting force and tool wear are reduced.

2.4.6. The Turning Mechanism of NMQL Using Carbon Nano-Enhancers

From the above research, it can be seen that the different morphology of C-group nano-enhancers have different processing properties (see Table 2). Especially the tubular CNTs show the best antifriction and antiwear resistance performance, which is related to the wetting film-forming mechanism and antifriction and antiwear mechanism in the turning area (see Figure 5).

Table 2. Conclusion of processing properties in turning

Working condition	Characterization paramter		Nano-enhancers	Conclusion
Turning	Cutting force		GR, GO, GNP, MoS_2	Order of decreasing degree: MoS_2<GNP=GR<GO
	CoF		GR, GNP, CNTs	Order of decreasing degree: GR< GNP< CNTs
	Chip morphology	Thickness	GO, CNTs	Better experimental results for carbon nano-enhancers
		Sawtooth	CNTs	
		Chip curl diameter	GR	
	Tool	Tool life	GR, GNP, MoS_2	Order of increasing degree: MoS_2<GNP<GR
		VB	GR, GNP, CNTs	GNP<GR<CNTs
		Tool morphology	GR, GNP, CNTs, Al_2O_3, MoS_2	Spherical nanoparticle material removal form--bulk spalling
		Tool adhesion	GR, GNP	Better experimental results for carbon nano-enhancers
	Surface integrity	*Ra*	GR, GNP, CNTs, Al_2O_3	Order of decreasing degree: GNP<Al_2O_3<CNTs
		SEM	GNP, Al_2O_3, MoS_2	MoS_2 is better than GNP

2.4.6.1. Mechanism of Wetting Film Formation in Turning

In turning process, the turning tool shows the working state of continuous material removal (see Figure 5a). For the wedge-shaped interface of rake face/chip and flank surface/fresh surface, the infiltration and film formation of lubricant is a continuous process. The cutting load is reduced from the maximum to zero from the tool nose to the non-contact area, so the tool nose is the most serious area of tool wear. Because of the weak load-carrying capacity of the lubricant, the contact surfaces of the rake face/chip and the flank/fresh surface are in the boundary lubrication state, so it is difficult for the lubricant to infiltrate into the tool nose. The wettability and load-carrying capacity of the lubricant will be greatly improved by adding nano-enhancers, but the wetting effect of different nano-enhancers is essentially different. For the layered nano-enhancers, with the continuous infiltration of nanofluid from the entrance of the wedge-shaped area to the tool nose, the layered nano-enhancers are self-spreading and infiltrating, which is easy to form a film at the entrance of the wedge-shaped area, and has a filling effect on the micro pits on the tool surface, which makes the surface contour of the two contact surfaces gradually smooth. However, the space for further infiltration of nanofluid is reduced, so the infiltration speed and depth of layered nano-enhancers are low (see Figure 5f, g, h). For the tubular and spherical nano-enhancers, the wetting is the rolling infiltration during the mutual movement of the two interfaces. It can not only infiltrate to the area closer to the tool nose, but also can carry the friction interface, which makes it change from boundary lubrication to oil film lubrication (see Figure 5b, c and 5d, e). Moreover, due to the gap between the friction pair interface with the tubular and spherical, which provides space for the subsequent infiltration of nano-enhancers. Therefore, tubular and spherical nano-enhancers not only have high infiltration speed, but also have large infiltration depth.

2.4.6.2. Mechanism of Tool/Workpiece Antifriction and Antiwear

Furthermore, due to the different antifriction and antiwear mechanisms of different morphologies of nano-enhancers, different processability will also be caused. Spherical nano-enhancers are pointing contact rolling friction reduction. Due to its hard texture, the tool surface easily produces stress concentration and cracks when rolling at high pressure interface, which is why it is found in the experiment that the tool wear is in the form of block spalling. Layered nano-enhancers are surface contact sliding friction reducing and play the role of solid lubricant at the tool/workpiece interface, so it will not form spalling wear of materials. Although graphene has super lubrication properties, its wettability is poor, and part of the surface of friction pair is still in the boundary lubrication state, which makes it easy to cause repeated changes of friction force and fluctuation of turning force, resulting in tool vibration and *Ra* increase. Tubular nano-enhancers are line contact rolling friction reduction, compared with the layered nano-enhancers, the tubular nano-enhancers have a stronger wettability and are easier to enter the tool nose area. Compared with the spherical nano-enhancers, the tubular nano-enhancers have larger bearing area and a more stable lubrication state. Therefore, nano-enhancers have the best antifriction and antiwear reducing properties in the turning process, which is fully confirmed by the experimental results.

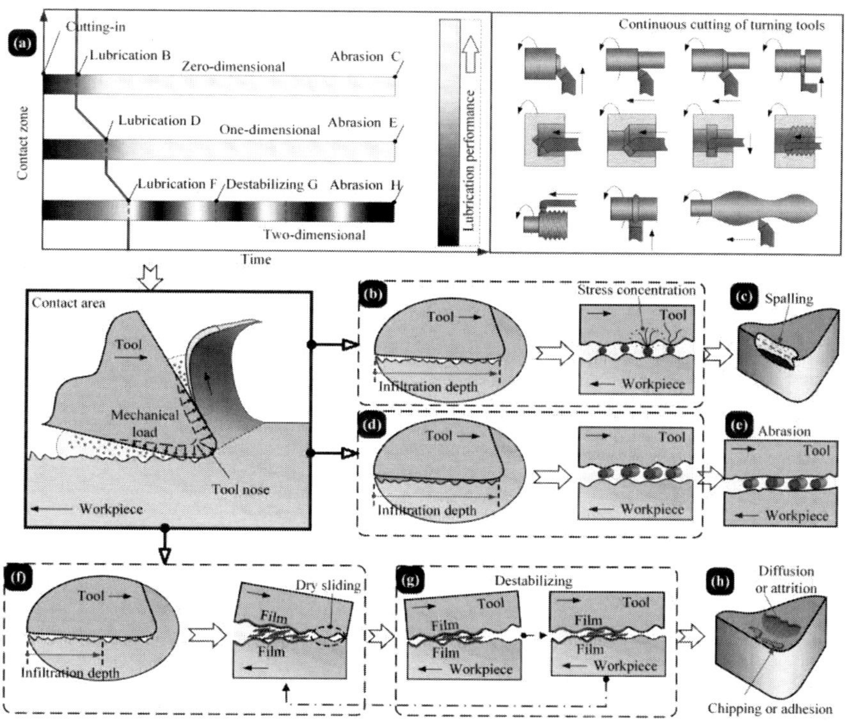

Figure 5. Turning mechanism of NMQL using carbon nano-enhancers. (a) Wetting and lubricating properties of different carbon group nanoparticles, (b) Lubrication behavior of spherical nanoparticles, (c) Ball nanoparticle working condition tool wear, (d) Lubrication behavior of tubular nanoparticles, (e) Tubular nanoparticles working condition tool/workpiece wear, (f) Lubrication behavior of layered nanoparticles, (g) Instability of layered nanoparticles under working conditions, (h) Layered nanoparticles work under tool wear.

2.5. The Application of CGNs in MQL Milling

Milling is one of the main processing forms in difficult to process materials, it is a semi-finishing method, mainly used for the processing of aero-engine blades, frame structural parts [33, 34]. Different from turning, milling is an intermittent material removal process of multi cutting edges, which changes tribological characteristics and influences the law of C-group nano-enhancers. Therefore, it's necessary to put forward a comprehensive review and conclusive milling mechanism in this section.

2.5.1. Milling Force

Compared with MQL, the large force reduction of 18.22% and 20.83% direction was observed when using graphene for TC4 and TC21 milling [35, 36]. Bai et al. [37] used CNTs and graphite to milling Ti-6Al-4V. The results showed that the milling forces were reduced by 22.4% and 8.43% respectively, and CNTs were reduced by 15.35% compared with graphite. Sirin and Kivak [38] used graphite to mill Inconel X-750, which reduced the milling force by 16.23% compared with MQL (see Figure 6). According to the above literature, the order of decreasing milling force by adding different kinds of nano-enhancers is: graphite<graphene< CNTs.

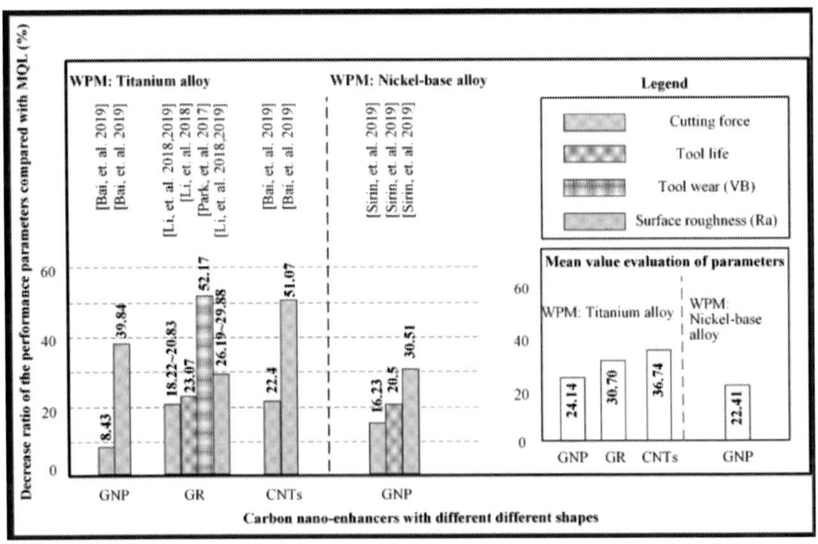

Figure 6. Analysis of experimental data in milling.

2.5.2. Chip Morphology

Bai et al. [37], Park et al. [39] and Yin et al. [40] carried out the experimental study of C-NMQL with employment graphene, graphite, CNTs, and other nano-enhancers, and observed the titanium alloy chips. Compared with MQL, the chip thickness decreased with the addition of nano-enhancers. However, different morphologies of nano-enhancers have different chip shapes. For the layered graphite, the chip shape is similar to that obtained by MQL, and the chip is C-shaped, and the fresh surface of the chip is rough and has irregular scratches. Moreover, there are obvious cracks on the surface of graphite nanofluid milling, which indicates that the chip has experienced a large shear plastic deformation, and the lubrication effect is the worst.

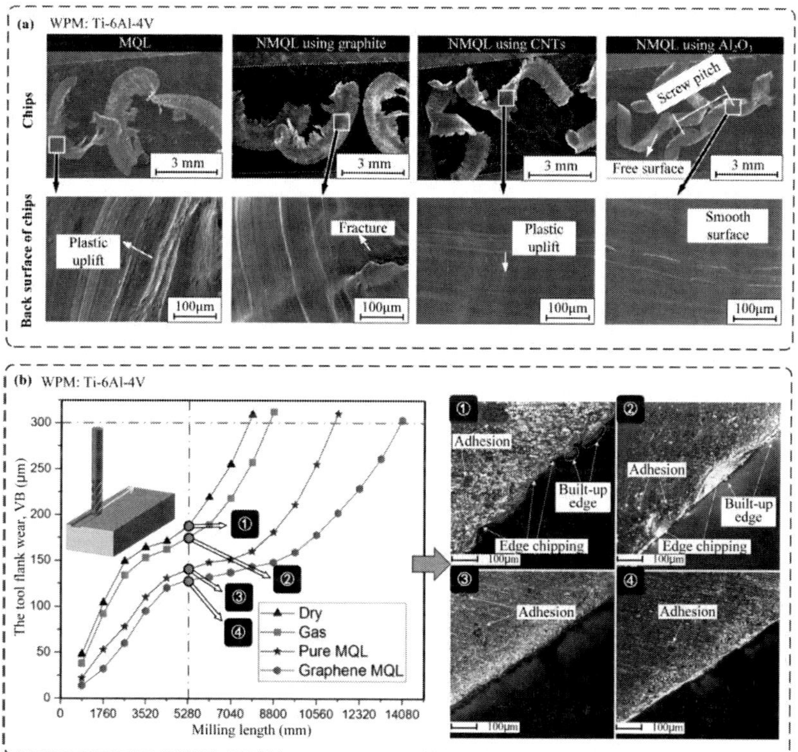

Figure 7. Chip and tool analysis of NMQL milling using C-group nano-enhancers. (a) Micromorphology of chip, (b) Tool wear analysis.

For tubular CNTs, the chip is spiral, the fresh surface of the chip is smooth without obvious spalling, but there are some scratches, and the lubrication effect is in the middle. When comparing CNTs with graphite, it is found that CNTs can obtain thinner chips, but both are larger than spherical Al_2O_3. The results show that the morphology of the chip milled with Al_2O_3 nanofluid is the best, which is thin and long spiral, and the chip shape is more slender than other conditions. Compared with other nano-enhancers, the morphology is significantly improved, the fresh surface is smoother, no obvious tool scratch is found. The metal deformation of the cutting layer is more uniform, the free surface of the chip has a small degree of sawtooth, and the chip thickness remains basically unchanged (see Figure 7a).

According to the previous paper, this is because different nano-enhancers have different lubrication effect, and the order of lubrication effect is: graphene<CNTs< Al_2O_3. This shows that different NMQL lubrication can improve the chip processing morphology to a certain extent, which is conducive to improving the surface quality and tool life.

2.5.3. Tool Life and Wear

2.5.3.1. Tool Life

Sirin and Kivak [38] milling Inconel X-750 with graphite increased by 20.5% compared with MQL tool life. When VB=300 μm was defined as tool failure, and the tool life of TC4 was prolonged by 23.07% by using graphene MQL (see Figure 6) [36].

2.5.3.2. Tool Wear (VB)

Park et al. [39] found that compared with MQL, the flank wear of Ti-6Al-4V alloy was reduced by 52.17%. While, when milling TC4, the flank wear decreased by only 8.57% with employment of graphene [36]. Sirin and Kivak [38] was used to mill Inconel X-750 with graphite. The maximum wear value VB was 0.32 mm under the condition of MQL, and the tool wear decreased significantly after adding nano-enhancers. With the increase of nano-enhancers concentration, the tool wear first decreased and then increased (see Figure 7b). This is due to the aggregation of nano-enhancers due to the increase of the concentration of nano-enhancers, which leads to poor stability of lubricants.

Under dry, high-pressure gas, MQL and NMQL conditions, tool wear will go through four stages: rapid wear, stable wear, rapid wear and tool failure (see Figure 7b) [36]. Compared with MQL, the tool wear curve with graphene had smaller slope and longer length in the first three stages, which indicated that the wear rate was slower under this condition. Compared with MQL, the tool surface adhesion under graphene condition was significantly reduced, and edge cutting, and build-up edge disappeared. Moreover, using graphene, the cutting edge keeps better straightness.

2.5.4. Surface Integrity

2.5.4.1. Surface Roughness

Compared with MQL, when graphene was used to milling TC21 and TC4, *Ra* decreased by 29.88% and 26.19%, respectively. While micro-hardness decreased by 8.84% (see Figure 8b) [35, 36]. Bai et al. [37] carried out milling experiments of Ti-6Al-4V. CNTs and graphite were selected as nano-enhancers to prepare nanofluids. Compared with MQL, the *Ra* values were reduced by 51.07% and 39.84% respectively. Sirin and Kivak [38] used graphite to milling Inconel X-750. Compared with MQL, *Ra* of the concentration of 0.50 vol% graphite nanofluid decreased by 30.51% (see Figure 6). The results of surface roughness experiments showed that CNTs with tubular structure can obtain better surface quality than those with layered graphite and graphene.

2.5.4.2. Morphology of Machined Surface

When graphene was used in titanium alloy milling experiment, compared with MQL lubrication condition, adhesion, surface pit and large fuse disappeared, and the workpiece surface was significantly improved [35, 36]. The experimental results of different nano-enhancers showed that graphite NMQL has the deepest furrow and plastic uplift, while CNTs NMQL was only slightly scratched. The improvement of the surface quality was also due to the film-forming lubrication of nano-enhancers. Yin et al. [41] analyzed the element composition of the workpiece surface by EDS, and found that the content of C element deposited on the workpiece surface by CNTs and graphite milling was the highest, reaching 4.72% and 5.31% respectively (see Figure 8a). However, it had to be admitted that the content of this element was less than that detected in turning, which also indicated that the wetting film-forming performance of nano-enhancers in milling was worse than that in turning.

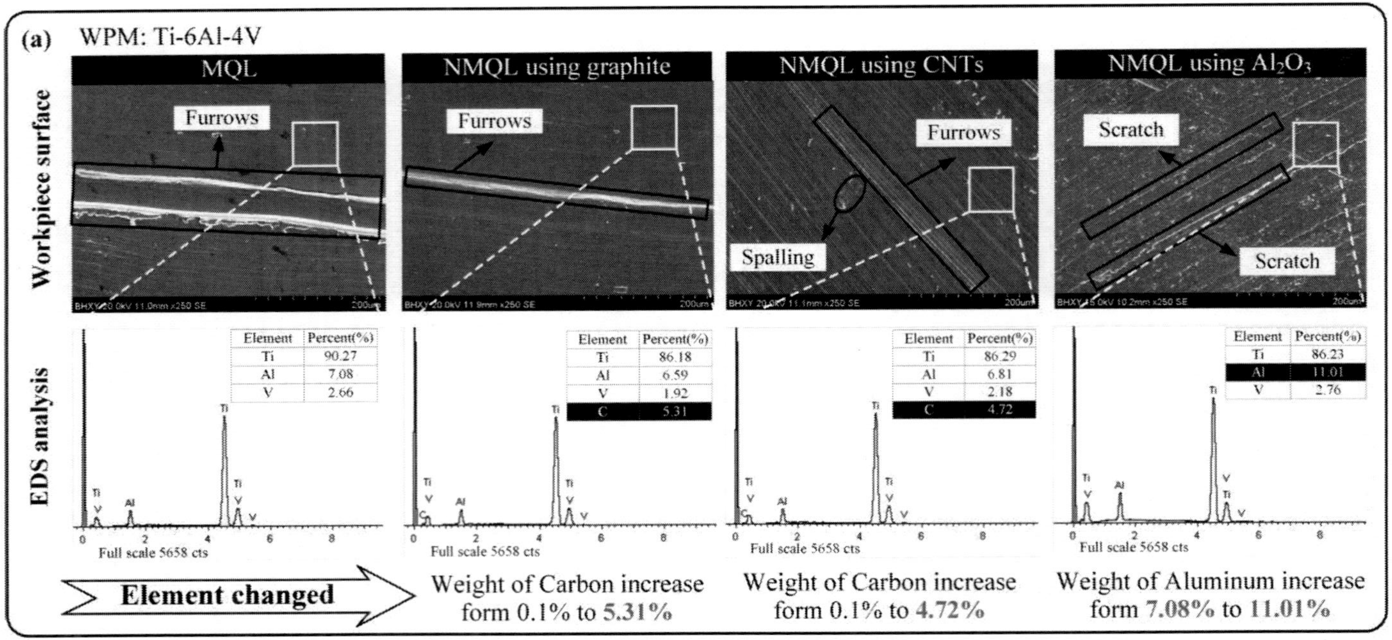

Figure 8. (Continued).

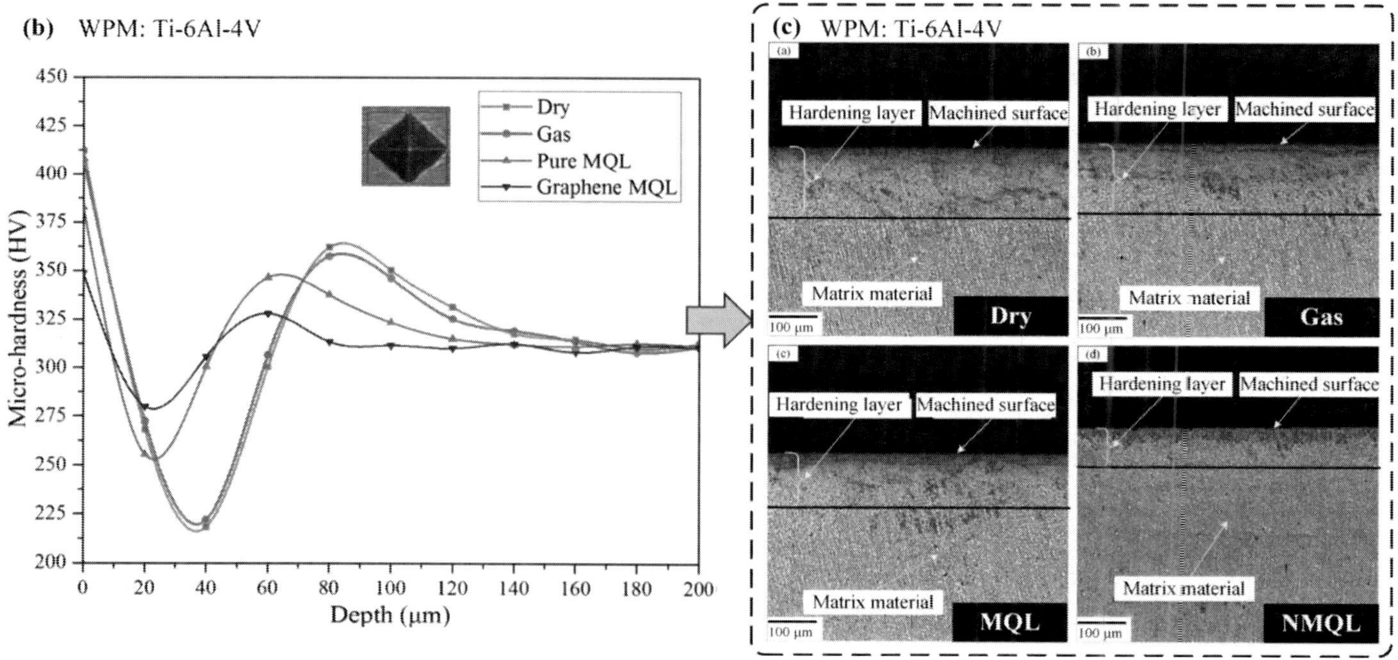

Figure 8. Workpiece analysis of NMQL milling using C-group nano-enhancers. (a) Micromorphology and EDS analysis, (b) Hardness measurement of workpiece surface.

2.5.5. The Milling Mechanism of NMQL Using Carbon Nano-Enhancers

Based on the above milling results, it can be seen that the order of antifriction and antiwear characteristics of C-group nano-enhancers with different morphology is: layered<tubular<spherical, which is different from the law obtained by milling (see Table 3). The reason is that after the change of cutting form, the infiltration film formation and antifriction and antiwear mechanism have also changed (see Figure 9).

Table 3. Conclusion of processing properties in milling

Working condition	Characterization parameter		Nano-enhancers	Conclusion
Milling	Cutting force		GR, GNP, CNTs	Order of decreasing degree: GNP<GR<CNTs
	Chip morphology	Thickness	GNP, CNTs, Al_2O_3	Order of decreasing degree: GR<CNTs< Al_2O_3
		Sawtooth		
		Chip curl diameter		
		Fresh surface		
	Tool	Tool life	GR, GNP	Better experimental results for carbon nano-enhancers
		VB	GR, GNP	
	Surface integrity	*Ra*	GR, GNP, CNTs	Tubular is better than layered, Milling has less lifting than turning
		SEM		

2.5.5.1. Mechanism of Wetting Film Formation in Milling

The milling process is the successive intermittent cutting of several edges to remove material (see Figure 9a). For a single cutting edge, the two wedge-shaped friction pairs of rake face/chip and flank surface/fresh surface do not exist all the time, but appear and disappear at a certain frequency (see Figure 9b). Compared with turning, the wetting boundary of nano-enhancers is essentially changed, which is the infiltration of a single edge in a very short time. This intermittent cutting is not conducive to nano-enhancers infiltration and film formation. On the one hand, when the cutting edge is not cutting, nano-enhancers are easily lost under the impact cleaning of high-pressure gas from the nozzle. On the other hand, it is difficult to form a more stable lubricating oil film quickly due to the extremely short wetting time of nano-enhancers and the continuous wetting like turning. Some nano-enhancers flow out with the cutting tool before they infiltrate into the cutting area, and even some nano-enhancers do not enter the cutting area and cannot play a cooling and lubricating role. This is why it is found that the reduction percentage of milling force with the same nano-enhancers is less than that of turning force, and the content of C element found on the surface of milling workpiece is far lower than that of turning.

2.5.5.2. Mechanism of Tool/Workpiece Antifriction and Antiwear

In addition, for different morphology of nano-enhancers, the wetting ability and antifriction and antiwear ability also have great differences. The self-spreading form of layered nano-enhancers is not conducive to the formation of film at the interface of rapid cutting. Compared with spherical and tubular, layered nano-enhancers have larger specific surface area, greater resistance to move in liquid, and poor wettability. It is not easy to enter the cutting area in the intermittent cutting process of cutting edge, and it is difficult to spread a large area of lubricating film in the cutting area in a very short time.

Therefore, the lubricity of the nano-enhancers is worse than that of the other two nano-enhancers. On the one hand, spherical nano-enhancers have less resistance to move in liquid, so they have better wettability. On the other hand, with the relative sliding of tool rake face and fresh surface, spherical nano-enhancers are more likely to roll into the interface depth and play a rolling lubrication role. Although the tubular CNTs are also rolling infiltration, they are easy to get entangled with each other, which increases the wetting resistance. In a very short time, the infiltration speed and depth are smaller than that of spherical nano-enhancers. Therefore, the spherical nano-enhancers are the best ones in the milling process (see Figure 9c). However, in past research, few scholars have carried out NMQL processing experiments for diamond or fullerene. The above rules are based on the experiment of Al_2O_3. Further research on the application of diamond or fullerene in milling must be done.

2.6. The Application of CGNs in MQL Grinding

Grinding, a finish machining method, is one of the main machining forms in hard machining materials, which could satisfy high precision and surface quality requirements for blade mortise, tenon and other important parts of the aero-engine [42]. The high requirements for machining precision and surface quality of the blade mortise and tenon and other important parts of the aero-engine determine that grinding occupies an important position in the manufacturing of aircraft engine parts mainly composed of hard machining materials. It determines that grinding occupies an important position in the manufacturing of aircraft engine parts mainly composed of hard machining materials. Compared with turning and milling, it is more different and meaningful to study the tribological characteristics of grinding, whose process is characterized by negative rake angle cutting of abrasives.

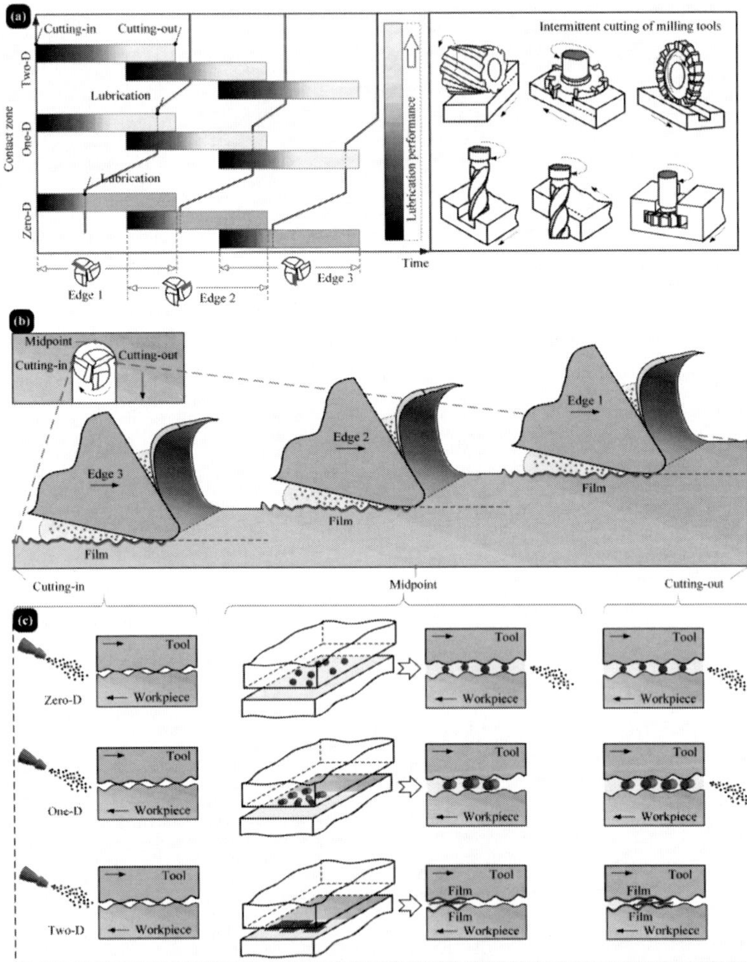

Figure 9. Milling mechanism of NMQL using carbon nano-enhancers. (a) Wetting and lubricating properties of different carbon group nanoparticles, (b) Kinematic analysis of material removal with multiple cutting edges, (c) Wetting properties of different nanometer phase.

2.6.1. Grinding Force

The NMQL grinding experiments were carried out with different nano-enhancers, and different conclusions were obtained (see Figure 10). (i) Layered nano-enhancers. When using graphene to grind Ti-6Al-4V material, compared with MQL, the F_n and F_t were reduced by 11.1% and 26.71%, respectively [43]. In the study of Ibrahim et al. [44], Fn and Ft decreased by 19.39% and 79.13% respectively. When Li et al. [45] grinds TC4 with graphene, F_n and F_t decreased by 22.28% and 42.79% respectively. The experiment of grinding Inconel 718 with graphene nanofluids by Pavan et al. [46] showed that Fn and Ft decreased by 25.26% and 27.22% respectively. (ii) Tubular nano-enhancers. Compared with MQL Inconel 718, the grinding force Ft decreased by 14.15% compared with MQL with CNTs [47]. Wang et al. [48] carried out the same experiment, and the grinding forces Fn and Ft decreased by 12.23% and 24.31% respectively. (iii) Spherical nano-enhancers. The experiment of grinding Inconel 718 with ND by Jia et al. showed that Ft decreased by 20.28% compared with MQL [47].

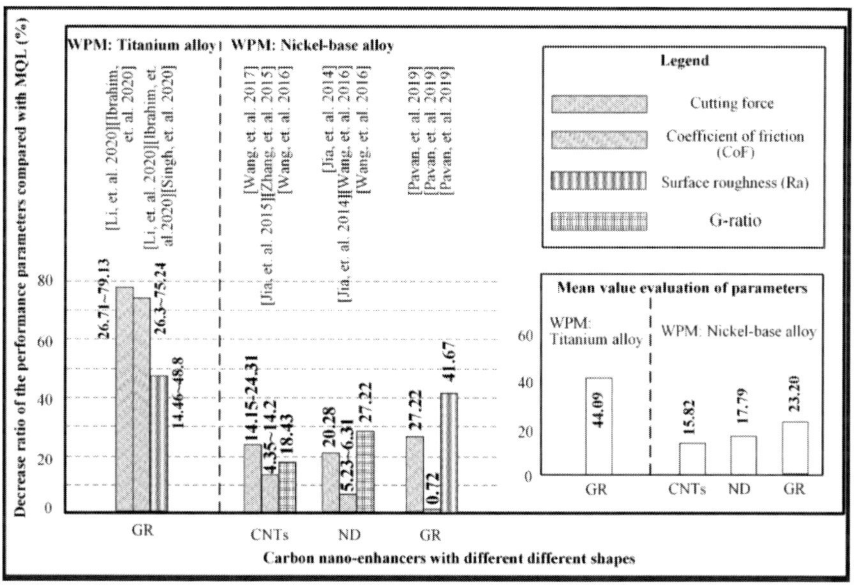

Figure 10. Analysis of experimental data in grinding.

Through the grinding force test of different morphologies of nano-enhancers, it is found that the order of reducing grinding force is:

tubular<spherical<layered. Further research on the application of different layered nano-enhancers was carried out to find the best process. Singh et al. [49] used different layered nano-enhancers to grind Ti-6Al-4V. On comparison with plain canola oil based MQL, it was found that 1.5 wt% graphene additives in canola oil based MQL had resulted in 9.8%, 25.6% reduced value of F_n, F_t. The grinding force decreased most under the lubrication of graphene. The order is: MoS_2<graphite<graphene. This also shows that graphene has great advantages in layered nano-enhancers.

Moreover, the effect of different nano-enhancers concentration on grinding force is also significant. In the experiment of grinding TC4 with graphene by Li et al. [45] further analyzed the influence of different graphene mass fraction on grinding force. With the increase of the mass fraction, the grinding force first increased and then decreased. It reached the minimum value at 0.1 wt%, which was due to the enhancement of the lubricating property of the oil film by appropriate amount of graphene. However, when the concentration is too low, graphene's function is weakened due to insufficient lubrication of grinding area. At the same time, when the concentration is too high, excessive graphene will easily block the grinding zone and hinder the formation of lubricating oil film, which will lead to the decrease of graphene strengthening effect due to insufficient lubrication.

2.6.2. Coefficient of Friction (CoF)

The index of CoF is very important in grinding. Compared with turning and milling, 90% of the grinding heat comes from the large area friction between the abrasive and the workpiece surface. Therefore, reducing the CoF is of great significance to reduce the grinding heat. The NMQL grinding experiments were carried out with different nano-enhancers, and different conclusions were obtained (see Figure 10). (i) Layered nano-enhancers. Li et al. [45] and Ibrahim et al. [44] respectively carried out the experimental research on grinding TC4 and Ti–6Al–4V with graphene, and their CoFs decreased by 26.3% and 75.24% respectively. The experimental study of grinding Inconel 718 with graphene by Pavan et al. [46] showed that the CoF was reduced by 0.72%. (ii) Tubular nano-enhancers. Jia et al. [47], Zhang et al. [50] and Wang et al. [48] carried out the comparative experiment of CNTs grinding Inconel 718. The results showed that the CoFs decreased by 4.35%, 14.29% and 10.75% respectively compared with MQL. (iii) Spherical nano-enhancers. Jia et al. [47] and Wang et al. [48] showed that the CoF of ND grinding Inconel 718 was reduced by 5.23% and 6.31% respectively compared with MQL.

It is found that the layered graphene has more potential to reduce the CoF, that is, to improve the lubrication performance compared with spherical diamond and tubular CNTs. For different layered nano-enhancers, they still have different lubricating properties. When graphene, graphite and MoS_2 were used to grind Ti-6Al-4V. The order of CoF was MoS_2>graphite>graphene [43, 49]. Graphene can reduce 21.9% compared with MoS_2 disulfide. This is due to the different interlayer bond energy of different nano-enhancers. Cui et al. [51] summarized the antifriction mechanism of layered nano-enhancers. The atomic binding force between the layers of the nano-enhancers is weak, thus forming a low shear plane. When the upper and lower layers move each other, the nano-enhancers extend along the fracture layer of the low shear plane and form a film. The weaker the interlayer bonding force (i.e., the weaker the interlayer bond energy), the easier it is to spread the film. The interlayer bond energy of nano-enhancers is proportional to the mass fraction. Among the three nano-enhancers, the order of mass fraction is MoS_2 disulfide>graphite>graphene. Therefore, it is proved that the lubrication performance of graphene is better than the other two.

2.6.3. Chip Morphology

The change rule of chip observed in grinding is similar to that in cutting. Compared with MQL, the surface of the chips with diamond was smoother and its shape became longer and finer [48]. It is proved that the diamond has excellent lubricating properties and reduce the friction angle of material removal, which plays an important role in reducing abrasive wear and reducing grinding force [51, 52].

2.6.4. Tool Life and Wear

2.6.4.1. Abrasive Wear

Ibrahim et al. [44] characterized the abrasive wear by friction and wear experiments. Using graphene NMQL as lubricating condition, the wear rate of ZrO_2 was reduced by 90% compared with MQL. Therefore, the abrasive can maintain good shape and size (i.e., sharpness), so that they can be processed with less grinding force and CoF.

2.6.4.2. Grinding Wheel Wear

Wang et al. [48] carried out a comparative experiment of diamond and CNTs grinding Inconel 718. The results showed that the G-ratio of CNTs and diamond increased by 18.43% and 27.22% respectively compared with MQL. This also showed the excellent properties of C-group nano-enhancers in antifriction and antiwear. This was initially confirmed by Kalita et al. [52] (see Figure 10). They found a lubricating oil film mainly composed of nano-enhancers by observing the surface of abrasive and scanning electron microscope (SEM) and energy dispersive spectrometer (EDS).

2.6.5. Surface Integrity

2.6.5.1. Surface Roughness

Ibrahim et al. [44], Li et al. [45] used graphene to grind Ti-6Al- 4V and TC4 alloys respectively. The results showed that the *Ra* values were reduced by 32.65% and 14.46% respectively. At the same time, in TC4 grinding experiment, the micro-hardness under graphene lubrication also decreased by 10.81%. Singh et al. [49] used graphene to grind Ti-6Al-4V, and the Ra value decreased by 48.8%. Pavan et al. [46] used graphene to grind Inconel 718. The results showed that the Ra value was reduced by 41.67% (see Figure 10).

2.6.5.2. Morphology of Machined Surface

For different layered nano-enhancers, Singh et al. [49] used MoS_2, graphite and graphene to grind titanium alloy. The results showed that the surface quality was improved significantly, and the chip/burr welding phenomenon disappeared. The order of surface quality was graphene>graphite>MoS_2. Li et al. [45] carried out an experimental study on grinding TC4 with graphene. Compared with the MQL, the surface quality of the workpiece was significantly improved after adding graphene, and large furrow, adhesion and grinding burn disappeared. Furthermore, Ibrahim et al. [44] and Li et al. [45] compared the surface quality of graphene with different concentrations, and the surface quality of graphene showed a trend of first improving and then decreasing with the concentration (see Figure 11a).

In order to verify the film-forming phenomenon on the workpiece surface, the SEM image and element distribution of the grinding surface were analyzed in the research of graphene grinding titanium alloy by Ibrahim et al. [44]. The C content on the surface of Ti-6Al-4V workpiece increased from 0.03% before grinding to 71.7% (see Figure 11a). Therefore, it was proved that there is a large amount of graphene on the surface of the workpiece to form the lubrication film, which plays a role in antifriction and antiwear (see Figure 11b and 11c).

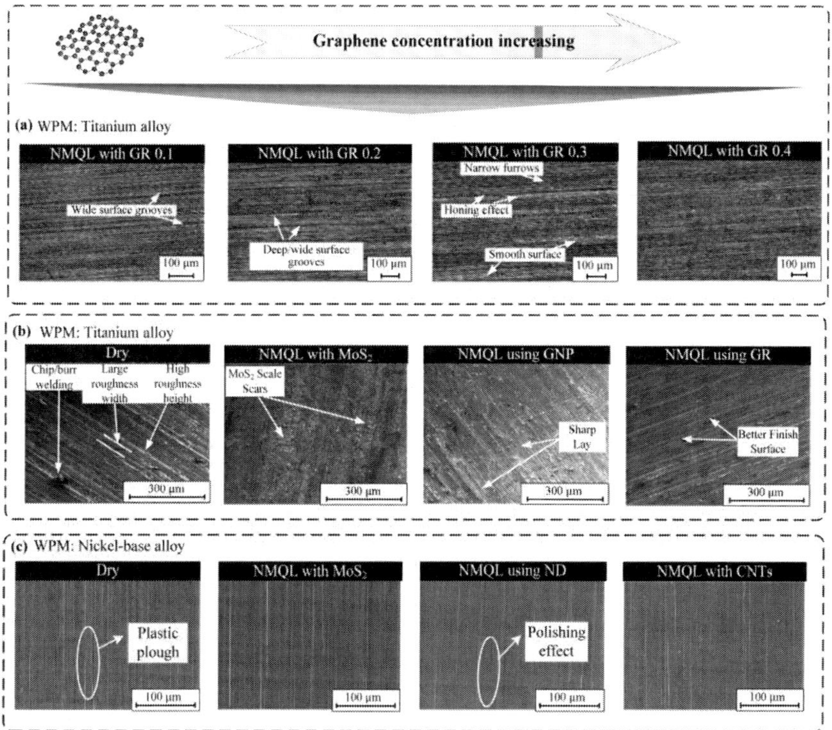

Figure 11. Workpiece analysis of NMQL grinding using C-group nano-enhancers. (a) Micromorphology and EDS analysis of workpiece in different graphene concentration, (b) Micromorphology analysis of titanium alloy, (c) Micromorphology analysis of nickel-base alloy.

2.6.6. The Grinding Mechanism of NMQL Using Carbon Nano-Enhancers

Based on the above grinding results, it can be seen that the order of antifriction and antiwear characteristics of C-group nano-enhancers with different morphology is tubular<spherical<layered, which is different from the rules obtained by turning and milling (see Table 4). The infiltration film formation and antifriction and antiwear mechanism in grinding is shown in Figure 12.

Table 4. Conclusion of processing properties in grinding

Working condition	Characterization parameter		Nano-enhancers	Conclusion
Grinding	Cutting force		GR, CNTs, GNP, ND, MoS_2	Order of decreasing degree:MoS_2<GNP<GR (Tubular< spherical<layered)
	CoF		GR, CNTs, ND, MoS_2	
	Chip morphology		ND	Better experimental results for carbon nano-enhancers
	Tool	Abrasive wear	GR	Better experimental results for carbon nano-enhancers
		Grinding wheel wear	CNTs, ND	Order of decreasing degree: CNTs<ND
	Surface integrity	*Ra*	GR	Better experimental results for carbon nano-enhancers
		SEM	GNP, GR, MoS_2	Order of decreasing degree: MoS_2<GNP<GR

2.6.6.1. Mechanism of Wetting Film Formation in Grinding

Grinding is similar to milling, in which a single abrasive/cutting edge successively and intermittently cuts to remove material (see Figure 12a). However, there are many abrasives and pores on the surface of the grinding wheel, and the abrasives are negative rake angle to remove the material, so the wetting film-forming mechanism is essentially different from milling.

From the aspect of cutting characteristics, the negative rake angle cutting characteristic is favorable to the nano-film formation. Grinding is a negative rake angle cutting, and the grinding process will go through obvious sliding and ploughing stages. In this process, nano-enhancers will be crushed by the abrasives on the surface of the workpiece to form an oil film. At the same time, due to the interaction, oil film will be formed on the top of abrasives. However, milling is a positive front angle cutting, and there is no obvious sliding and ploughing stage in the cutting process, that is, the nanofluid is not rolled on the surface of the workpiece to form an oil film, so the surface of the workpiece cannot form or form an unstable oil film. In this case, the self-spreading infiltration of layered nano-enhancers can form a lubricating oil film with larger coverage and stronger antifriction and antiwear ability, and it is easier to combine with the surface of friction pairs to improve the stability of oil film.

Considering the material removal mechanism, the lubricating oil film on the workpiece surface can play a greater role in the sliding and ploughing stages of grinding process. In grinding process, the machining track of the latter abrasive is based on the former one, that is, the surface processed by the latter abrasive is the fresh surface processed by the former abrasive. Since a stable lubricating oil film is formed on the surface of the workpiece after the former abrasive is processed. The oil film on the workpiece surface plays a good lubricating effect when the latter abrasive is in the stage of sliding and ploughing, which greatly reduces the friction heat between the tool and the workpiece. However, for milling, the last cutting edge will not cut along the track of the latter one, so whether the cutting edges form oil film on the surface of the workpiece has no effect on the latter edge (see Figure 12b).

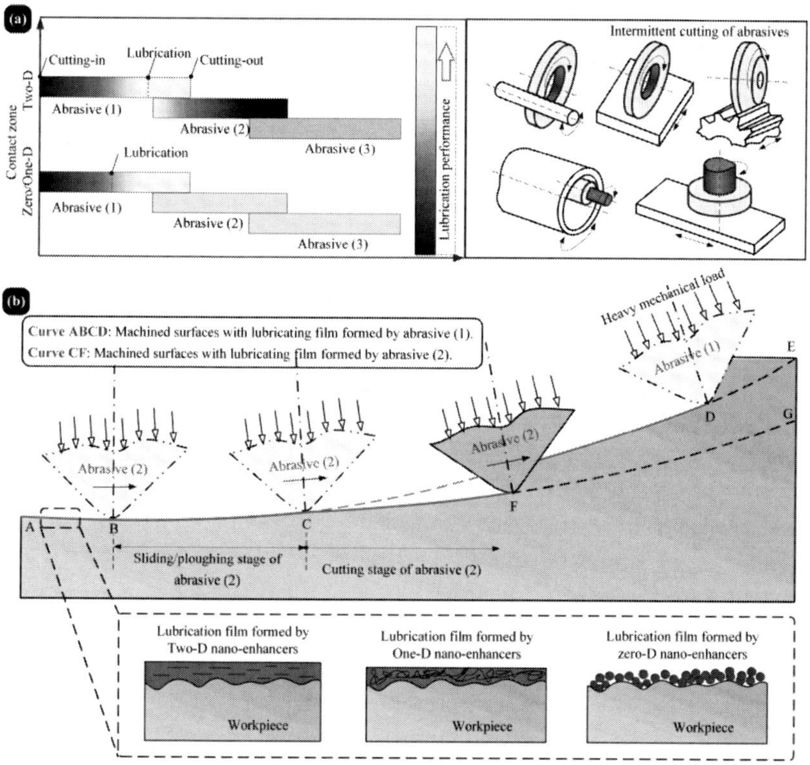

Figure 12. Grinding mechanism of NMQL using carbon nano-enhancers. (a) Wetting and lubricating properties of different carbon group nanoparticles, (b) Kinematic relationship of different abrasive particles and antifriction and antiwear mechanism of different nano-enhancers.

Considering the tool structure, the structure of grinding wheel is beneficial to the infiltration of lubricating oil and the formation of oil film. For milling cutter, the distance between cutting edges is large, which is not conducive to the storage and full infiltration of lubricant. Grinding is different. In the grinding zone with large cutting arc length, there are dozens or even more than 100 abrasives cutting at the same time. Because of the pores on the surface of grinding wheel, capillary/microchannel network is formed in the grinding zone. On the one hand, the dynamic pumping effect of capillary provides space and power for lubricant infiltration. On the other hand, the lubricant can be stored in the pores of the grinding wheel and always be filled, and sufficient nanofluid will be rolled into film repeatedly during the repeated grinding of the grinding wheel. Therefore, under this favorable boundary condition, the infiltration speed and depth are no longer the standard to measure different kinds of nano-enhancers, and the spreading area becomes the key performance, and the layered nano-enhancers are more advantageous than the other two.

2.6.6.2. Mechanism of Tool/Workpiece Antifriction and Antiwear

Compared with the spherical and tubular nano-enhancers, the interlayer atomic binding force of the layered nano-enhancers is weak, thus forming a low shear plane. When the upper and lower layers move mutually, the nano-enhancers extend along the fracture layer of the low shear plane and form a film. Due to the repeated grinding of the grinding wheel, the nano-enhancers spread repeatedly, and the coverage area increases, so the lubrication performance is enhanced, and the continuous and stable lubricating oil film is easy to form. Moreover, the weaker the interlayer bonding strength is, the easier it is to spread the film. While it is difficult for spherical and tubular nano-enhancers to form a stable lubricating oil film under the condition of sufficient wetting, but they always work in the way of rolling friction reduction, which results in the poor antifriction and antiwear performance of the particles in the repeated cutting process (see Figure 12b).

Conclusion

The application of C-group nano-enhancers in MQL is reviewed. In this review, the force, CoF, material removal, tool wear and workpiece surface integrity in turning, milling and grinding of aerospace difficult-to-machine materials in the aerospace industry are reviewed and evaluated. The conclusions are as follows.

i. In the MQL machining of aerospace difficult-to-machine materials in the aerospace industry, the lubrication state of tool/workpiece interface changes essentially after adding nano-enhancers. The addition of nano-enhancers avoids the direct contact between the tool and the workpiece during boundary lubrication and reduces the friction between the tool and the workpiece. For the tool, a dynamic or static protective oil film is formed at the cutting edge, which delays the tool failure, increases the sharpness of the tool tip, reduces the tool wear and the tool surface adhesion. For the chip, the addition of nano-enhancers makes the chip thickness smaller, the sawtooth height and chip curl diameter and chip surface scratch decreases. For the surface integrity of the workpiece, the addition of nano-enhancers will improve the cooling and lubrication performance of the tool/workpiece contact area and form a nano lubrication film between the tool flank and the workpiece surface, reducing the friction and heat generation, and protecting the newly formed surface of the workpiece.

ii. For turning titanium alloy and nickel-base alloy, C-group nano-enhancers have better lubrication properties than MQL. The results show that the turning force is reduced by 10.64~45.65%, the CoF is reduced by 27.54~68.63%, the tool wear VB is reduced by 2.38~63.63%, the tool life is reduced by 2.38~27.27%, and the surface roughness Ra value is reduced by 28.54~79.03%. In addition, abrasion, diffusion, chipping, attrition are avoided in the tool. Adhere, small micro burrs and surface pitting are avoided on the workpiece. Compared with spherical and layered nano-enhancers, tubular nano-enhancers are easier to enter the continuous cutting zone, which avoids the vibration of the tool, stabilizes the lubrication state of the cutting zone, and greatly reduces the friction. Therefore, tubular nano-enhancers are the best choice in turning.

iii. For milling titanium alloy and nickel-base alloy, C-group nano-enhancers have better lubrication properties than MQL. The milling force decreased by 8.43~22.4%, tool wear VB decreased by 8.57~52.17%, tool life decreased by 20.5~23.07%, and surface roughness Ra decreased by 26.19~51.07%. Moreover, adhesion, edge chipping, build-up edge are avoided on the tool, and adhesion, surface pit, scratch and large fuse are avoided on the workpiece. The spherical nano-enhancers are easier to roll into the depth of the dynamic intermittent cutting interface and play a rolling lubrication role. The moving resistance in the liquid is smaller, so it has better wettability. It has better cooling and lubricating properties than tubular and layered materials. Therefore, spherical nano-enhancers are the best choice in milling.

iv. For grinding titanium alloy and nickel-base alloy, C-group nano-enhancers have better lubrication properties than MQL. The grinding force is reduced by 9.8~79.13%, the CoF is reduced by 4.35~75.24%, the tool wear rate is reduced by 18.43~71.92%, and the surface roughness Ra value is reduced by 14.46~48.8%. The surface of the workpiece is improved. Compared with tubular and spherical particles, layered nano-enhancers are more beneficial to antifriction and antiwear at the interface of high load abrasive/workpiece under sufficient wetting conditions. Therefore, layered nano-enhancers are the best choice in grinding.

References

[1] Zhang XT, Li CH, Zhou ZM, et al. (2023). Vegetable Oil-Based Nanolubricants in Machining: From Physicochemical Properties to Application. *Chin. J. Mech. Eng.*, *36*(1), 39. 10.1186/s10033-023-00895-5.

[2] Song YX, Xu ZL, Li CH, et al. (2023). Research progress on the grinding performance of nanobiolubricant minimum quantity lubrication. *Surface Technology*, *52*(12), 1-20.

[3] Hu SG, Li CH, Zhou ZM, et al. (2023). Nanoparticle-enhanced coolants in machining: mechanism, application, and prospects. *Front. Mech. Eng.*, *18*(4), 47. 10.1007/s11465-023-0769-8.

[4] Shi Z, Guo SM, Liu HJ, et al. (2021). Experimental evaluation of minimum quantity lubrication of biological lubricant on grinding properties of GH4169 nickel-base alloy. *Surface Technology*, *50*(12), 71-84.

[5] Joly-Pottuz L, Matsumoto N, Kinoshita H, et al. (2008). Diamond-derived carbon onions as lubricant additives. *Tribol. Int.*, *41*(2), 69-78. 10.1016/j.triboint.2007.05.001.

[6] Stankovich S, Dikin DA, Piner RD, et al. (2007). Synthesis of graphene-based nanosheets via chemical reduction of exfoliated graphite oxide. *CARBON*, *45*(7), 1558-1565.

[7] Li S, Wang CA, Yang FQ, et al. (2021). Hollow-grained "Voronoi foam" ceramics with high strength and thermal superinsulation up to 1400°C. *Materials Today*, *46*(6), 35-43.

[8] Wang YG, Li CH, Zhang YB, et al. (2017). Comparative evaluation of the lubricating properties of vegetable-oil-based nanofluids between frictional test and grinding experiment. *J. Manuf. Process.*, *26*, 94-104. 10.1016/j.jmapro.2017.02.001.

[9] Yildirim ÇV. (2020). Investigation of hard turning performance of eco-friendly cooling strategies: Cryogenic cooling and nanofluid based MQL. *Tribol. Int.*, *144*, 16. 10.1016/j.triboint.2019.106127.

[10] Tiwari JN, Tiwari RN, Kim KS. (2012). Zero-dimensional, one-dimensional, two-dimensional and three-dimensional nanostructured materials for advanced electrochemical energy devices. *Prog. Mater. Sci.*, *57*(4), 724-803. 10.1016/j.pmatsci.2011.08.003.

[11] Kharlamova MV. (2016). Advances in tailoring the electronic properties of single-walled carbon nanotubes. *Prog. Mater. Sci.*, *77*, 125-211. 10.1016/j.pmatsci.2015.09.001.

[12] Liu L, Fang Z, Gu A, et al. (2011). Lubrication Effect of the Paraffin Oil Filled with Functionalized Multiwalled Carbon Nanotubes for Bismaleimide Resin. *Tribology Letters*, *42*(1), 59-65.

[13] Rümmeli MH, Rocha CG, Ortmann F, et al. (2011). Graphene: Piecing it Together. *Adv. Mater.*, *23*(39), 4471-4490. 10.1002/adma.201101855.

[14] Novoselov KS. (2011). Nobel Lecture: Graphene: Materials in the Flatland. *Rev. Mod. Phys.*, *83*(3), 837-849. 10.1103/RevModPhys.83.837.

[15] Berman D, Erdemir A, Sumant AV. (2013). Few layer graphene to reduce wear and friction on sliding steel surfaces. *Carbon*, *54*, 454-459. 10.1016/j.carbon.2012.11.061.

[16] Kinoshita H, Nishina Y, Alias AA, et al. (2014). Tribological properties of monolayer graphene oxide sheets as water-based lubricant additives. *Carbon*, *66*, 720-723. 10.1016/j.carbon.2013.08.045.

[17] Chan CH, Tang SW, Mohd NK, et al. (2018). Tribological behavior of biolubricant base stocks and additives. *Renew. Sust. Energ. Rev.*, *93*, 145-157. 10.1016/j.rser.2018.05.024.

[18] Wu WT, Li CH, Yang M, et al. (2019). Specific Energy and G ratio of Grinding Cemented Carbide under Different Cooling and Lubrication Conditions. *Int. J. Adv. Manuf. Technol.*, *105*(1-4), 67-82. 10.1007/s00170-019-04156-5.

[19] Singh R, Dureja JS, Dogra M, et al. (2019). Influence of graphene-enriched nanofluids and textured tool on machining behavior of Ti-6Al-4V alloy. *Int. J. Adv. Manuf. Technol.*, *105*(1-4), 1685-1697. 10.1007/s00170-019-04377-8.

[20] Mishra SK, Ghosh S, Aravindan S. (2020). Machining performance evaluation of Ti6Al4V alloy with laser textured tools under MQL and nano-MQL environments. *J. Manuf. Process.*, *53*, 174-189. 10.1016/j.jmapro.2020.02.014.

[21] Dhar NR, Islam MW, Islam S, et al. (2006). The influence of minimum quantity of lubrication (MQL) on cutting temperature, chip and dimensional accuracy in turning AISI-1040 steel. *Journal of Materials Processing Technology*, *171*(1), 93-99.

[22] Wang XM, Li CH, Zhang YB, et al. (2022). Tribology of enhanced turning using biolubricants: A comparative assessment. *Tribol. Int.*, *174*, 30. 10.1016/j.triboint.2022.107766.

[23] Li GX, Yi S, Li N, et al. (2019). Quantitative analysis of cooling and lubricating effects of graphene oxide nanofluids in machining titanium alloy Ti6Al4V. *Journal of Materials Processing Technology*, *271*, 584-598. 10.1016/j.jmatprotec.2019.04.035.

[24] Nguyen D, Lee PH, Guo Y, et al. (2019). Wear Performance Evaluation of Minimum Quantity Lubrication With Exfoliated Graphite Nanoplatelets in Turning Titanium Alloy. *J. Manuf. Sci. Eng.-Trans. ASME*, *141*(8), 8. 10.1115/1.4043899.

[25] Marques A, Suarez MP, Sales WF, et al. (2019). Turning of Inconel 718 with whisker-reinforced ceramic tools applying vegetable-based cutting fluid mixed with solid lubricants by MQL. *Journal of Materials Processing Technology*, *266*, 530-543. 10.1016/j.jmatprotec.2018.11.032.

[26] Hegab H, Kishawy HA, Gadallah MH, et al. (2018). On machining of Ti-6Al-4V using multi-walled carbon nanotubes-based nano-fluid under minimum quantity lubrication. *Int. J. Adv. Manuf. Technol.*, *97*(5-8), 1593-1603. 10.1007/s00170-018-2028-4.

[27] Singh R, Dureja JS, Dogra M, et al. (2020). Wear behavior of textured tools under graphene-assisted minimum quantity lubrication system in machining Ti-6Al-4V alloy. *Tribol. Int.*, *145*, 13. 10.1016/j.triboint.2020.106183.

[28] Hegab H, Umer U, Deiab I, et al. (2018). Performance evaluation of Ti-6Al-4V machining using nano-cutting fluids under minimum quantity lubrication. *Int. J. Adv. Manuf. Technol.*, *95*(9-12), 4229-4241. 10.1007/s00170-017-1527-z.

[29] Hegab H, Darras BM, Kishawy HA. (2018). Sustainability Assessment of Machining with Nano-Cutting Fluids. *Journal of Procedia Manufacturing*, *26*, 245-254.

[30] Gutnichenko AQ, Bushlya AV, Bihagen BS, et al. (2018). Influence of GnP additive to vegetable oil on machining performance when MQL-assisted turning Alloy 718. *Procedia Manufacturing*, *25*, 330-337.

[31] Gupta MK, Mia M, Pruncu CI, et al. (2020). Modeling and performance evaluation of Al₂O₃, MoS₂ and graphite nanoparticle-assisted MQL in turning titanium alloy: an intelligent approach. *J. Braz. Soc. Mech. Sci. Eng.*, *42*(4), 21. 10.1007/s40430-020-2256-z.

[32] Revuru RS, Zhang JZ, Posinasetti NR, et al. (2018). Optimization of titanium alloys turning operation in varied cutting fluid conditions with multiple machining performance characteristics. *Int. J. Adv. Manuf. Technol.*, *95*(1-4), 1451-1463. 10.1007/s00170-017-1299-5.

[33] Liu DW, Li CH, Dong L, et al. (2022). Kinematics and improved surface roughness model in milling. *Int. J. Adv. Manuf. Technol.*, 22. 10.1007/s00170-022-10729-8.

[34] Duan ZJ, Li CH, Zhang YB, et al. (2023). Mechanical behavior and semiempirical force model of aerospace aluminum alloy milling using nano biological lubricant. *Front. Mech. Eng.*, *18*(1), 1-15. 10.1007/s11465-022-0720-4.

[35] Li M, Yu T, Li H, et al. (2019). Research on surface integrity in graphene nanofluid MQL milling of TC21 alloy. *International Journal of Abrasive Technology* *9*(1).

[36] Li M, Yu TB, Zhang RC, et al. (2018). MQL milling of TC4 alloy by dispersing graphene into vegetable oil-based cutting fluid. *Int. J. Adv. Manuf. Technol.*, *99*(5-8), 1735-1753. 10.1007/s00170-018-2576-7.

[37] Bai XF, Li CH, Dong L, et al. (2019). Experimental evaluation of the lubrication performances of different nanofluids for minimum quantity lubrication (MQL) in milling Ti-6Al-4V. *Int. J. Adv. Manuf. Technol.*, *101*(9-12), 2621-2632. 10.1007/s00170-018-3100-9.

[38] Sirin S, Kivak T. (2019). Performances of different eco-friendly nanofluid lubricants in the milling of Inconel X-750 superalloy. *Tribol. Int.*, *137*, 180-192. 10.1016/j.triboint.2019.04.042.

[39] Park KH, Suhaimi MA, Yang GD, et al. (2017). Milling of titanium alloy with cryogenic cooling and minimum quantity lubrication (MQL). *Int. J. Precis. Eng. Manuf.*, *18*(1), 5-14. 10.1007/s12541-017-0001-z.

[40] Yin QG, Li CH, Dong L, et al. (2021). Effects of Physicochemical Properties of Different Base Oils on Friction Coefficient and Surface Roughness in MQL Milling AISI 1045. *Int. J. Precis Eng Manuf-Green Technol.*, *8*(6), 1629-1647. 10.1007/s40684-021-00318-7.

[41] Yin QA, Li CH, Dong L, et al. (2018). Effects of the physicochemical properties of different nanoparticles on lubrication performance and experimental evaluation in the NMQL milling of Ti-6Al-4V. *Int. J. Adv. Manuf. Technol.*, *99*(9-12), 3091-3109. 10.1007/s00170-018-2611-8.

[42] Cui X, Li CH, Zhang YB, et al. (2022). Grindability of titanium alloy using cryogenic nanolubricant minimum quantity lubrication. *J. Manuf. Process.*, *80*, 273-286. 10.1016/j.jmapro.2022.06.003.

[43] Singh H, Sharma VS, Dogra M. (2020). Exploration of graphene assisted vegetables oil based minimum quantity lubrication for surface grinding of TI-6AL-4V-ELI. *Tribol. Int.*, *144*, 13. 10.1016/j.triboint.2019.106113.

[44] Ibrahim AMM, Li W, Xiao H, et al. (2020). Energy conservation and environmental sustainability during grinding operation of Ti–6Al–4V alloys via eco-friendly oil/graphene nano additive and Minimum quantity lubrication. *150*, 106387.

[45] Li M, Yu TB, Zhang RC, et al. (2020). Experimental evaluation of an eco-friendly grinding process combining minimum quantity lubrication and graphene-enhanced plant-oil-based cutting fluid. *J. Clean Prod.*, *244*, 13. 10.1016/j.jclepro. 2019.118747.

[46] Pavan RB, Gopal AV, Amrita M, et al. (2019). Experimental investigation of graphene nanoplatelets-based minimum quantity lubrication in grinding Inconel 718. *Proc. Inst. Mech. Eng. Part B-J. Eng. Manuf.*, *233*(2), 400-410. 10.1177/0954405417728311.

[47] Jia DZ, Li CH, Zhang DK, et al. (2014). Experimental verification of nanoparticle jet minimum quantity lubrication effectiveness in grinding. *J. Nanopart. Res.*, *16*(12), 15. 10.1007/s11051-014-2758-7.

[48] Wang Y, Li C, Zhang Y, et al. (2016). Experimental evaluation of the lubrication properties of the wheel/workpiece interface in MQL grinding with different nanofluids. *Tribol. Int.*, *99*, 198-210.

[49] Singh H, Sharma VS, Singh S, et al. (2019). Nanofluids assisted environmental friendly lubricating strategies for the surface grinding of titanium alloy: Ti6Al4V-ELI. *J. Manuf. Process.*, *39*, 241-249. 10.1016/j.jmapro.2019.02.004.

[50] Zhang DK, Li CH, Jia DZ, et al. (2015). Specific grinding energy and surface roughness of nanoparticle jet minimum quantity lubrication in grinding. *Chin. J. Aeronaut.*, *28*(2), 570-581. 10.1016/j.cja.2014.12.035.

[51] Cui X, Li CH, Zhang YB, et al. (2019). Tribological properties under the grinding wheel and workpiece interface by using graphene nanofluid lubricant. *Int. J. Adv. Manuf. Technol.*, *104*(9-12), 3943-3958. 10.1007/s00170-019-04129-8.

[52] Kalita P, Malshe AP, Kumar SA, et al. (2012). Study of specific energy and friction coefficient in minimum quantity lubrication grinding using oil-based nanolubricants. *14*(2), 160-166.

Chapter 3

How the Physicochemical Properties of Nanolubricants Affect the Grinding Performance of Difficult-to-Machine Materials

Abstract

As the base oil of nanolubricant, the properties of biolubricant have significant influence on the grinding performance of aerospace material. However, the primary cause for this condition is the unknown and complex influencing mechanisms of the biolubricant physicochemical properties on grindability. In this chapter, a comparative assessment of grindability is performed using titanium alloy, nickel-based alloy, and high-strength steel. Firstly, this chapter considers the physicochemical properties as the main factors, and the antifriction and heat dissipation behaviours of biolubricant in a high temperature and pressure interface are comprehensively analyzed. Secondly, the comparative assessment of force, temperature, wheel wear and workpiece surface for titanium alloy, nickel-based alloy, and high-strength steel confirms that biolubricant is a potential replacement of traditional cutting fluids because of its improved lubrication and cooling performance. Biolubricant with high viscosity and high fatty acid saturation characteristics should be used to overcome the bottleneck of wheel wear and nickel-based alloy surface burn. The nano-enhancers with high hardness and spherical characteristics are better choices. Furthermore, a different option is available for high-strength steel grinding, which needs low-viscosity biolubricant to address the debris breaking difficulty and wheel clogging. Finally, the current challenges and potential methods are proposed to promote the application of biolubricant.

Keywords: grinding, aerospace, difficult-to-machine material, biolubricant, physicochemical property, grindability

3.1. Introduction

Grinding is one of the most important methods of precision forming parts, and the use of grinding fluids exceeds that of cutting fluids [1, 2]. In addition, grinding is a crucial process for ensuring manufacturing accuracy and surface quality, especially for aerospace materials, and it accounts for more than 60% of all processing procedures [3, 4]. Accordingly, MQL should be applied in grinding. However, the MQL technology is still constrained by insufficient heat transfer capacity when it comes to aerospace material grinding. These difficult-to-machine materials are more likely to experience thermal damage in the high temperature grinding zone due to their characteristics of low thermal conductivity, high specific grinding energy and fracture toughness [5, 6]. The grinding process of titanium alloy, superalloy, high-strength steel, composite material, and other difficult-to-machine materials has the serious problems of heavy grinding force, high temperature, and severe wheel wear or clogging [7, 8]. The afore-mentioned issues in MQL grinding must be addressed as soon as possible [9]. Nano-enhanced biolubricants, which are prepared by adding nanoparticles in the biolubricant, are expected to improve grindability and meet the demand for green manufacturing [10, 11].

Biolubricant, as a vital medium in the manufacturing system, is essential for ensuring machining performance and environmental friendliness [12, 13]. The core problem is to solve heat dissipation, antifriction and antiwear problems through the preparation and use of biolubricant or nano-enhanced biolubricants under extreme grinding conditions. Scholars have carried out a series of experimental studies using vegetable oil as biological cooling medium to evaluate the grinding performance parameters [14, 15]. However, the influencing law and knowledge system amongst (i) physicochemical property of lubricants, (ii) antifriction and heat transfer behavior, and (iii) machining performance parameters remain unknown. There is no basis for scientific research and industrial application. On this basis, this work conducted a systematic review of biolubricant MQL for machining aerospace difficult-to-machine materials and summarized the influencing mechanism of the physicochemical properties on grindability. The structure of this paper is shown in Figure 1. Furthermore, the machining performance of titanium alloy, nickel-based alloy, and high-strength steel was reviewed by quantitative evaluation. Finally, this work aims to provide theoretical references for further research by improving the influencing law of biolubricant physicochemical properties on heat dissipation, antifriction, and antiwear.

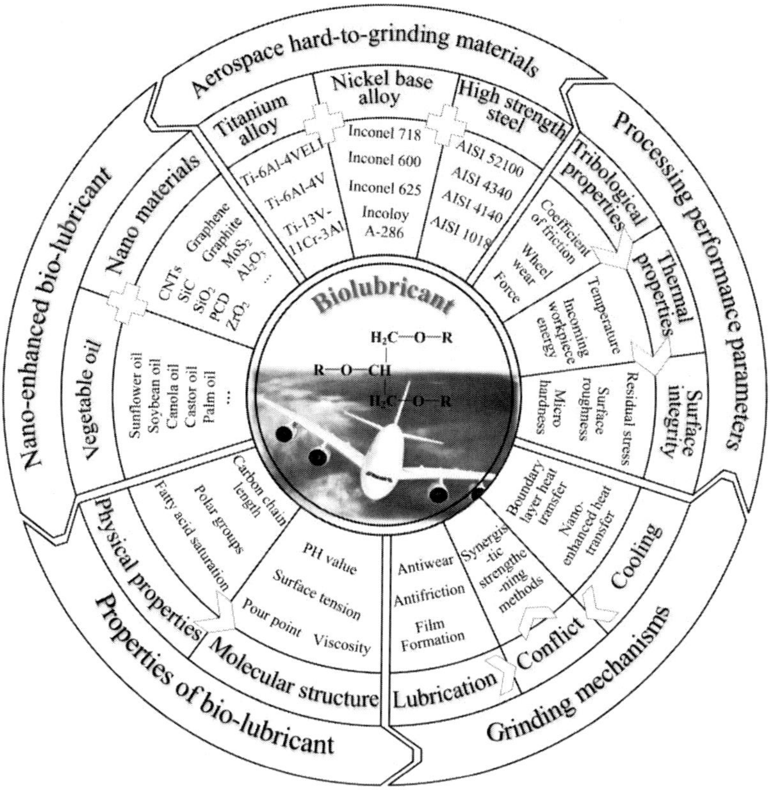

Figure 1. Paper structure.

3.2. Nanolubricant Physical Properties and Grinding Performance

In the early days of MQL, mineral oils (e.g., liquid paraffin) were often used as micro-lubricants. At present, some research uses mineral synthetic oils (e.g., Castrol limited). Although mineral oil as a micro-lubricant can meet the cooling and lubrication needs, it is still dependent on non-renewable resources and poses a threat to the environment and workers' health. Kelly and Cotterell [16] used vegetable oil as micro-lubricant in drilling experiments for the first time in 2002, and the processing temperature, torque and surface roughness were superior to flood with a certain drilling speed, which opened the research exploration of biolubricant applied in MQL. Biolubricant is both naturally degradable and environmentally friendly, making it perfectly compatible with the original purpose of MQL. Biolubricants currently used for MQL include various vegetable oils (e.g., soybean, palm, peanut, castor, sunflower, rapeseed, sesame, corn and coconut oil, etc.) and synthetic esters (made from degradable components). Scholars have conducted a comprehensive study on the processing performance of vegetable oil as a lubricant [17, 18]. Including a verifiable comparison of its application performance under different cooling and lubrication methods, analysis of the optimality of different vegetable oils for different processing processes and workpiece materials, and revelation of cooling and lubrication mechanisms.

The main components of vegetable oil are triacylglyceride, a small amount of free fatty acid, part of glyceride, phosphorus ester (0.1%~0.5%), tiny amounts of sterols, tocopherol and vitamin E. Each triacylglyceride molecule contains three C12~C22 fatty acid chains, including saturated fatty acids such as palmitic, stearic and oleic acids, and a variety of unsaturated acids such as erucic acid, jatropha, ricinoleic acid and so on. The main parameters that affect the cooling and lubrication performance include: fatty acid carbon chain length, fatty acid saturation (number of C=C double bonds), functional group type, viscosity, surface tension, PH value, etc. The Grinding mechanisms with usage of biolubricant was showed in Figure 2.

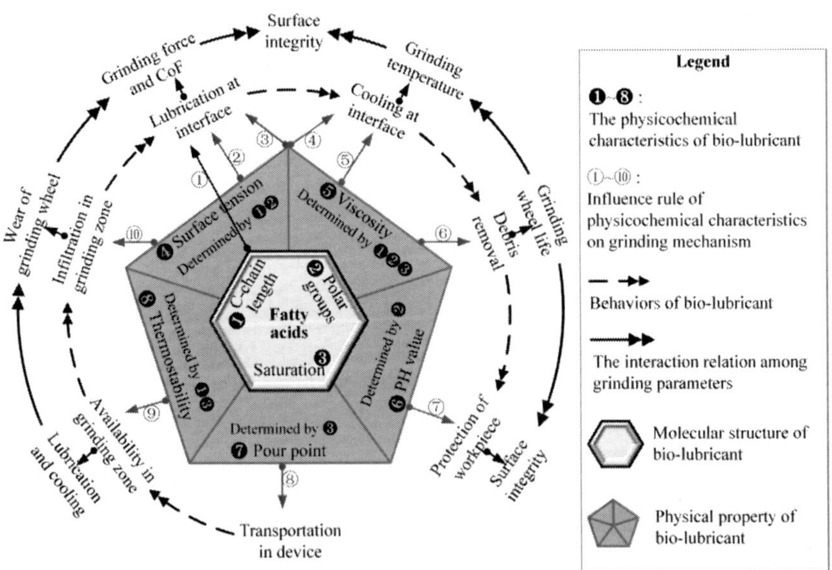

Figure 2. Grinding mechanisms with usage of biolubricant.

3.2.1. The Molecular Structure of Fatty Acids

The molecular structure of biolubricants is composed of one glyceride and three straight-chain fatty acids. The carbon chain length of fatty acid molecule ranges from 14 to 22, including saturated and unsaturated fatty acids. The different fatty acids and their contents in biolubricants determine their different physicochemical properties, which affect cooling and lubrication performance in grinding (mechanism ①).

The polar atoms (S, O, N, etc.) and polar groups (-OH, -COOH, -COOR, -CoR, -CN, etc.) contained in the lubricant have excellent affinity. Physical adsorption occurs with metal material molecules through van der Waals force, forming a physical adsorption film with antifriction and antiwear, thus reducing the grinding force [19]. However, both fatty acids and triglycerides in biolubricants contain polar groups, thus forming a lubricating oil film with high strength, which improves the lubrication performance of the grinding zone. In addition, fatty acid compounds tend to undergo metal saponification reaction on metal surface during high temperature grinding process, forming semi-chemical bonded oily film with single molecular layer with vertical adsorption characteristics. Compact adsorption on the workpiece material surface under intermolecular force, thus increasing tool life. Regardless of physical adsorption or chemical reaction, the factors affecting the strength of lubricating film are fatty acid saturation and carbon chain length.

It is well known that the strength of the physical adsorption film formed by biolubricants containing saturated fatty acids is higher than that formed by unsaturated fatty acids, and the less C=C double bonds in unsaturated fatty acids, the higher the strength of the lubrication film. This is due to the easy oxidation of the polar C=C double bond leading to degradation of biolubricants, reducing their thermal stability and oxidation resistance, resulting in the failure of the physical adsorption film. In addition, fatty acids without C=C double bond and linear molecular chain can form a tighter fatty acid molecular structure with compact and uniform shape, which improves the strength and lubrication performance of film.

For carbon chain length, when the number of C atoms in saturated fatty acids is greater than 16, the increase of C atom number will not affect its frictional properties. In unsaturated fatty acids, the density of the adsorbed film is reduced due to the adsorption effect of the oleic bond on the polar unsaturated bond, thus reducing the strength and lubrication performance of the oil film. Therefore, the number of C atoms in saturated and unsaturated fatty acids is the same, the lubrication film performance of unsaturated fatty acids is poor. On the other hand, the cohesion between molecules is proportional to the number of C atoms, so the strength and lubrication performance of the adsorption film of unsaturated fatty acids with long carbon chain length are stronger than those with short carbon chain length.

3.2.2. Viscosity

Viscosity is the variable exchange and adhesion generated by regular intermolecular motion and is one of the main factors affecting the cooling and lubrication performance of biolubricants. The viscosity of lubricant mainly affects its wetting performance, lubrication performance and heat transfer performance in grinding. (i) Influence of viscosity on debris removal (mechanism ⑥): After the lubricant enters the grinding zone at a certain speed and angle, the fluidity of the lubricant with high viscosity becomes poor due to the viscous force, and it cannot fully penetrate into the wheel/workpiece interface and take away the debris. (ii) Influence of viscosity on cooling performance (mechanism ⑤): The lubricant entering the grinding zone flows at a certain speed, and in the process of the relative motion of the lubricating oil film formed on the wheel/workpiece interface, the relative motion of the liquid layer is also carried out. Therefore, the heat transfer process of lubricant microdroplet conforms to the convective heat transfer theory of flowing liquid, and the heat transfer coefficient is related to viscosity. When the Reynolds number of lubricant is much larger than 2300, the oil film in the grinding zone carries out convection heat transfer in the way of turbulence. According to the heat transfer theory, the temperature gradient of the thermal boundary layer is the largest in the viscous bottom layer, and changes gently in the turbulent branch layer. Therefore, the heat transfer performance of lubricant is determined by the viscous bottom layer. Thus, the greater the viscosity of lubricant, the thicker the viscous bottom layer, the smaller the temperature rise per unit time, the lower the heat transfer capacity. (iii) Influence of viscosity on lubrication performance (mechanism ③): Reducing the interfacial friction between the abrasive rake face/debris and the back face/fresh surface of the workpiece can reduce the heat generated by the friction, lower the temperature and improve the lubrication performance. After lubricant enters the grinding zone, because of the difference of wettability, the lubricant with higher viscosity has better lubrication performance.

3.2.3. Surface Tension

When lubricant microdroplets enter the grinding zone, the smaller the surface tension is, the smaller the particle size of droplets is and the more uniform the distribution of droplets is. At the same time, the contact angle of the droplet is smaller so that it has a larger infiltration area per unit particle size volume. In conclusion, the wettability and lubrication of biolubricant was improved (mechanism ②).

On the other hand, surface tension further affects the cooling performance through the influence of contact angle. It is mainly reflected in two aspects: (i) The smaller the surface tension is, the smaller the contact angle of the microdroplet is, which increases the wetting area and further provides the cooling performance of lubricant (mechanism ⑩). (ii) The residence time of lubricant microdroplets is very short after entering the grinding zone, and they are carried out of the grinding zone rapidly with the wheel. According to the principle of convective heat transfer, there are two parts in convective heat transfer: thermal boundary layer and mainstream region, and the thickness of the thermal boundary layer is basically unchanged. However, in the very short cooling time, the lubricant in the mainstream region has not fully absorbed enough heat away and does not play a good heat transfer role. When the contact angle of microdroplet is reduced, the thermal boundary layer area increases and the proportion of grinding fluids in the mainstream region decreases, which improves the cooling performance (mechanism ④).

3.2.4. PH Value

As lubricating medium in cutting, the influence of acid-base property on the surface quality of parts is reflected after parts are processed [20]. Alkaline lubricants tend to passivate metal surfaces or form insoluble hydroxide or oxide film layers. The acid lubricant will promote the single displacement reaction to corrosion of metal, but also make the synthetic esters have incomplete hydrolysis reaction, affecting the processing effect. Biolubricants are generally acidic, especially in the preparation of synthetic ester biolubricants, in order to increase the antiwear performance, a variety of polar functional groups (-OH, COOH, R-O-R', etc.) are usually added to the carbon chain. The carboxy group ionizes hydrogen ions in an aqueous solution, making the lubricant acidic. Therefore, the preparation of synthetic esters usually requires the addition of some amines (amine is alkaline in aqueous solution, the representative substance is diethylene glycol amine) for neutralization to alkaline (mechanism ⑦).

3.2.5. Pour Point

The proper pour point of biolubricant is the premise of its effectiveness. Pour point is the lowest flow temperature of biolubricant, which is a conventional index to characterize the low temperature fluidity. The pour point of vegetable oil is generally between -15°C and -19°C, which is higher than that of mineral oil, because of the large amount of C=C double bond in vegetable oil. Biolubricants stored in micro lubricants supply units will lose their fluidity and cannot be transported properly if the ambient temperature is too low. The influence of vegetable oil pour point should be considered locally. For example, in northeast of China, Russia and other countries, vegetable oil cannot be used when the ambient temperature is lower than 20°C. Therefore, the reduction of pour point has become one of the research topics of biolubricants (mechanism ⑧). Guo [21] from Beijing University of Chemical Technology proposed a "two-step method for preparing polyol ester lubricating oil," with the pour point reduced to -50°C.

3.2.6. Thermal Stability

Thermal stability of biolubricants at high temperature boundary in grinding zone is the premise of effective cooling and lubrication, and flash point is an important parameter to characterize its thermal stability. The flash point of vegetable oil is generally about 280°C, such as soybean and cottonseed oil are 280°C, rapeseed oil is 275°C and peanut oil is 285°C. When the flash point is lower than the grinding temperature, the vegetable oil will fail. In addition, the density of β-H and electron cloud on the carbon framework of vegetable oil is large, and it is easy to be decomposed by heat at high temperature. There are C=C double bonds in branched fatty acids, which decompose easily at high temperature. For C=C double bond, epoxidation + ring-opening reaction, selective hydrogenation and hydrogenation cracking are commonly used in vegetable oil modification. For β-H, polyols without β-H can be esterified or transesterified with fatty acid or fatty acid ester to prepare polyol ester lubricating base oil. Guo [21] from Beijing University of Chemical Technology raised the opening flash point up to 305°C through "two-step method for preparing polyol ester lubricating oil" (mechanism ⑨). However, for grinding applications, the improvement of thermal stability of vegetable oil is still a difficult problem to be solved.

3.3. Titanium Alloy Grinding

The types of titanium alloys mainly include Ti-6Al-4V, Ti-6Al-4VELI, Ti-13V-11CR-3Al, etc., which are used to engine blades, compressor discs, landing gear, heat shields, etc. The chemical composition and mechanical properties of the commonly used Ti-6Al-4V are shown in Table 1 and Table 2. The grinding problems of titanium alloys come from three aspects: the enhanced mechanical activity, extremely low thermal conductivity and low elastic modulus [22, 23]. Therefore, the current research focuses on the workpiece surface integrity and grinding force/heat performance parameters.

Table 1. Chemical composition of Ti-6Al-4V (%)

Material	Ti-6Al-4V
Al	6.18
V	4.19
Fe	0.3
O	0.15
N	0.05
C	0.1
H	0.015
Ti	Balance
Si	0.15

Table 2. Mechanical properties of Ti-6Al-4V

Material	Ti–6Al–4V
Hardness	30 HRC
Yield strength	861 MPa
Elongation	14%
Tensile strength	993 MPa
Elasticity modulus	114 GPa
Density	4.43 g/cm^3
Thermal conductivity	5.44 W·m/K
Specific heat	526.3 J/Kg·K

3.3.1. Force and Coefficient of Friction (CoF)

The grinding process is a complex interference between abrasive and workpiece. The grinding force includes material removal force, plough force and interface friction force. Due to the negative rake cutting characteristics of the abrasive, the proportion of friction is more than 80%. Therefore, the reduction of grinding force under different lubrication conditions is mainly the reduction of friction. Friction is the main source of grinding heat. The value of grinding force is the main parameter to analyze not only the tribological characteristics of the interface, but also the grinding heat output and dissipation. Many experimental studies have been carried out on the grinding force of titanium alloy.

As shown in Figure 3, compared with soluble oil, a large normal and tangential force respectively reduction of 72.2% and 61.9% were directly observed by Sadeghi et al. [24] when using biolubricant in MQL for Ti-6Al-4V. However, in the synthetic ester MQL, the normal and tangential force were reduced by 16.7% and 20.5% compared with biolubricant MQL, respectively. It confirms the feasibility of replacing water-soluble cutting fluids with biolubricants. The grinding force that can be significantly improved lies in the excellent lubrication performance of biolubricants, which produces low friction and better debris removal performance. On the other hand, with certain chemical modifications, synthetic esters have better grinding performance than biolubricants. The reason is that synthetic esters have good lubricity and film strength, and maintain their integrity at high temperature, which can reduce friction more effectively than biolubricants. Li et al. [25] used biolubricant MQL in grinding TC4 and obtained the normal and tangential force decreased by 16.6% and 42.9% compared with dry, respectively. Further, the advantage of biolubricant can be reflected by CoF, which decreased from 0.562 to 0.384. Ibrahim et al. [26] also obtained lower CoF compared with flood and dry when using palm oil and synthetic esters as lubrication medium. Based on the above verification results, it can be made known that compared with mineral fluids, biolubricants and vegetable synthetic esters are more conducive to the formation of lubricating oil film due to their unique molecular structure (long molecular chain, polar group), thus obtaining better interface tribological properties.

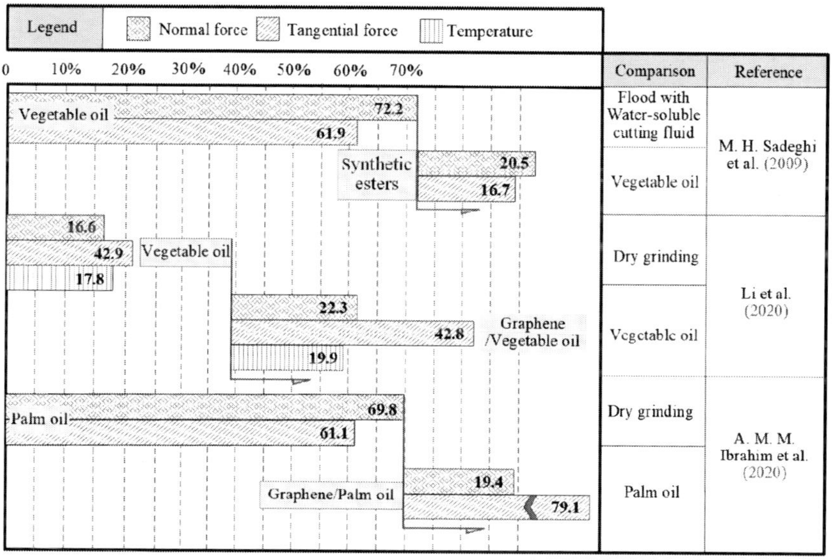

Figure 3. The improvement ratio of biolubricant MQL compared with traditional flood or dry grinding.

Research on grinding performance of different biolubricants is the premise of process optimization. Singh et al. [27] compared the different results of canola, soybean and olive oil as lubricating media in grinding Ti-6Al-4V-ELI. Among the three oils, canola oil obtained the lowest grinding force (F_t=4.68 N, F_n=15.25 N), CoF (0.307) and specific grinding energy (17.16 J/mm^3). In terms of viscosity, the viscosity of canola oil reaches 38 mPa·s at 40°C, while the soybean and olive oil are 26 mPa·s and 32 mPa·s respectively. It indicated that viscosity plays a key role in improving tribological characteristics.

For molecular structure, the main fatty acid composition of canola oil is erucic acid (C22:1), which has high straightness and a long carbon chain. While the main fatty acid composition of soybean and olive oil are linoleic acid (C18:2), with low straightness and a short carbon chain, and oleic acid (C18:1), with high straightness and a short carbon chain, respectively. As a result, canola oil achieves the best lubrication performance.

In addition, when graphene nanoparticle was introduced to prepare nano-enhanced biolubricants by Li et al. [25], the normal and tangential grinding force were 22.3% and 42.8% lower than those of biolubricants due to their excellent antifriction and enhanced heat transfer characteristics, and the CoF was also reduced to 0.283. Ibrahim et al. [26] also found that a significant improvement in lubrication when graphene was added, with tangential grinding force reduced by 79.1% and normal grinding force reduced by 19.4% compared to pure palm oil, and CoF was reduced from 0.247 to 0.061.

A lubricating oil film was found on the surface of the workpiece. EDS analysis was carried out at three different points on the workpiece surface, and it can be seen that the content of carbon element at A, B and C positions are 18.22%, 34.43% and 19.79%, respectively, while the carbon content in Ti-6Al-4V sample is less than 0.03%. Therefore, the presence of a large amount of carbon content is attributed to the formation of graphene lubrication films during the lubrication process of graphene nano-enhancer with biolubricants. It was also found that the use of Al_2O_3 nano-enhanced biolubricant significantly reduced the CoF as the grinding time increased by Setti et al. [28], while dry grinding and flood did not have this gain effect. This advantage was explained by the discovery of lubricating oil film on the workpiece surface, and EDS analysis fully confirmed the contribution of nanoparticles.

For nano-enhanced biolubricants, the suitability of nano-enhancers and biolubricants has a significant effect on grinding performance, which is mainly reflected in different types and mass fractions of nano-enhancer. Singh et al. [27] used layered nano-enhancer graphene, graphite and MoS_2 to prepare nano-enhanced biolubricants with sunflower oil. The results showed that graphene nano-enhancer achieved the best grinding performance, the CoF and specific grinding energy are 0.253 and 12.76 J/mm^3, respectively. When the mass fraction of graphene is increased, the grinding performance parameters increase first and then decrease, and the optimal mass fraction of graphene and canola oil was 1.5wt.%. The increase of concentration, on the one hand, improves the number of nano-enhancer for effective lubrication at the interface, and on the other hand, increases the viscosity of the biolubricant. However, when the concentration is too high, nano-enhancer will aggregate, precipitate and thus fail. How to prevent agglomeration and improve the optimal mass fraction is also a current research focus.

3.3.2. Wheel Wear

CBN and silicon carbide wheels are generally used in titanium alloy grinding, and the lubrication performance of biolubricant has a significant effect on the wear and life of wheel. Ibrahim et al. [26] carried out friction and wear experiments using zirconia balls and titanium alloy workpiece to simulate grinding conditions. Experimental results showed that the wear rate of synthetic ester (LB2000) and palm oil are significantly reduced compared with dry lubrication. Compared with palm oil, the wear rate decreased by an order of magnitude to 3.65×10^{-4} when graphene was selected as nano-enhancer. Setti et al. [28] observed the wheel wear under different lubrication conditions. An obvious wear plane was formed on the abrasive surface under poor lubrication performance in dry grinding. When nano-enhancers were added to biolubricants, the wear was significantly controlled, and good abrasive sharpness was maintained. Under high pressure impact, the nano-enhancers enter the pore of wheel for cooling and lubrication. On the other hand, the sliding/rolling of nano-enhancers prevents direct contact between wheel and workpiece. In addition, nano-enhancers promote the discharge of debris from the pore of the wheel and thus keep the wheel clean with the effect of MQL pressure.

3.3.3. Grinding Temperature

In titanium alloy grinding, it is a puzzle that the burn of the workpiece surface because the heat generated by friction gathers on the workpiece surface. The total heat flux density generated per unit time in the grinding zone is related to the grinding force. There are four transfer outlets of grinding heat, namely, the heat flux density of workpiece, the heat flux density of abrasive, the heat flux density of debris and the heat flux density of cooling medium. Increasing heat flux density of cooling medium is one of the feasible methods to reduce the maximum surface temperature of the workpiece (i.e., grinding temperature).

As shown in Figure 3, the grinding temperature was obtained in vegetable oil MQL is 229.18°C by Li et al. [25], which is 17.8% lower than that obtained by dry grinding. After further adding graphene nano-enhancer, the temperature decreased to 183.61°C, 34.2% lower than that in dry grinding. Liu et al. [29] carried out a multi-factor experimental study of Ti-6Al-4V grinding, among which cooling lubrication (four levels: dry, flood, MQL and NMQL) was one of the factors. Compared with dry and MQL grinding, NMQL can obtain higher S/N value, which indicates that nano-enhanced biolubricants are most effective in reducing grinding temperature. What`s more, the lowest temperature (133.8°C) in NMQL can be obtained.

The addition of nano-enhancers makes the biolubricants become a two-phase fluid, and the heat transfer performance of micro element, with the nanomaterial as the core and the biological lubricant as the coating layer, will be significantly improved under the Brownian movement effect of the nanomaterial. In addition, the viscosity and surface tension of biolubricants can change in a beneficial direction. The reduction of grinding temperature depends on the improvement of heat transfer performance of lubricating medium, which conforms to the convection heat transfer mechanism [30]. Convective heat transfer coefficient is a key factor affecting convective heat transfer capacity. Yang et al. [31] established a convective heat transfer coefficient formula under NMQL. According to the formula, the main factors affecting convective heat transfer coefficient include: specific heat capacity, viscosity, and thermal conductivity of biological lubricants, thermal conductivity, specific heat capacity, content of nanomaterials, and Brownian movement activity of nanoparticles, etc. The above factors can be used as the basis for heat exchange performance optimization of NMQL.

3.3.4. Debris Morphology

Due to the negative rake cutting in the process of abrasive removal, the material removal strain rate is much higher than that of cutting. The shape of debris is generally C-shaped, and the free surface is shearing band shape. The tribological and heat transfer characteristics of the rake face/debris interface during material removal can be analyzed by the degree of finishing of the debris side and the fresh surface and the regularity of the shear band on the free surface.

Liu [32] observed the debris morphology of grinding Ti-6Al-4V, as shown in Figure 4. The lubrication concludes dry condition with cryogenic air (-5°C) and NMQL with synthetic ester based Al_2O_3 nano-enhanced biolubricant with volume fraction of 2%. In dry grinding, there is no lubricant in the abrasive/debris interface, resulting in the fresh surface of the abrasive with obvious furrow marks without sufficient lubrication performance (due to the abrasive wear too fast). In addition, the fresh surface of debris is accompanied by material adhesion and accumulation, which is the material flow accumulation caused by the increase of plasticity and interfacial friction of debris under high temperature. Secondly, the side of debris is not smooth and has the zigzag defect formed after the material pulling fracture, which is caused by the material strong local plasticity at high temperature. Finally, the distribution of free surface shear bands is chaotic. This is due to the poor tribological properties of the interface, which leads to the stuttering and crawling phenomenon in the material removal process. In NMQL, the above phenomenon has been significantly improved. However, there are still some cases where the fresh surface is not smooth, the side has a serrated defect, and the free surface is not uniform. Due to the high mechanical and thermal loads, the thermal softening effect in the material removal process results in plastic enhancement and fracture stress reduction. This increases the critical cut depth of the material resulting in debris width, plastic accumulation on both sides of the furrow, and material adhesion between the workpiece and debris surface.

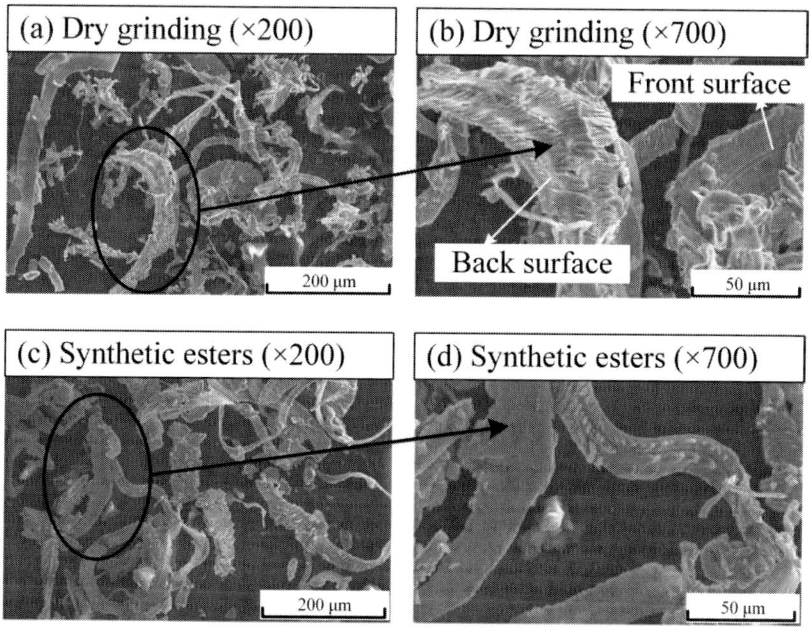

Figure 4. The debris morphology under different lubrication conditions. (a) Dry grinding (×200), (b) Dry grinding (×700), (c) Synthetic esters (×200), (d) Synthetic esters (×700).

3.3.5. Surface Integrity

The surface roughness of the workpiece (*Ra*) reflects the average height of the micro-convex peak and micro-concave valley on the workpiece surface. Due to the change of lubrication conditions, thermal boundary removal and material constitutive changes are caused, which have a significant effect on surface roughness. In the study of Sadeghi et al. [24], the surface roughness was reduced to 0.2~0.31 μm by using vegetable oil and synthetic esters as lubricating media. The surface roughness of synthetic ester is lower than that of vegetable oil under different lubricating medium usage and flow rate. Li et al. [25] found that the *Ra* value of MQL with vegetable oil could be reduced to 0.438 μm, 28.4% lower than that with dry grinding. Furthermore, it was reduced to 0.295 μm in the condition of graphene nano-enhanced biolubricant, which was 51.8% lower than that of dry grinding. When palm oil was used, *Ra* value reached 0.325 μm, which was 43.97% lower than that of flood. Similarly, when graphene is mixed with palm oil, *Ra* value reached 0.278 μm, 52.07% lower than that of flood [26]. However, when using water-based nanofluids, the surface roughness ranges from 0.6 μm to 0.7 μm, which is at the same level as that of dry grinding [28]. It is also fully confirmed that biolubricants have a significant effect on improving surface roughness. When comparing different vegetable oils, the surface roughness of canola, soybean, olive and sunflower oil fluctuated in the range of 0.41~0.44 μm, and canola oil had a slight advantage. As expected, *Ra* value increased significantly after nanoparticle was introduced, and the lowest value of was 0.21 μm. With the change of nanoparticle concentration, *Ra* value and grinding force change in the same rule, also exist an optimal mass fraction.

The microhardness of the workpiece surface reflects the change of material properties under the influence of heat/force during machining. In the study of Li et al. [25], the microhardness value reached 431.53 HV in dry grinding, while it decreased to 391.69 HV and 349.37 HV when vegetable oil and nano-enhanced biolubricant were used, which decreased by 9.2% and 19.0% compared with dry, respectively. The decrease of force and temperature results in the decrease of the thickness of the thermal affected layer on the workpiece surface and microhardness.

The surface morphology of the workpiece can directly reflect the machining defects, and the suppression of surface defects formed by machining is the most important method to improve the surface integrity. For titanium alloy grinding, common surface defects include adhesion, burn, plastic accumulation and scratch. As shown in Figure 5, compared with flood, plastic deformation, thermal damage (such as debris adhesion, surface deposition, etc.) significantly reduced in vegetable oil and synthetic ester MQL condition through the conclusion of Sadeghi et al. [24]. When Li et al. [25] used biolubricant, grinding burns were avoided. Yet it still existed adhesion and wide furrows. While, by using 0.1 wt.% concentration of graphene nano-enhanced biolubricant, the optimal workpiece surface was obtained without obvious defects. Ibrahim et al. [26] also obtained the same conclusion. Singh et al. [27] also found that plastic uplift existed in flood and large grinding burns occurred in dry. The workpiece surface of MQL with canola oil was still rough despite the reduction of plastic uplift defects. Canola oil-based graphene NMQL achieves a smooth workpiece surface. Meanwhile, different kinds of vegetable oil and mass fraction of nanoparticle also had a significant effect on the inhibition of surface defects.

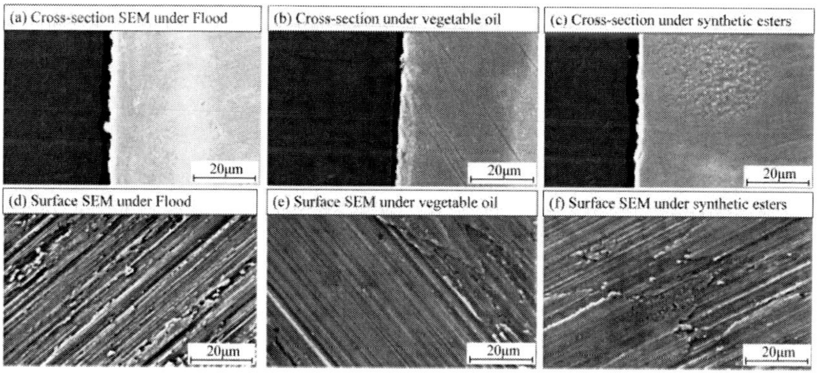

Figure 5. The surface and cross-section morphology under different lubrication conditions. (a) Cross-section SEM under Flood, (b) Cross-section under vegetable oil, (c) Cross-section under synthetic esters, (d) Surface SEM under Flood, (e) Surface SEM under vegetable oil, (f) Surface SEM under synthetic esters.

3.3.6. An Analysis of Grinding Defects and Solutions

According to the research results, the grinding defects and bottlenecks of titanium alloy can be summarized as follows:

i. Deterioration of tribological properties due to low elastic modulus. The elastic modulus of titanium alloy is 114 GPa, which is about 60% of that of nickel-based alloy and high strength steel. The abrasive removes the material with negative rake cutting, which makes the grinding force appear as "high pressure and low shear" state, resulting in low elastic modulus becoming the bottleneck. The high pressure at the interface between the abrasive tip and the machined surface caused the titanium alloy material to rebound and retreat, resulting in vibration and surface roughness reduction during the grinding. In this case, the ratio of friction energy to grinding energy of the interface is higher than that of other materials, which not only produces more friction heat, but intensifies abrasive wear.

ii. Deterioration of workpiece surface integrity due to low thermal conductivity. The transfer outlet of grinding heat in the grinding zone includes: workpiece, abrasive, lubricants, debris and air. In the titanium alloy grinding, the energy transmitted into the workpiece can reach more than 58% of the total energy. However, due to the low thermal conductivity of titanium alloy, the heat cannot be transmitted quickly through the matrix, thus accumulating on the surface of the workpiece. Now, research can confirm that the grinding defects caused by this bottleneck include the following aspects: ① Grinding burn and crack. The black area that can be observed on the workpiece surface is grinding burn; ② Material adhesion on the workpiece. The material is crushed to the surface of the workpiece by the abrasive under the effect of thermal softening at high temperature is adhesion. ③ Ploughing and plastic uplift. The plastic ploughing phenomenon intensified under the action of thermal softening, and the material removal rate decreased, thus accumulating on both sides of the furrow.

According to the grinding experimental results, the bottleneck in (ii) has a stronger impact on the machining results than that in (i). Currently, scholars focus on how to improve the heat transfer capacity of lubricants, so as to reduce the proportion of incoming energy to the workpiece. Accordingly, the current research methods and conclusions can be summarized as follows:

i. Graphene with high thermal conductivity was used as nano-enhancer to prepare nano-enhanced biolubricant with palm oil, which had better heat exchange and antifriction performance, and achieved the best grinding experimental results compared with other lubricants.
ii. Cryogenic air cooling coupled with nano-enhanced biolubricant can significantly improve the heat transfer performance, which has been verified by preliminary experiments. This is one of the powerful techniques to solve the problem of titanium alloy grinding

3.4. Nickel-Based Alloy Grinding

Nickel-based alloy is one of the superalloys, which has high strength and certain oxidation corrosion resistance at 650 ~ 1000°C. Nickel-based alloy types for aerospace applications include the Inconel series (718, 600, 625, etc.)

and the Incoloy series (825, A-286, etc.). It is mainly used in heat resistant parts of aeroengines, turbine blades, space shuttle engines, gas turbines and so on. In addition to the low thermal conductivity, the high hardness of nickel-based alloys is a source of grinding bottlenecks that distinguishes them from titanium alloys and high strength steels. The chemical composition and mechanical properties of the commonly used Inconel 718 are shown in Table 3 and Table 4.

Table 3. Chemical composition of Inconel 718 (%)

Material	Inconel 718
Al	0.95
Mn	0.35
Fe	Balance
Cr	18.8
Cu	0.3
Ni	53.4
Co	1
Mo	2.99
Si	0.35
C	0.08

Table 4. Mechanical properties of Inconel 718

Material	Ti–6Al–4V
Hardness	100 HRC
Yield strength	550 MPa
Elongation	45%
Tensile strength	965 MPa
Elasticity modulus	199.9 GPa
Density	8.24 g/cm^3
Thermal conductivity	14.7 W·m/K
Specific heat	435 J/Kg·K

3.4.1. Force and Coefficient of Friction (CoF)

When grinding Inconel 718, sunflower, rice bran, palm and peanut oil were used as MQL lubricants [33, 34]. It was found that the use of biolubricant significantly reduced the CoF compared with flood, while the force, specific grinding energy and temperature are inferior to flood, as shown in Figure 6. Compared with the four biolubricants, palm oil obtained the best experimental results, with the specific grinding energy and CoF reduced by 2.5% and 25%, respectively, compared with flood (as shown in Figure 6). Based on the research it can be found that palm oil has achieved excellent lubrication performance among various vegetable oils, and the reasons can be summarized into two aspects: ① Palmitic acid (C16:0, 35%~48%), the main fatty acid component of palm oil, is saturated fatty acid. While the component of sunflower, peanut and rice bran oil (oleic acid, linoleic acid, etc.) is unsaturated fatty acid; ② Palm oil showed a higher viscosity value, reaching 38 mPa·s at 40°C. Wang et al. [35] compared the MQL grinding performance of various vegetable lubricants (castor, soybean, rapeseed, corn, peanut, palm and sunflower oil) for grinding of Inconel 718. Castor oil achieved the minimum normal and tangential grinding forces (F_n=76.83 N, F_t=24.33 N). The unique advantages over other vegetable oils as follows: ① Ricinoleic acid (C18:1), the main component of castor oil, contains polar group -OH, which significantly enhances the adsorption strength of lubricating oil film at the interface; ② The viscosity of castor oil is one order of magnitude higher than that of other vegetable oils, reaching 260 mPa·s at 40°C. It is beneficial to the improvement of tribological characteristics. Zhang et al. [36] also found that in the grinding of nano-enhanced biolubricant prepared by Al_2O_3/SiC nanoparticles with palm oil and castor oil, the grinding force obtained with palm oil as base oil is less than that obtained with castor oil.

For nano-enhanced biolubricant, compared with biolubricant, Virdi et al. [34] obtained the reduction of CoF and specific grinding energy with 17.7% and 29.4% respectively in grinding Inconel 718 with palm oil/Al_2O_3 NMQL. In sunflower oil/Al_2O_3 NMQL, the normal and tangential grinding force decreased by 4.15% and 17.5%. The CoF decreased from 0.142 to 0.117, and the specific grinding energy from 39.542 to 32.901 J/mm^3 [33]. In the NMQL experiment conducted by Wang et al. [37], different performances of Al_2O_3 and MoS_2 were compared. The CoF decreased from 0.45 to 0.4 in palm oil/MoS_2 NMQL and 0.38 in palm oil/Al_2O_3 NMQL respectively compared with MQL. Because spherical Al_2O_3 nanoparticle has high hardness, it plays a supporting role between the wheel and workpiece, reducing the actual contact area of friction pair. When it enters the grinding zone, the rolling effect is produced on the uneven surface of the workpiece so that the sliding friction is transformed into rolling friction to produce antiwear effect, which can reduce tangential force and CoF in the grinding process. Zhang et al. [38] used synthetic ester, mixed with CNTs, MoS_2 and CNTs/MoS_2 hybrid nanomaterials to prepare lubricating medium. The experiment of grinding Inconel 718 showed that hybrid nanoparticles showed stronger cooling and lubrication performance compared with a single nanoparticle.

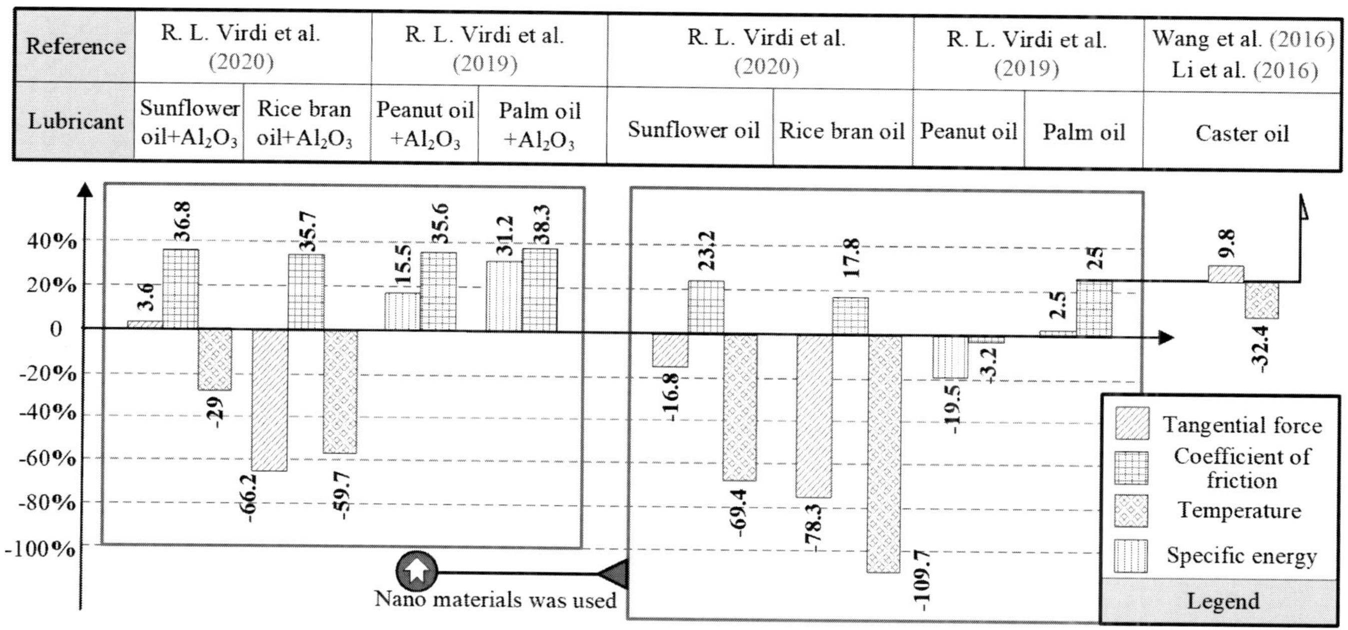

Figure 6. The improvement ratio of biolubricant MQL compared with traditional lubrication conditions.

The two nanoparticles with different structures can improve the lubrication state of the grinding zone through physical synergy, and the grinding force is significantly reduced. The addition of different forms of nanoparticles has different abilities to improve cooling and lubrication performance. Spherical and layered nanoparticles are more favorable for antifriction and antiwear but have lower thermal conductivity. While tubular CNTs have higher thermal conductivity and heat transfer capacity, but are not conducive to lubrication.

The concentration experiment of Wang et al. further found that with the increase of Al_2O_3 concentration, the specific grinding energy showed a trend of decreasing first and then increasing. At 1.5 vol. %, it reached the minimum value (64.95 J/mm^3), which was 34.1% lower than that of pure oil [39]. Therefore, proper concentration can improve the tribological properties, while when too high will lead to aggregation, which is not conducive to the tribological properties.

3.4.2. Wheel Wear

In the grinding process of nickel-based alloy, severe wear and short service life of wheel are the main technical bottlenecks. The forms of wheel wear include abrasive loss, adhesion blockage and even abrasive shedding. In addition, the material volume removed per unit volume of wheel wear (G ratio) is also one of the parameters reflecting the life of wheel, and the larger the G ratio is, the smaller the degree of wheel wear is and the longer the service life is. Virdi et al. [34] compared wheel wear of MQL grinding Inconel 718 with peanut and palm oil, and compared with peanut oil, the G ratio increased by 50.6% with palm oil. In the other experiment, the G ratio in sunflower oil increased by 3.61% compared with rice bran oil [33]. This is because the viscosity of sunflower oil is higher than that of rice bran oil, which reduces the fluidity and prolongs the lubrication time of grinding interface, thus preventing wheel wear. The life rule of wheel with different vegetable oil is consistent with grinding force, which depends on lubrication performance. When Al_2O_3 was added to sunflower oil, the G ratio increased from 0.688 to 1.532, and the stable lubricating oil film formed by nano-enhanced biolubricant protected the abrasive from wear. In addition, the G ratio was significantly increased by increasing the concentration of nanoparticle. Hegab et al. [40] found in grinding Inconel 718 that wheel wear was significantly improved with the addition of nanoparticles. This is mainly attributed to the addition of nanoparticles, which improves the lubrication and wettability of the rake face/debris and back face/machined workpiece, thus improving the wear behavior. Moreover, compared with Al_2O_3, the wear decreases by 26.3% using MWCNT.

When Inconel 718 was machined with palm oil/MoS_2 and palm oil/Al_2O_3, the addition of nanoparticles increased the G ratio. Compared with palm oil, NMQL increased by 23.9% and 32.9%, respectively. Different kinds of nanoparticles had different degrees of improvement in G ratio [37]. However, in the study of the effect of different volume fractions of Al_2O_3 nanoparticle on wheel wear, G ratio shows a trend of increasing first and then decreasing with the increase of volume fraction and reaches the maximum at 2.5 vol.% [39].

In addition, SEM and EDS analysis of abrasive before and after grinding shows that, as shown in Figure 7, nano-enhanced biolubricant forms a micro-storage layer similar to capillary on the surface of wheel, which can temporarily store nano-enhanced biolubricant to ensure that nanoparticles can effectively enter the grinding zone. On the other hand, spherical Al_2O_3 plays a good bearing role in the grinding zone, changing the sliding friction between the abrasive and the workpiece into rolling friction, and reducing the wear plane of the abrasive by reducing the contact area between the abrasive and the workpiece.

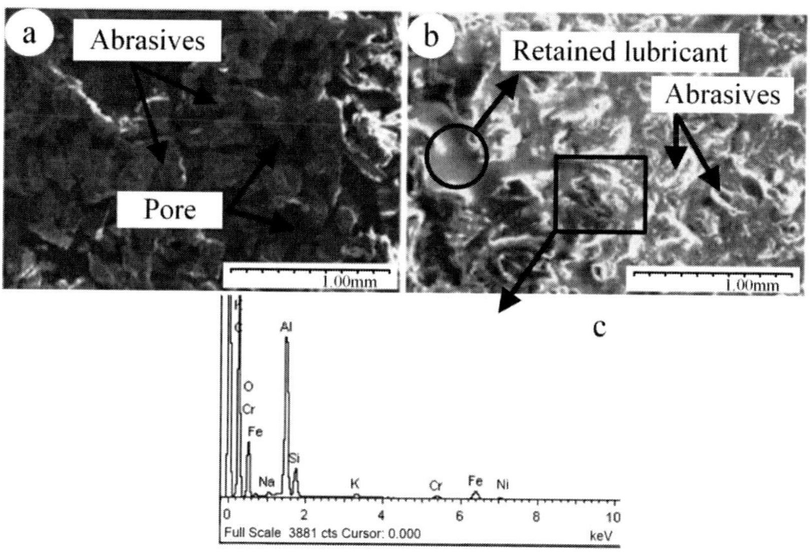

Figure 7. The morphology and EDS analysis of grinding wheel. (a) before grinding, (b) after grinding, (c) EDS analysis.

3.4.3. Grinding Temperature

Corundum wheel and CBN super hard abrasive wheel are mostly used in grinding of nickel-based alloy. However, the biggest problem is adhesion blockage in the process of grinding. The nickel-based alloy with low thermal conductivity generates a large amount of heat and concentrates on the workpiece surface during material removal, resulting in an increase in grinding temperature. Thermal damage caused by high grinding temperature seriously affects workpiece quality and limits production efficiency. Therefore, the effect of effective cooling and heat transfer of lubricant is one of the important factors.

During the grinding of Inconel 718 by Virdi et al. [33], the grinding temperature of sunflower oil MQL was 19.2% lower than that of rice bran oil MQL (as shown in Figure 6). However, after adding Al_2O_3 nanoparticles into sunflower oil, the grinding temperature was further reduced from 105°C to 80°C (as shown in Figure 6). In the study of grinding Inconel 718 by Li et al. [41], grinding temperatures of castor, soybean, rapeseed, corn, peanut, palm and sunflower oil MQL were compared, and it was found that palm oil MQL reached the minimum value (119.6°C) and castor oil MQL reached the maximum value (176°C). The result of grinding temperature is different from that of grinding force and the fundamental reason is that the physicochemical properties of vegetable oil have different influence trends on lubrication and cooling performance: (i) Because castor oil contains -OH and its viscosity is too high, the molecular fluidity of vegetable oil molecules is poor in the heat transfer process, resulting in reduced heat transfer efficiency; (ii) Although castor oil obtained the lowest grinding force (i.e., the minimum heat output) in the study of Wang et al. [37], the reduced heat transfer efficiency still resulted in the energy transmitted to the workpiece reaching the maximum value among various oils; (iii) Palm oil is used as a lubricant with moderate viscosity to achieve a minimum grinding temperature. In addition, Zhang et al. [42], also found that surface tension of different vegetable oils had a significant impact on heat transfer performance, mainly because it affected the particle size and spread area of microdroplets, thus changing the effective heat transfer area of lubricant.

Furthermore, Li et al. [43] added nanoparticles into palm oil to significantly improve the heat transfer performance of lubricant. The effects of different nano-enhancers (MoS_2, SiO_2, PCD, CNT, Al_2O_3 and ZrO_2) on grinding temperature were compared. The thermal conductivity of CNT was the highest, resulting in the lowest grinding temperature under the condition of CNT nano-reinforced biolubricants. In the palm oil/CNT experiment with different concentrations of nanoparticles, it can be seen that the concentration should not be too high or too low, and an appropriate concentration can reduce the grinding zone temperature.

3.4.4. Debris Morphology

Wang et al. [44] observed the debris morphology in the Inconel 718 grinding experiment (as shown in Figure 8). Compared with flood, the debris morphology in MQL condition was longer and narrower, and the surface was smoother. Furthermore, compared with palm oil MQL, the debris under palm oil/Al2O3 NMQL condition are thinner, which is determined by the excellent lubrication performance of nanoparticles. Nanoparticles can be well adsorbed in the texture of the debris surface or the grooves of the workpiece surface, acting as rolling bearings to reduce the friction between the abrasive and the debris [37]. Moreover, when the volume concentration of Al2O3 nano-enhanced biolubricant reaches 2 vol.%, the surface of the debris is smooth and the strip debris with uniform width are presented. The change of debris morphology reflects the material removal mechanism, and the excellent lubrication performance plays an important role [39].

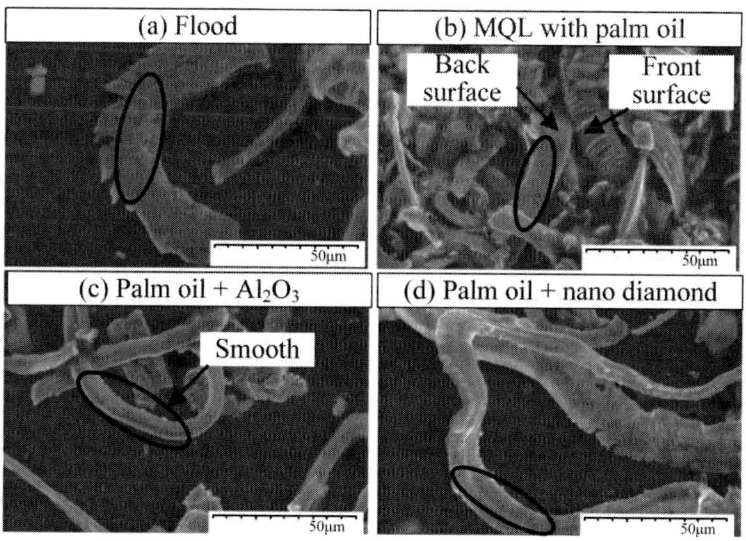

Figure 8. The debris morphology under different lubrication conditions. (a) Flood, (b) MQL with palm oil, (c) Palm oil+Al_2O_3, (d) Palm oil + nano diamond.

3.4.5. Surface Integrity

According to the experiment of grinding Inconel 718 with sunflower and rice bran oil, as shown in Figure 9, the surface roughness of both biolubricants can be reduced compared with that of flood [33]. However, compared with rice bran oil MQL, *Ra* in sunflower oil MQL condition also decreased by 5.49%. When Al_2O_3 nanoparticles were added, sunflower oil/Al_2O_3 NMQL decreased by 41.9% compared with MQL.

Wang et al. [35] compared the surface quality of the workpiece when grinding Inconel 718 with different biolubricants. Castor oil, as a lubricant, benefited from its high viscosity, and the surface roughness reached the lowest value of 0.36 μm, and the optimal workpiece surface was obtained. Furthermore, MoS_2 and Al_2O_3 were used to prepare nano-enhanced biolubricants with palm oil. Compared with palm oil, the addition of both nanoparticles reduced *Ra* values, where The *Ra* value of Al_2O_3 nanoparticles decreases by 13.3% compared with MoS_2 [37]. Li et al. [43] studied the surface quality of workpiece under various nanomaterial conditions and found that nano diamond has more advantages than other materials.

Wang et al. [39] found that when the volume concentration reaches 2%, the surface quality of the workpiece is the best (as shown in Figure 9) in the study of different concentrations of nanoparticles (palm oil/Al_2O_3 NMQL), and EDS analysis also found an increase in Al element content on the surface of the workpiece. In the experiment of Virdi et al. [33], by observing the surface topography of the workpiece, it can be seen that sliding and ploughing exist in almost all conditions, and some of them also have deep gullies and adhesion. In the castable lubrication mode, severe plough and adhesion lead to poor workpiece surface, and adhesion further aggravates the friction between grinding wheel and workpiece. In the flood lubrication, severe plough and adhesion lead to poor workpiece surface, and adhesion further aggravates the friction between wheel and workpiece. After the use of vegetable oil, the surface quality of the workpiece improved, but there was also slight stacking of materials, adhesion and wear marks caused by plough accumulation. This is due to the fact that the viscosity of vegetable oil declines more slowly than mineral oil at higher temperatures, and the higher viscosity ensures that they provide more stable lubricity at the operating temperature range. Further, the smoother surfaces, small number of narrower furrows, and slight accumulation of plough are revealed under NMQL. EDS analysis also confirmed the existence of lubricating oil film on the workpiece surface.

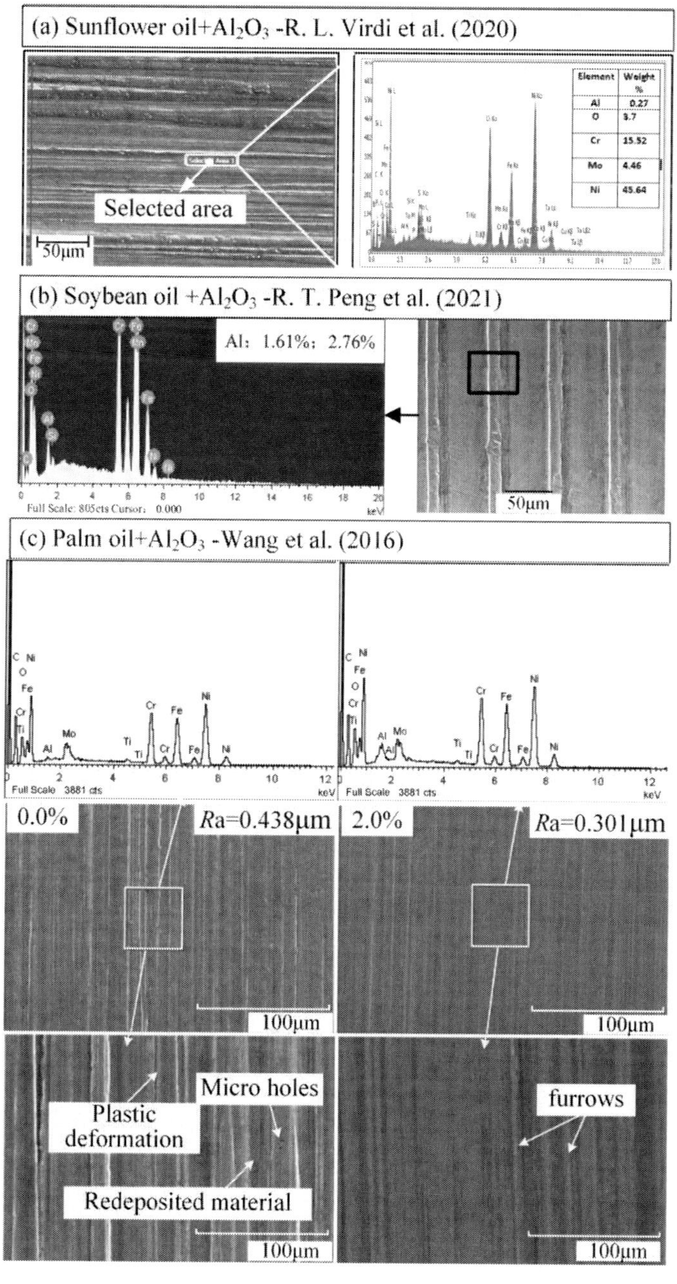

Figure 9. The surface SEM and EDS analysis of Ni-based alloy with usage of Al_2O_3 biolubricant. (a) Sunflower oil+Al_2O_3-R.L. Virdi et al. (2020), (b) Soybean oil + Al_2O_3-R.T. Peng et al. (2021), (c) Palm oi1+ Al_2O_3-Wang et a1. (2016).

In the experiment of grinding Inconel 718 with soybean oil/Al_2O_3 NMQL, Peng et al. [45] found O element on the workpiece surface, which was attributed to the chemical reaction and the formation of lubricating oil film, which effectively improved the lubrication condition of grinding.

3.4.6. Ann Analysis of Grinding Defects and Solutions

According to the research results, the grinding defects and bottlenecks of nickel-based alloy can be summarized as follows:

i. Wheel wear is aggravated by high hardness. The hardness of nickel-based alloys can reach about 100 HRC, 3 to 5 times that of titanium alloys and high strength steels. The high hardness comes from the following reasons: ① The high melting point alloying elements of nickel-based alloys (nickel, iron, chromium, etc.) combine with other elements to form high hardness materials, such as austenitic alloys, high hardness compounds, high hardness intermetallic compounds, and hard particles; ② Nickel-based alloy is generally treated with solution treatment before machining, and work hardening occurs in the process, leading to the increase of hardness; ③ After aging treatment, the hard phase in the solid solution is precipitated and the lattice is distorted, resulting in aging strengthening. In the grinding, bottlenecks caused by high hardness are as follows: ① Due to the low plastic deformation degree of the workpiece in the abrasive material removal, the interference area between abrasive and debris decreases, which leads to the increase of the unit mechanical/thermal load of effective abrasive region and the aggravation of abrasive wear; ② The interference between hard phase and abrasive also aggravates abrasive wear. Therefore, scholars also focus on the wheel wear, G ratio and other parameters.

ii. Surface burns caused by low thermal conductivity. At 100°C, the thermal conductivity of nickel-based alloy is 14.7 W·m/K. Despite it is higher than that of titanium alloy, there are still some problems such as burn, furrow and plastic accumulation on the workpiece surface. However, the experimental results show that such defects do not reach the severity of titanium alloy.

Therefore, the lubricant should first meet the characteristics of antiwear and antifriction, while having a certain heat exchange capacity. The current research methods and conclusions can be summarized as follows:

i. Base oil of biolubricant should have high viscosity, fatty acid saturation, containing polar groups and other characteristics (e.g., palm oil, castor oil), which is conducive to the formation of strong adsorption of lubricating oil film. Nano-enhancers have high hardness and spherical characteristics (e.g., Al_2O_3), so it plays an excellent role of rolling antifriction and antiwear in high pressure grinding interface.
ii. The recommended nano-enhancers have low thermal conductivity, in order to give consideration to heat transfer capacity and solve the grinding burn phenomenon, the use of hybrid nano-enhancers is one of the feasible solutions. Commonly used nano-enhancers with high thermal conductivity include CNTs, SiC, etc.

3.5. High Strength Steel Grinding

High strength steel, the main models are AISI 4140, 4340, 52100 and etc., plays an important role in the aerospace field because of its high tensile strength. The applications include rocket engine housing, aircraft landing gear, bulletproof steel plate and other parts with special performance requirements. As the yield strength and tensile strength of high strength steel are higher than that of titanium alloy and nickel-based alloy, it is difficult to cut debris in the grinding, the wheel is blocked, and the grinding temperature increases, which leads to the surface layer of the workpiece changes, which is directly reflected in the changes of microhardness and residual stress. Chemical composition and physicochemical properties of commonly used high strength steels are shown in Table 5 and Table 6.

Table 5. Chemical composition of high strength steels (%)

Material	AISI 4140	AISI 4340	AISI 52100
C	0.38-0.43	0.38-0.43	0.95-1.05
Mn	0.70-1.00	0.6-0.8	0.25-0.45
P	0.035	0.035	0.025
S	0.04	0.04	0.025
Cr	0.80-1.1	0.7-0.9	1.4-1.65
Mo	0.15-0.25	0.2-0.3	0.08
Si	0.17-0.37	0.15-0.35	0.15-0.35
Ni	0.3	1.65-2.00	0.3

Table 6. Mechanical properties of high strength steels

Material	AISI 4140	AISI 4340
Hardness	22.2 HRC	27.8 HRC
Yield strength	930 MPa	835 MPa
Elongation	12%	12%
Tensile strength	1080 MPa	980 MPa
Elasticity modulus	185 GPa	208 GPa
Density	7.85 g/cm^3	7.83 g/cm^3
Thermal conductivity	46 W·m/K	--
Specific heat	0.27-0.3 J/Kg·K	0.3 J/Kg·K

3.5.1. Force and Coefficient of Friction (CoF)

Sadeghi et al. [46] used biolubricant and synthetic ester to conduct AISI 4140 grinding experiment, which significantly improved the performance compared with flood. Taking tangential grinding force for example, vegetable oil and synthetic ester were reduced by 57.2% and 67.5%, respectively, compared with flood. While compared with biolubricants, synthetic ester produced less grinding force, with a 22.8% and 22.9% reduction in normal and tangential force (as shown in Figure 10). In grinding AISI 1018 experiment of Shao et al. [47], commercial vegetable oil was used as MQL lubrication medium, and the CoF was reduced from 0.299 to 0.271 compared with flood. The above conclusions are consistent with titanium alloy grinding, and also verify the advantages of synthetic ester. This is due to the introduction of polar groups in the preparation of synthetic ester, thus improving the adsorption capacity of lubricating oil film. Mao et al. [48] also used water and water-based Al_2O_3 nanofluids as lubricants to conduct AISI 52100 grinding experiments. Compared with water-based MQL, the tangential force decreased by 10.92% while the force ratio decreased by 11.11% with the addition of nanoparticles. ManojKumar and Ghosh [49] added MWCNT into sunflower oil to form NMQL lubrication conditions when they carried out the experiment of grinding AISI52100. Compared with MQL, the specific grinding energy of sunflower oil/MWCNT NMQL decreased by 11.9%, and the force ratio decreased from 0.37 to 0.35. The addition of nano-enhancers increases the viscosity of the lubricant from 44.2 mPa·s to 47.96 mPa·s, thus enhancing the lubrication performance. When grinding AISI 52100 with pure water MQL and NMQL by Molaie et al. [50], smaller results were also obtained with the addition of nano-enhancers than under MQL conditions.

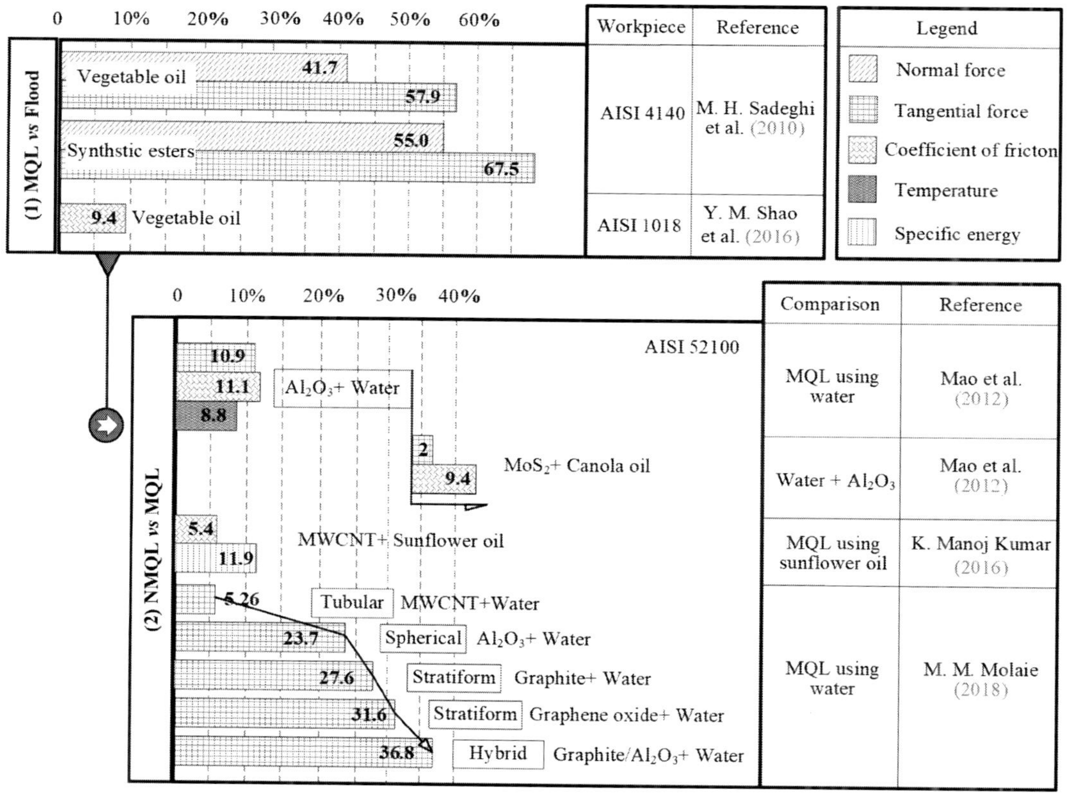

Figure 10. Grinding performance of high strength steel with usage of biolubricant.

Molaie et al. [50] further compared the grinding performance of nano-enhanced biolubricants prepared by pure water and four kinds of nano-enhancers, namely Al_2O_3, graphite, graphene oxide and CNTs, respectively. The results show that the grinding force of water/graphene oxide was the smallest (F_n= 74.2 N, F_t= 26.3 N). In the subsequent experiments, scholars mixed Al_2O_3 and graphite to prepare hybrid nano-enhanced biolubricant, and the grinding performance was improved after mixing, which is due to the different tribological mechanism of the two kinds of nano-enhancers. The layered structure of graphene oxide allows easy access to the sliding zone and reduces the direct contact between particles with the metal, forming a protective and durable friction film. While spherical Al_2O_3 nanoparticles with high hardness can produce ball effect between friction surfaces.

3.5.2. Wheel Wear

Belentani et al. [51] used flood, vegetable oil and vegetable oil/water mixture (mixing ratio 1:5) as lubricants to grind AISI 4340, as shown in Figure 11. Compared with flood, the wheel wear rate was reduced under MQL. The wheel wear degree depends on the thermal/force boundary conditions in the grinding zone: ① The lack of lubrication ability under flood leads to the aggravation of abrasive wear; ② As a lubricant, vegetable oil still has a high wear rate, which is due to the high viscosity lubricant does not have the wheel washing performance, resulting in porosity blockage; ③ Using vegetable oil/water mixture is the solution to the above problems. With the increase of water proportion, the washing performance of the lubricant on the wheel is also enhanced, and the wheel wear rate is lower. Furthermore, Javaroni et al. [52] used a cooled wheel cleaning jet (CWCJ) in combination with MQL to significantly optimize wheel wear rate by using CWCJ's cleaning effect on the wheel. Moreover, with the decrease of high-pressure gas temperature, wheel wear rate further decreases, as shown in Figure 12. Garcia et al. [53] used vegetable oil/water mixture (mixing ratio 1:5) as lubricant and in combination with wheel cleaning jet (WCJ) technology in AISI 52100 grinding experiment, the wear rate was reduced by 33.3%. It can be seen that the wheel cleaning performance of the lubrication process is very important to the life of wheel.

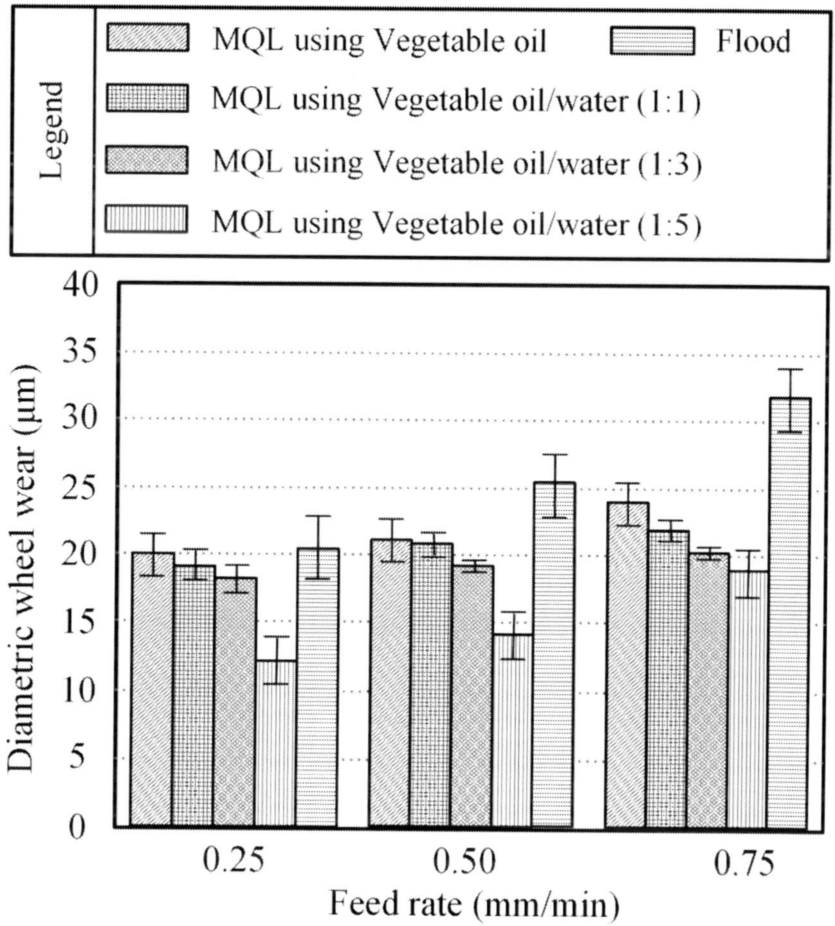

Figure 11. Diametric wheel wear with usage of vegetable oil+water hybrid lubricant.

ManojKumar and Ghosh [49] employed soluble synthetic oil, sunflower oil and sunflower oil/MWCNT nano-enhanced biolubricant as lubrication media in the grinding experiment of AISI52100. Compared with flood, the G ratio of the three lubricants increased by 23.1%, 26.9% and 69.2% respectively. The addition of nano-enhancers has a more prominent effect on the wheel life.

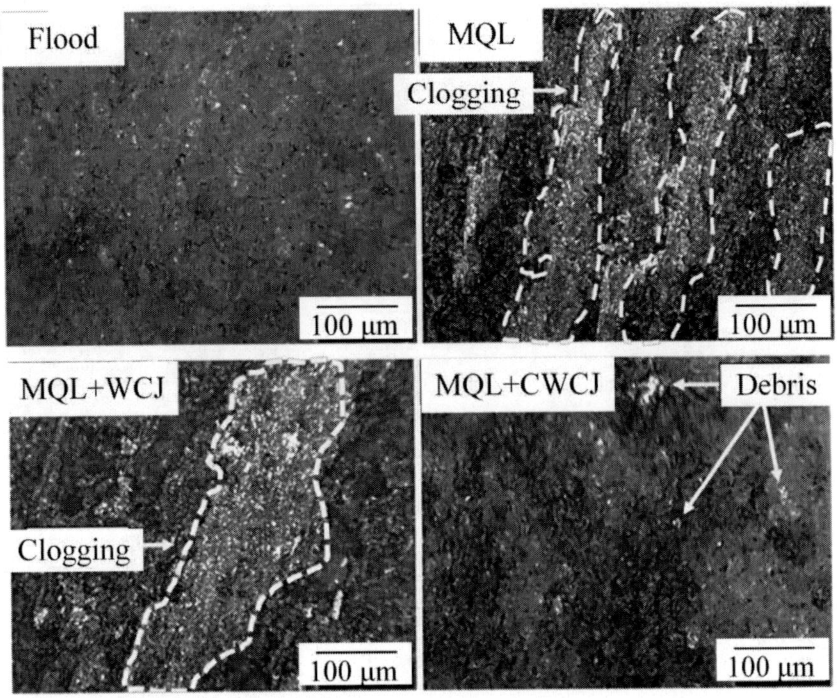

Figure 12. SEM of grinding wheel under different lubrication conditions.

3.5.3. Grinding Temperature

Mao et al. [48] analyzed the change of grinding temperature in AISI 52100 grinding experiment. Compared with pure water MQL, the grinding temperature after adding nanoparticle (Al_2O_3) decreases from 484°C to 442°C, which decreases by 8.68%, as shown in Figure 13. The reduction in grinding temperature can also be verified by the workpiece microstructure distribution. Due to the influence of grinding heat, the grinding surface after machining is generally composed of white layer and dark layer, which have an impact on the service life and surface quality of the workpiece. It is found that the thickness of white layer and dark layer is 8 μm and 32 μm respectively when water is used as lubrication medium, while 4 μm and 21 μm when water based nanofluids is used. Therefore, the reduction of grinding temperature has a significant effect on the microstructure change of high strength steel.

Figure 13. Cross-section morphology under different lubrication conditions. (a) Dry grinding, (b) MQL using water, (c) MQL using water based nanolubricant.

3.5.4. Debris Morphology

Javaroni et al. [52] explored the morphology of debris under flood, MQL and MQL+WCJ technologies. In flood process, debris are strip and banding, indicating that debris formation is a ductile flow process (as shown in Figure 14). While in MQL condition, the debris is spherical. In MQL+WCJ condition, the number of spherical debris is significantly lower than that of MQL, which indicates that the cooling performance is stronger than that of MQL. Under the condition of MQL+WCJ (high-pressure gas with -15°C), the shape is similar to that of flood.

ManojKumar and Ghosh [49] observed the debris morphology in the grinding experiment of AISI52100 and found that the biolubricant MQL optimizes material removal process. As shown in Figure 15, a large number of elongated debris ① and a small number of weakly plastic deformation debris ② are present in sunflower oil condition; While, in the case of sunflower oil /MWCNT, the number of weakly plastic deformation debris is small, and the debris is formed by cutting of sharp abrasive, indicating that the nanofluids environment has better sharpness maintenance ability. Further, when the lubricant flow is reduced, the number of strip debris increases.

However, spherical debris were produced at the flow of 50 ml/h, indicating that the temperature in the grinding zone increased, which led to the melting and re-solidification of the debris material into a spherical shape. Therefore, the flow of lubricant should be controlled in an appropriate range.

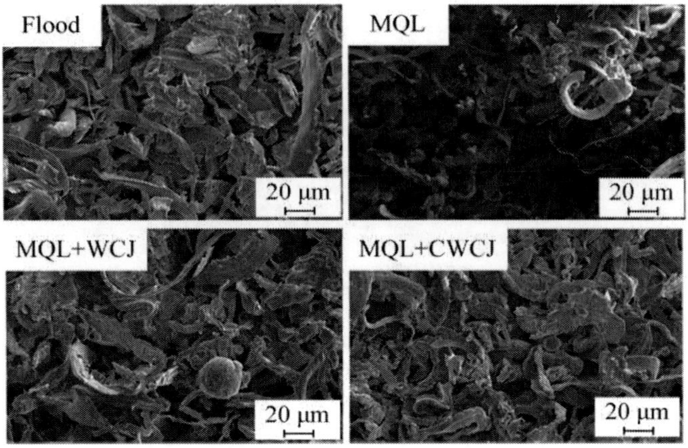

Figure 14. Debris morphology under different lubrication conditions.

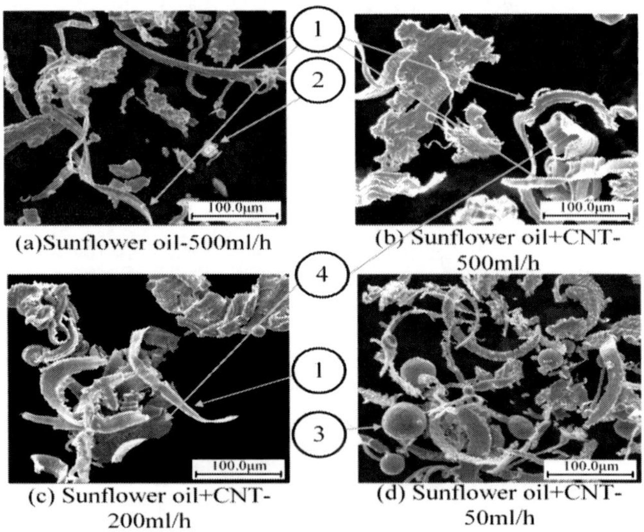

Figure 15. Debris morphology under different lubricant and usage amount. (a) Sunflower oil-500ml/h, (b) Sunflower oil+CNT-500ml/h, (c) Sunflower oil+CNT-200ml/h, (d) Sunflower oil+CNT-50ml/h.

3.5.5. Surface Integrity

Shao et al. [47] explored the distribution of residual stress of workpiece in AISI 1018 grinding and found that residual compressive stress would be generated on the workpiece surface in MQL and flood conditions, while higher residual tensile stress would be generated in dry conditions. As shown in Figure 16, ManojKumar and Ghosh [49] also obtained the same rule in their experiments. Residual compressive stress was generated on the workpiece surface, which was due to the better reduction of grinding zone temperature by nanofluids through reducing friction and enhancing thermal conductivity. The formation mechanism of residual stress in grinding process is complex, which is influenced by grinding heat and grinding force: ① The grinding force results in local plastic deformation of workpiece and residual compressive stress on the surface; ② Thermal expansion and cold contraction also have a certain effect on the residual tensile stress, and the extremely high grinding temperature may induce phase transition and introduce residual stress.

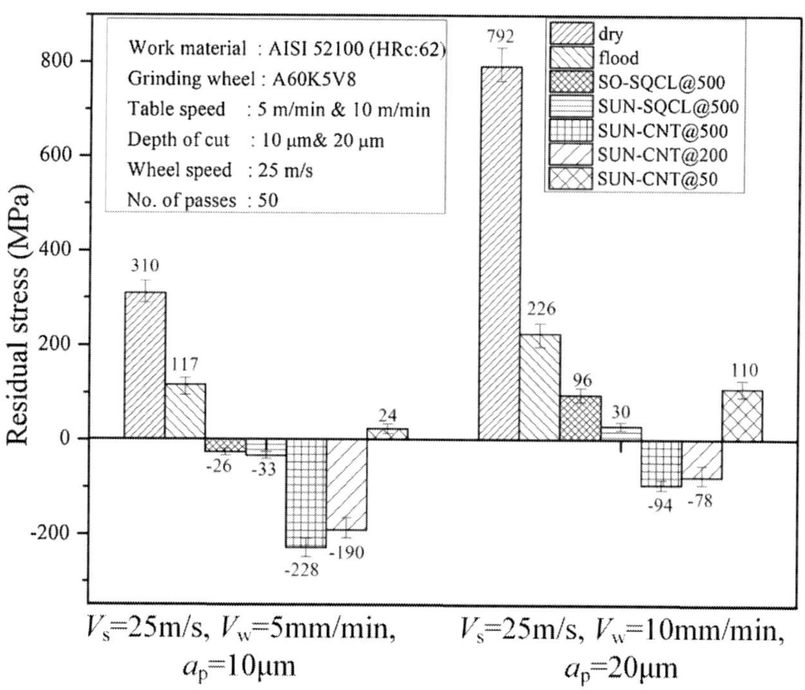

Figure 16. Residual stress of the workpiece under different lubrication conditions.

Sadeghi et al. [46] compared the workpiece surface quality when grinding AISI 4140 with vegetable oil and synthetic ester MQL, which obtained a reduction of 19.0% in surface roughness under synthetic ester compared with vegetable oil MQL. The *Ra* value of the two biolubricants is lower than that of flood when the grinding depth is increased to 15 μm. This also fully proves that biol-lubricants are more advantageous in the case of large removal (high grinding force). As shown in Figure 17, by further observation of the surface and cross section of the workpiece, it can be seen that the surface using biol-lubricant is smoother, while burr and obvious plastic uplift are observed on the cross section. Mao et al. [48] added A12O3 nanoparticles in water when grinding AISI 52100, *Ra* value decreased by 38.71% compared with MQL. Nevertheless, a large number of plastic uplifts were found on the workpiece surface. Belentani et al. [51] used vegetable oil mixed with water as lubricant (mixing ratio 1:5) to carry out the grinding experiment of AISI 4340, and the surface roughness decreased from 0.7 μm to 0.35 μm under the condition of vegetable oil. Molaie et al. [50] used four water-based NMQL (Al_2O_3, graphite, graphene oxide, MWCNT) conditions in grinding AISI 52100. Compared with pure water MQL (*Ra*=1.18 μm), the surface roughness obtained by graphene oxide nano-enhanced biolubricants was the smallest (*Ra*=0.65 μm).

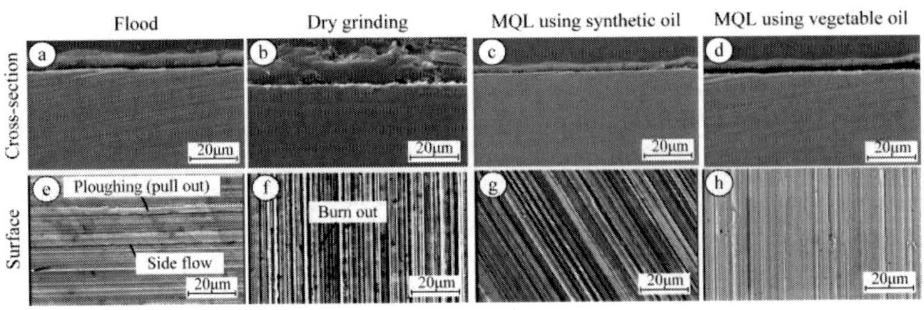

Figure 17. The surface and cross-section morphology under different lubrication conditions. (a) Cross-section-Flood, (b) Cross-section-Dry grinding, (c) Cross-section-using synthetic oil, (d) Cross-section-MQL using vegetable oil, (e) Surface-Flood, (f) Surface-Dry grinding, (g) Surface -using synthetic oil, (h) Surface -MQL using vegetable oil.

Considering the flow rate of biol-lubricant, Javaroni et al. [54] found in the grinding experiment of AISI 4340 that the workpiece surface quality was optimized with the increase of flow rate, and the plastic uplift and debris adhesion on the workpiece surface were effectively avoided at 160 ml/h flow rate. As the biolubricant MQL has limited performance for wheel cleaning, the workpiece surface quality is further optimized when MQL and WCJ are used together, as shown in Figure 18.

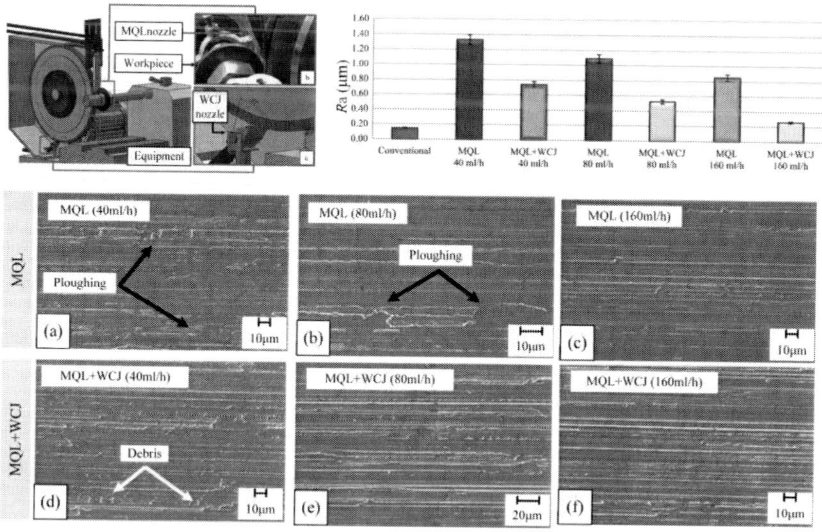

Figure 18. Surface integrity under various usage amount and with or without WCJ method. (a) MQL (40 ml/h), (b) MQL (80 ml/h), (c) MQL (160 ml/h), (d) MQL+WCJ (40 ml/h), (a) MQL+WCJ (80 ml/h), (a) MQL+WCJ (160 ml/h).

3.5.6. An Analysis of Grinding Defects and Solutions

According to the research results, the grinding defects and bottlenecks of high strength steel can be summarized as follows:

i. Wheel blockage caused by high strength. The tensile strength of high strength steel is higher than that of titanium alloy and nickel-based alloy, which is difficult for the material to deform after it is removed and formed into debris. The experimental results also show that the deformation curvature of the debris of high strength steel is less than that of the other two materials. Therefore, the discharge direction of debris is toward the wheel, and it is easy to accumulate in the wheel pore and cause the blockage, which is consistent with the observation that the wheel blockage is more serious than that of titanium alloy and nickel-based alloy. In addition, spherical debris is also found in the grinding of high strength steel, which is due to the material melting caused by high temperature in the grinding zone. Part of the melted material is re-solidified as spherical debris and discharged from the cutting zone, while most will flow into the pore of the grinding wheel and cause serious blockage.

ii. Microstructure transformation of workpiece caused by high grinding temperature. Due to the blockage of wheel and abrasive wear, the tribological characteristics deteriorate and the grinding heat output increases, which makes the grinding zone temperature of high strength steel too high. Therefore, it has been confirmed that the material structure changes under the influence of high grinding temperature. Furthermore, residual stress and microhardness of workpiece will increase.

Therefore, the way to solve the grinding bottleneck of high strength steel is to improve the cleaning performance of wheel and heat transfer performance of grinding zone. The current research methods and conclusions can be summarized as follows:

i. Low viscosity biolubricants (canola oil, sunflower oil) have good fluidity and cleanliness, compared with high viscosity lubricants are more suitable for high strength steel. The use of nano-enhancers can significantly improve the heat transfer performance, but there is no special requirement on the type and characteristics.
ii. The wheel cleanness of NMQL still needs to be improved. At present, there are two feasible schemes: ① Using vegetable oil/water mixed lubricant to improve grinding performance by reducing lubricant viscosity and improving thermal conductivity; ② The use of WCJ or CWCJ with MQL, also has obvious effect.

Conclusion

A systematic review of MQL grinding of aerospace materials with biolubricants was carried out, revealing the mechanism of physicochemical property of biolubricants on antifriction and antiwear, heat dissipation, wheel cleanness, and analyzing the unique grinding bottlenecks of typical materials (titanium alloy, nickel-based alloy and high strength steel). The physicochemical property and compatibility of biolubricants under different grinding conditions were proposed. The specific conclusions are as follows:

i. The long carbon chain, high fatty acid saturation and polar groups of biolubricants are the key characteristics to improve the adsorption capacity and strength of lubricating oil film in grinding zone. High viscosity and surface tension are beneficial to lubrication performance, while the cooling performance and wheel cleaning performance will be limited due to too high viscosity. Pour point, thermal stability and PH value are the limiting conditions of thermal/force boundary of biolubricants, and the thermal stability of grinding process should be considered emphatically.
ii. For titanium alloy, workpiece burn, and material plastic accumulation was discovered in grinding due to its enhanced mechanical activity, low thermal conductivity and high elastic modulus. Compared with traditional cutting fluids, the normal force can be reduced by up to 72.2% with biolubricant, while the tangential force can be reduced by 79.1% and *Ra* value can be reduced by 52.07% using graphene + palm oil compared with pure oil. The advantage of biolubricant can be reflected by CoF and grinding temperature, which decreased from 0.562 to 0.384 and 278.9°C to 229.18°C, respectively. At the same time, the microhardness of workpiece surface, the accumulation of material plough and the adhesion of debris were improved significantly.
iii. For nickel-based alloy, the bottleneck of wheel wear and workpiece surface burn was caused by its high hardness and low thermal conductivity. Using palm oil as biolubricant, it has better grinding performance than other kinds of vegetable oil. When Al_2O_3+ palm oil is used, the specific energy and friction coefficient are 31.2% and 38.3% lower than those of flood, respectively. The G ratio increased from 0.688 to 1.532 due to the addition of Al_2O_3 nano-enhancer. Moreover, the use of biolubricants can significantly reduce wheel wear, increase the G ratio, and control the surface defects caused by high grinding temperature.
iv. For high strength steel, the phenomenon of debris breaking difficulty, wheel blockage and high grinding temperature as the high tensile strength, which is directly reflected in the increase of microhardness and residual stress compared with the cutting fluids, the tangential grinding force was reduced by 67.5% when using biolubricant, and

the G ratio was increased by 69.2% when using MWCNT + sunflower oil. At the same time, the residual stress and microhardness caused by high grinding temperature were significantly reduced with employment nano-enhanced biolubricant.

v. Due to the different grinding bottlenecks of difficult-to-machine materials, the physical and chemical characteristics of biolubricants are also different. Titanium alloy grinding needs lubricants with strong lubrication and cooling capabilities and requires nano-enhancers with high thermal conductivity (e.g., graphene) and biolubricants with high viscosity (e.g., castor oil and palm oil). Nickel-based alloy grinding needs lubricants with modeling antiwear characteristics, base oil should have high viscosity and fatty acid saturation, containing polar groups and other characteristics (e.g., palm oil, castor oil), nano-enhancers corresponding with high hardness, spherical characteristics (e.g., Al_2O_3). High strength steel grinding needs lubricants with wheel cleaning performance, and low viscosity physical lubricants (e.g., canola oil, sunflower oil) are more suitable. The use of nano-enhancers can significantly improve the heat transfer performance, but there is no need to make special requirements on their types and properties.

References

[1] Gao T, Li CH, Yang M, et al. (2021). Mechanics analysis and predictive force models for the single-diamond grain grinding of carbon fiber reinforced polymers using CNT nano-lubricant. *J. Mater. Process. Technol.*, *290*, 16. 10.1016/j.jmatprotec.2020.116976.

[2] Yang M, Li CH, Luo L, et al. (2021). Predictive model of convective heat transfer coefficient in bone micro-grinding using nanofluid aerosol cooling. *Int. Commun. Heat Mass Transf.*, *125*, 1-12. 10.1016/j.icheatmasstransfer.2021.105317.

[3] Zhu SW, Xiao GJ, He Y, et al. (2022). Tip vortex cavitation of propeller bionic noise reduction surface based on precision abrasive belt grinding. *Journal of Advanced Manufacturing Science Technology*, 2(1), 2022003.

[4] Liu MZ, Li CH, Yang M, et al. (2023). Mechanism and enhanced grindability of cryogenic air combined with biolubricant grinding titanium alloy. *Tribol. Int.*, *187*, 1-22. 10.1016/j.triboint.2023.108704.

[5] Nadolny K, Rokosz K, Kaponek W, et al. (2017). SEM-EDS-based analysis of the amorphous carbon-treated grinding wheel active surface after reciprocal internal cylindrical grinding of Titanium Grade $2^{A®}$ alloy. *Int. J. Adv. Manuf. Technol.*, *90*(5-8), 2293-2308. 10.1007/s00170-016-9555-7.

[6] Sinha MK, Setti D, Ghosh S, et al. (2016). An investigation on surface burn during grinding of Inconel 718. *J. Manuf. Process.*, *21*, 124-133. 10.1016/j.jmapro.2015.12.004.

[7] Duan ZJ, Li CH, Ding WF, et al. (2021). Milling Force Model for Aviation Aluminum Alloy: Academic Insight and Perspective Analysis. *Chin. J. Mech. Eng.*, *34*(1), 35. 10.1186/s10033-021-00536 9.

[8] Gao T, Zhang YB, Li CH, et al. (2021). Grindability of carbon fiber reinforced polymer using CNT biological lubricant. *Sci Rep*, *11*(1), 1-14. 10.1038/s41598-021-02071-y.

[9] Xi XX, Yu TY, Ding WF, et al. (2018). Grinding of Ti_2AlNb intermetallics using silicon carbide and alumina abrasive wheels: Tool surface topology effect on grinding force and ground surface quality. *Precis. Eng.-J. Int. Soc. Precis. Eng. Nanotechnol.*, *53*, 134-145. 10.1016/j.precisioneng.2018.03.007.

[10] Liu MZ, Li CH, Zhang YB, et al. (2021). Cryogenic minimum quantity lubrication machining: from mechanism to application. *Front. Mech. Eng.*, *16*(4), 649-697. 10.1007/s11465-021-0654-2.

[11] Li HG, Zhang YB, Li CH, et al. (2022). Extreme pressure and antiwear additives for lubricant: academic insights and perspectives. *Int. J. Adv. Manuf. Technol.*, *120*(1-2), 1-27. 10.1007/s00170-021-08614-x.

[12] Gupta MK, Khan AM, Song QH, et al. (2021). A review on conventional and advanced minimum quantity lubrication approaches on performance measures of grinding process. *Int. J. Adv. Manuf. Technol.*, *117*(3-4), 729-750. 10.1007/s00170-021-07785-x.

[13] Tang LZ, Zhang YB, Li CH, et al. (2022). Biological Stability of Water-Based Cutting Fluids: Progress and Application. *Chin. J. Mech. Eng.*, *35*(1), 24. 10.1186/s10033-021-00667-z.

[14] Singh AK, Kumar A, Sharma V, et al. (2020). Sustainable techniques in grinding: State of the art review. *J. Clean Prod.*, *269*, 16. 10.1016/j.jclepro.2020.121876.

[15] Pimenov DY, Mia M, Gupta MK, et al. (2021). Improvement of machinability of Ti and its alloys using cooling-lubrication techniques: a review and future prospect. *J. Mater. Res. Technol-JMRT*, *11*, 719-753. 10.1016/j.jmrt.2021.01.031.

[16] Kelly J, Cotterell MJJoMPT. (2002). Minimal lubrication machining of aluminium alloys. *120*(1-3), 327-334.

[17] Song Y-X, Li C-H, Zhou Z-M, et al. (2024). Nanobiolubricant grinding: a comprehensive review. 1-42.

[18] Wang XM, Li CH, Yang M, et al. (2024). Research progress on the physical mechanism of minimum quantity lubrication machining with nano-biolubricants. *Journal of Mechanical Engineering*, *60*, 1-37.

[19] Debnath S, Reddy MM, Yi QS. (2014). Environmental friendly cutting fluids and cooling techniques in machining: a review. *J. Clean Prod.*, *83*, 33-47. 10.1016/j.jclepro.2014.07.071.

[20] Chawaloesphonsiya N, Guiraud P, Painmanakul P. (2018). Analysis of cutting-oil emulsion destabilization by aluminum sulfate. *Environ. Technol.*, *39*(11), 1450-1460. 10.1080/09593330.2017.1332101.

[21] Guo Z, Study on the chemical modification of vegetable oil to prepare low pour point polyol ester lubricant base oil, Beijing University of Chemical Technology, Beijing, 2020.

[22] Yang M, Kong M, Li CH, et al. (2023). Temperature field model in surface grinding: a comparative assessment. *Int. J. Extreme Manuf.*, *5*(4), 60. 10.1088/2631-7990/acf4d4.

[23] Liu MZ, Li CH, Zhang YB, et al. (2023). Heat transfer mechanism and convective heat transfer coefficient model of cryogenic air minimum quantity lubrication grinding titanium alloy. *Journal of Mechanical Engineering*, *59*(23), 343-357.

[24] Sadeghi MH, Haddad MJ, Tawakoli T, et al. (2009). Minimal quantity lubrication-MQL in grinding of Ti-6Al-4V titanium alloy. *Int. J. Adv. Manuf. Technol.*, *44*(5-6), 487-500. 10.1007/s00170-008-1857-y.

[25] Li M, Yu TB, Zhang RC, et al. (2020). Experimental evaluation of an eco-friendly grinding process combining minimum quantity lubrication and graphene-enhanced plant-oil-based cutting fluid. *J. Clean Prod.*, *244*, 13. 10.1016/j.jclepro. 2019.118747.

[26] Ibrahim AMM, Li W, Xiao H, et al. (2020). Energy conservation and environmental sustainability during grinding operation of Ti–6Al–4V alloys via eco-friendly oil/graphene nano additive and Minimum quantity lubrication. *150*, 106387.

[27] Singh H, Sharma VS, Singh S, et al. (2019). Nanofluids assisted environmental friendly lubricating strategies for the surface grinding of titanium alloy: Ti6Al4V-ELI. *J. Manuf. Process.*, *39*, 241-249. 10.1016/j.jmapro.2019.02.004.

[28] Setti D, Sinha MK, Ghosh S, et al. (2015). Performance evaluation of Ti-6Al-4V grinding using chip formation and coefficient of friction under the influence of nanofluids. *Int. J. Mach. Tools Manuf.*, *88*, 237-248. 10.1016/j.ijmachtools. 2014.10.005.

[29] Liu GT, Li CH, Zhang YB, et al. (2018). Process parameter optimization and experimental evaluation for nanofluid MQL in grinding Ti-6Al-4V based on grey relational analysis. *Mater. Manuf. Process.*, *33*(9), 950-963. 10.1080/10426914.2017.1388522.

[30] Wang XM, Zhang JC, Wang XP, et al. (2021). Temperature field model and verification of titanium alloy grinding under different cooling conditions. *China Mechanical Engineering*, *32*(05), 572-578.

[31] Yang M, Li CH, Zhang YB, et al. (2018). Theoretical analysis and experimental research on temperature field of microscale bone grinding under nanoparticle jet mist cooling. *Journal of Mechanical Engineering*, *54*(18), 194-203.

[32] Liu GT, The heat transfer enhancement mechanism and experimental research of cryogenic air atomizing nanofluids for minimal quantity lubrication titanium alloy, Qingdao University of Technology, Qing Dao, 2017.

[33] Virdi RL, Chatha SS, Singh H. (2020). Machining performance of Inconel-718 alloy under the influence of nanoparticles based minimum quantity lubrication grinding. *J. Manuf. Process.*, *59*, 355-365. 10.1016/j.jmapro.2020.09.056.

[34] Virdi RL, Chatha SS, Singh H. (2019). Experiment evaluation of grinding properties under Al_2O_3 nanofluids in minimum quantity lubrication. *Mater. Res. Express*, *6*(9), 8. 10.1088/2053-1591/ab301f.

[35] Wang Y, Li C, Zhang Y, et al. (2016). Experimental evaluation of the lubrication properties of the wheel/workpiece interface in minimum quantity lubrication (MQL) grinding using different types of vegetable oils. *J. Clean Prod.*, *127*, 487-499.

[36] Zhang XP, Li CH, Jia DZ, et al. (2019). Spraying parameter optimization and microtopography evaluation in nanofluid minimum quantity lubrication grinding. *Int. J. Adv. Manuf. Technol.*, *103*(5-8), 2523-2539. 10.1007/s00170-019-03642-0.

[37] Wang YG, Li CH, Zhang YB, et al. (2018). Processing Characteristics of Vegetable Oil-based Nanofluid MQL for Grinding Different Workpiece Materials. *Int. J. Precis Eng Manuf-Green Technol.*, *5*(2), 327-339. 10.1007/s40684-018-0035-4.

[38] Zhang Y, Li C, Jia D, et al. (2015). Experimental evaluation of the lubrication performance of MoS2/CNT nanofluid for minimal quantity lubrication in Ni-based alloy grinding. *International Journal of Machine Tools Manufacture*, *99*, 19-33.

[39] Wang Y, Li C, Zhang Y, et al. (2017). Experimental evaluation on tribological performance of the wheel/workpiece interface in minimum quantity lubrication grinding with different concentrations of Al_2O_3 nanofluids. *J. Clean Prod.*, *142*, 3571-3583.

[40] Hegab H, Darras BM, Kishawy HA. (2018). Sustainability Assessment of Machining with Nano-Cutting Fluids. *Journal of Procedia Manufacturing*, *26*, 245-254.

[41] Li BK, Li CH, Zhang YB, et al. (2016). Grinding temperature and energy ratio coefficient in MQL grinding of high-temperature nickel-base alloy by using different vegetable oils as base oil. *Chin. J. Aeronaut.*, *29*(4), 1084-1095. 10.1016/j.cja.2015.10.012.

[42] Zhang YB, Li CH, Yang M, et al. (2016). Experimental evaluation of cooling performance by friction coefficient and specific friction energy in nanofluid minimum quantity lubrication grinding with different types of vegetable oil. *J. Clean Prod.*, *139*, 685-705. 10.1016/j.jclepro.2016.08.073.

[43] Li BK, Li CH, Zhang YB, et al. (2017). Heat transfer performance of MQL grinding with different nanofluids for Ni-based alloys using vegetable oil. *J. Clean Prod.*, *154*, 1-11. 10.1016/j.jclepro.2017.03.213.

[44] Wang Y, Li C, Zhang Y, et al. (2016). Experimental evaluation of the lubrication properties of the wheel/workpiece interface in MQL grinding with different nanofluids. *Tribol. Int.*, *99*, 198-210.

[45] Peng RT, He XB, Tong JW, et al. (2021). Application of a tailored eco-friendly nano fluid in pressurized internal-cooling grinding of Inconel 718. *J. Clean Prod.*, *278*, 14. 10.1016/j.jclepro.2020.123498.

[46] Sadeghi MH, Hadad MJ, Tawakoli T, et al. (2010). An investigation on surface grinding of AISI 4140 hardened steel using minimum quantity lubrication-MQL technique. *Int. J. Mater. Form.*, *3*(4), 241-251. 10.1007/s12289-009-0678-3.

[47] Shao YM, Fergani O, Ding ZS, et al. (2016). Experimental Investigation of Residual Stress in Minimum Quantity Lubrication Grinding of AISI 1018 Steel. *J. Manuf. Sci. Eng.-Trans. ASME*, *138*(1), 7. 10.1115/1.4029956.

[48] Mao C, Tang XJ, Zou HF, et al. (2012). Investigation of Grinding Characteristic using Nanofluid Minimum Quantity Lubrication. *Int. J. Precis. Eng. Manuf.*, *13*(10), 1745-1752. 10.1007/s12541-012-0229-6.

[49] ManojKumar K, Ghosh A. (2016). Assessment of cooling-lubrication and wettability characteristics of nano-engineered sunflower oil as cutting fluid and its impact on SQCL grinding performance. *J. Mater. Process. Technol.*, *237*, 55-64. 10.1016/j.jmatprotec.2016.05.030.

[50] Molaie MM, Zahedi A, Akbari J. (2018). Effect of water-based nanolubricants in ultrasonic vibration assisted grinding. *Journal of Manufacturing Materials Processing*, *2*(4), 80.

[51] Belentani RD, Funes H, Canarim RC, et al. (2014). Utilization of Minimum Quantity Lubrication (MQL) with Water in CBN Grinding of Steel. *Mater. Res.-Ibero-am. J. Mater.*, *17*(1), 88-96. 10.1590/s1516-14392013005000165.

[52] Javaroni RL, Lopes JC, Diniz AE, et al. (2020). Improvement in the grinding process using the MQL technique with cooled wheel cleaning jet. *Tribol. Int.*, *152*, 13. 10.1016/j.triboint.2020.106512.

[53] Garcia MV, Lopes JC, Diniz AE, et al. (2020). Grinding performance of bearing steel using MQL under different dilutions and wheel cleaning for green manufacture. *J. Clean Prod.*, *257*, 13. 10.1016/j.jclepro.2020.120376.

[54] Javaroni RL, Lopes JC, Garcia MV, et al. (2020). Grinding hardened steel using MQL associated with cleaning system and cBN wheel. *Int. J. Adv. Manuf. Technol.*, *107*(5-6), 2065-2080. 10.1007/s00170-020-05169-1.

Chapter 4

Principles and Equipment Designs of Magnetic Traction Nanolubricant (MTN) Grinding

Abstract

The application of nanolubricant MQL in general grinding and point grinding has been successful, but it has not been explored in deep grinding, which is characterized by a large arc length grinding zone. This is because under the complex interface between the gas barrier layer of high-speed rotating grinding wheel, the nanolubricant presents lower infiltration performance in grinding zone, resulting in grinding wheel wear intensification. To solve this problem, the magnetic traction nanolubricant (MTN) grinding method was proposed. In this chapter, the principle, device and magnetic nanolubricant were introduced. Firstly, the adsorption and suction effect were used to improve infiltration force of nanolubricant, and magnetic traction devices were designed. Secondly, the preparation method and physical property characterization of magnetic nanolubricant were carried out. Finally, the saturation magnetization and relative permeability were measured and calculated for characterizing magnetic response performance. The MTN grinding method could be used in deep and form grinding of aerospace materials, which are in line with sustainable manufacturing.

Keywords: magnetic traction nanolubricant, deep grinding, magnetic traction device, magnetic nanolubricant, physical characteristics

4.1. Introduction

Magnetorheological Finishing (MRF) generates collimating hardening jet beams by locally applying magneto-rheological jet containing abrasive materials for axial magnetic field solidification to make deterministic precise polishing at a relatively far distance. After full mechanical stirring in the container, the magneto-rheological fluid mixed with fine polishing abrasives is taken into a pump by a low-pressure system and then transformed into high-pressure liquid which is then transported through pipelines to the ferromagnetic nozzle installed in the electromagnet, forming high-speed jets. The magnetic jet beams which are stabilized and hardened by locally axial magnetic field are sprayed to the workpiece surface at some distance for polishing. The used polishing liquid is finally filtered by the recycling device and returns to the container for reuse.

With reference to the MRF technology and device, a new magnetic traction nanolubricant (MTN) method is developed. It is expected to solve technological bottlenecks of poor infiltration and insufficient lubrication ability: ① the buffing wheel for polishing different from grinding wheel and it is a rag wheel. MRF wheel bears small stresses during processing and it can be designed with hollow structure, filled with a magnetic field. Influenced by magnetic field, the magneto-rheological fluid is adsorbed onto the buffing wheel for polishing of the workpiece surface. However, the grinding wheel, a high-speed rotating cutter, bears great grinding forces during material removal and its structure cannot be changed, or it will cause explosive cracking of the grinding wheel under large cutting forces due to stress concentration. Moreover, there are abundant pores on the grinding wheel surface, which are more beneficial for adsorption of magnetic nanofluids. Under the action of magnetic field, the magnetic nanoparticles not only are adsorbed on the grain surface, but also may be stored in pores. Metal is used for binding of grinding wheel (binding agent is generally cast iron). Cast iron has good magnetism. The magnetic field generator is installed on the grinding wheel side above the grinding zone and magnetization of binding agent can be achieved by applying the magnetic field, thus realizing magnetism of grinding wheel without changing the grinding wheel structure. The grinding wheel base is usually made of aluminium alloy, medium carbon steel (e.g., 45# steel) and low-carbon steel (e.g., Q235 steel). Designing the grinding wheel base as soft magnetic materials can achieve guidance and changes of magnetic induction line directions. ② Additionally, the buffing wheel can carry a lot of magneto-rheological fluid under the action of magnetic field to remove workpiece surface materials. Magnetic nanolubricant has a small dosage and low weight, and it can be adsorbed by a small magnetic force.

Carbon nanolubricant, especially graphene (GR), shows excellent cooling and lubrication performances in grinding [1, 2]. To make carbon nanolubricant adsorbed by magnetized grinding wheel, magnetic functionalization of carbon nanofluids is needed. Common magnetofluid is commonly prepared by Fe_3O_4 which doesn't have good cooling and lubrication performances. Therefore, this study explores the preparation technology of magnetic nanolubricant and tests its physical properties.

4.2. Magnetic Traction Method Based - The Adsorption Effect

4.2.1. Principle

The processing principle of the new magnetic traction nanolubricant grinding is shown in Figure 1. Firstly, a gradient magnetic field is produced on the grinding wheel surface by applying a permanent magnet and magnetic field strength presents gradient changes along radial direction of the grinding wheel. This provides a power source to adsorption and infiltration of magnetic nanolubricant. Existing nanolubricants are mainly prepared by nanoparticles without magnetic conduction. For magnetic functionalization of nanolubricant, Fe_3O_4 and nanoparticles with excellent cooling and lubrication performances were mixed to prepare magnetic nanolubricant, which assured excellent cooling and lubrication performances during magnetic functionalization. Furthermore, magnetic nanolubricant is sprayed by the nozzle under the action of high-pressure air and then atomized into micro-droplet groups. The micro-droplet groups run through the airflow field on the grinding wheel surface and impact onto the grinding wheel surface as a response to the suction of magnetic field forces, forming a layer of lubricating oil film. Since magnetic field forces can increase adsorption rate of micro-droplets on the grinding wheel surface, the efficiency of lubricant into the grinding zone is increased. When lubricant arrives at entrance of the grinding zone, its shear flowing speed and rate in micro and nano channels within the grinding zone are increased due to the magnetic field forces of the grinding wheel, thus improving the infiltration efficiency of lubricant in the long space with a large grinding arc.

Magnetic Fe_3O_4 nanoparticles can form an oriented arrangement under the effect of magnetic field, through which the spatial distribution of lamellar nanolubricant (e.g., graphene) can be controlled. Hence, it can improve friction reduction and wear resistance of grinding zone effectively. Based on study of the above technology, it can decrease the grinding force and improve workpiece surface quality during large-arc grinding of difficult-to-machine materials.

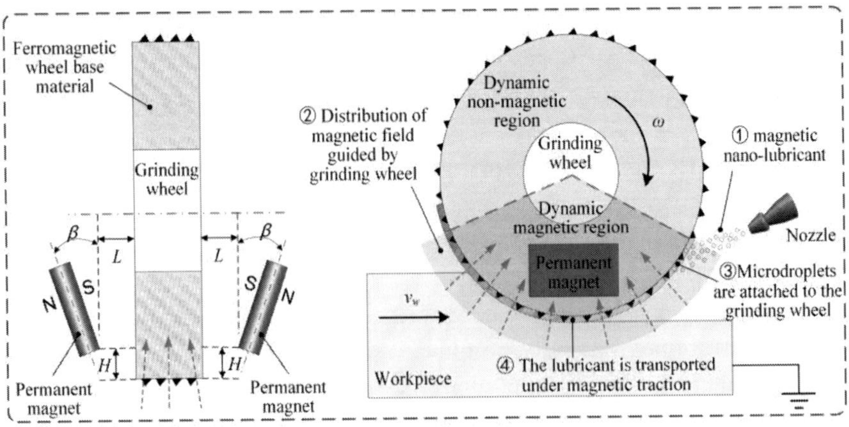

Figure 1. Principle of NMQL grinding assisted by magnetic.

4.2.2. Device Design

Due to changes of permanent magnet parameters and its relative positions with the grinding wheel, it will cause changes of magnetic field strength below the grinding wheel (defined as the dynamic magnetic region), thus adjusting size and distribution of magnetic field strength in the grinding zone. The blueprint of magnetic traction device is shown in Figure 2.

The device is designed for magnetization of grinding wheel and then controls the magnetic field strength. Therefore, the device not only requires positioning and tight clamping of the permanent magnet, but also adjusts the position of the permanent magnet, thus adjusting the magnetic field strength. The scheme is mainly composed of the following structures: 1. Magnet fixing plate×2; 2. Conical headed bolt×2; 3. Nut I×2; 4. grinding wheel; 5. Magnet (100×50×20mm)×2; 6. Fit bolts×8; 7. Connection block I×2; 8. Connection block II×2; 9. Pressing plate×2; 10. hexagon socket screw I×6; 11. Full-flight screw; 12. Nut II×8; 13. hexagon socket screw×4; 14. screw connection block×2; 15. Bolt I×3; 16. Left side plate; 17. Nut III×4; 18. positioning bolt; 19. Upper cover plate; 20. Bolt IV×4; 21. grinding wheel cover.

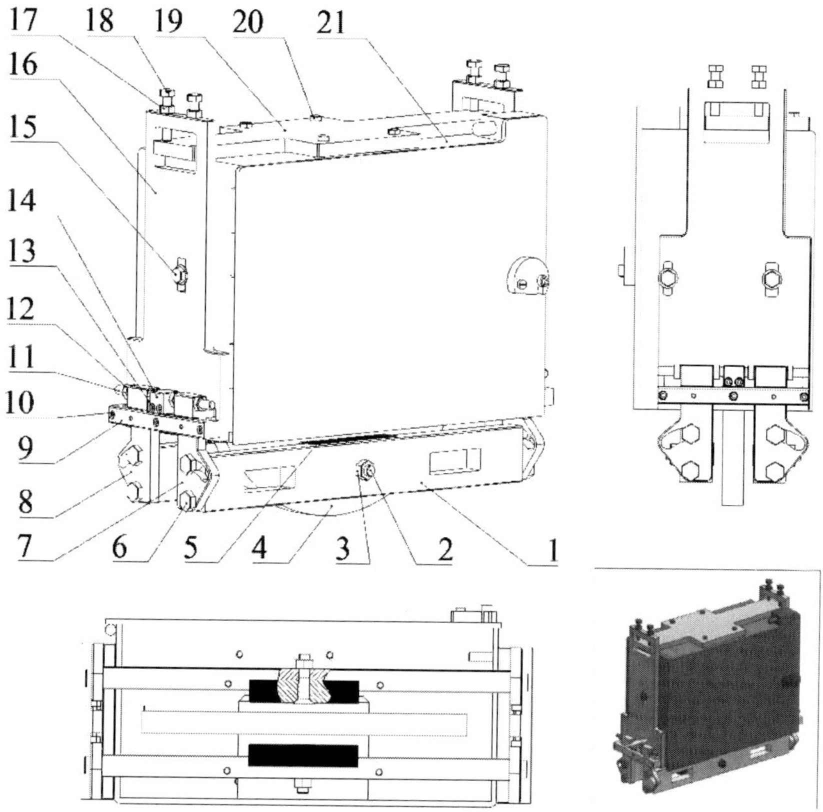

Figure 2. Design drawing of magnetic traction device.

The permanent magnet is fixed onto the magnet fixing plate (1) through the conical headed bolt (2) and nut (3). There's a square groove corresponding to the permanent magnet size on the magnet fixing plate for positioning of the permanent magnet. There are two pieces of fixing plates, which are at the right and left sides of the grinding wheel to apply two pieces of magnet. Two ends of the magnet fixing plate (1) are positioned onto the connection blocks I and II (7 and 8) through the fit bolt (6). Specifically, the lower nut fits the round hole on the connecting plate, while the upper nut fits the arc hole on the connecting plate. The angle of the fixing plate can be adjusted through the arc hole, thus driving rotation of the permanent magnet and forming an included angle (β: 0°~40°) with the side surface of the grinding wheel. Furthermore, there are slider structures above the connection plates I and II (7 and 8), which fit with the groove structures on the left plate (17) to realize moving along the axial direction of grinding wheel. The connection plates are clamped tightly through the pressing plate (9) and it realizes adjustment of distance (the distance L between the permanent magnet and grinding wheel side: 2 mm~20 mm) and positioning through the fixed full-flight screw (11) and nut II (12). The screw connection block (14) serves for positioning of full-flight screw. The left side plate (16) has a square groove, and it is fixed onto the grinding wheel cover (21) through the bolt (15). The distance (H) between the surface below the grinding wheel and surface below the permanent magnet can be adjusted within the range of 10 mm~30 mm through the bolt (18). Later, the permanent magnet is clamped tightly through the nut (17). The right plate and left plate (16) have symmetric structures and it is not introduced in the following text. The upper cover plate (19) is fixed onto the grinding wheel cover (21) by the bolt (20) and it is the basic support of the whole device. The structure of magnetic field assisted grinding device is shown in Figure 3.

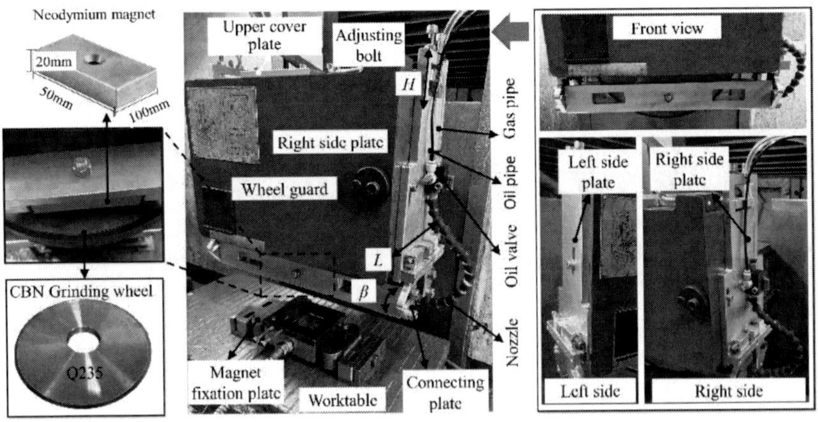

Figure 3. Magnetic field assisted grinding device diagram.

4.3. The Magnetic Traction Method - The Suction Effect

4.3.1. Principle

When the magnet is set in front of the direction of lubricant infiltration, the magnetic nanolubricant could be attracted by the magnetic force, therefore the increased the flow rate. As shown in Figure 4, in the present example, the model of the permanent magnet I-1 is N42, the remanence magnetic induction intensity is about 1.30 T, the maximum magnetic energy product is about 320 KJ/m^3, the arc angle is 20°, the width of the end face is 20 mm, and the length of the end face is 30 mm, and the magnetic field density of the magnetic field region generated by the permanent magnet I-1 and the magnetic traction force to the magnetic nanoparticles are as follows:

According to the "effective" dipole moment method, the magnetic force acting on the magnetic microstructure can be modeled by replacing a magnetic object with an "equivalent" point dipole with moment *m*, the force acting on the dipole being defined by F_m:

$$F_m = \mu_f (m_{peff} \cdot \nabla) B \tag{1}$$

Wherein, μ_f is the permeability of the medium, m_{peff} is the "effective" dipole moment of the object, and *B* is the magnetic field generated by an external source at the center of the object, where the dipole of the equivalent point is located. The dipole moment *m* is related to the volume and magnetic properties of an object and can be described as:

$$m = MV \tag{2}$$

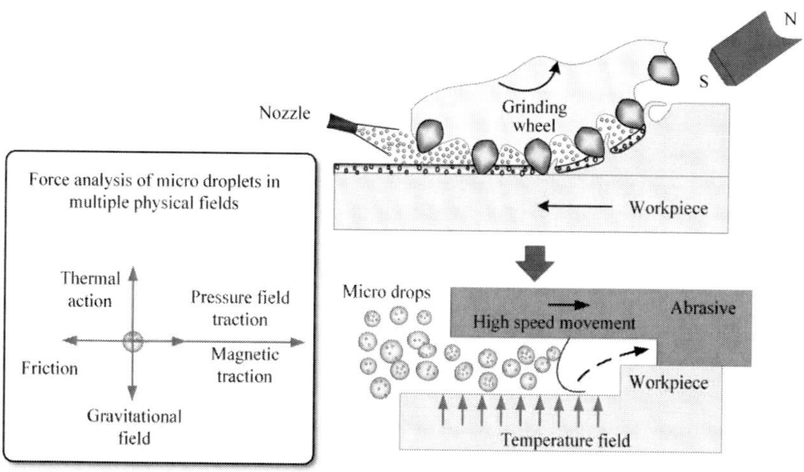

Figure 4. Schematic diagram of a force applied to a single droplet of the magnetic nanolubricant.

Wherein, *M* and *V* are the magnetization and the dipole volume, respectively. The force exerted on such a dipole changes the characteristics of the magnetic field source. It also depends on the distance between the source and target objects. Considering the permanent magnet in the present example, the magnetic field density of the grinding zone can be defined by:

$$B = \frac{\mu_0}{4\pi}\left(\frac{3(m \cdot r) \cdot r}{|r|^5} - \frac{m}{|r|^3}\right) \tag{3}$$

Wherein, μ_0 is the permeability of vacuum ($4\pi \times 10^{-7}$T·m/A), r is the distance vector from the source to the object. Under the situation of dispersing the gas barrier layer by the wind deflector, the range of the horizontal distance from the permanent magnet I-1 to the nozzle II-1 is about 10 – 15 cm, and the range of the horizontal distance from the workpiece V-4 to the permanent magnet I-1 is about 5 – 8 cm.

The arc angle of the permanent magnet is 20°, the width of the end face is 20 mm, and the length of the end face is 30 mm. Therefore, the volume of the permanent magnet is calculated as 31.42 cm^3 according to the formula $\pi(r_1^2 - r_2^2)L/18$, wherein r_1 is the radius of the large arc and is of 16 cm, and r_2 is the radius of the small arc and is of 14 cm. Since the volume of the hole required for fixation and recovery is about 3.04 cm^3, the residual volume of the permanent magnet is about 28.38 cm^3, and the magnetization is converted to SI units of 10342.61A/cm according to the maximum magnetic energy product, and B at the nozzle is 2.96×10^{-3} T by preliminary calculation. When the magnetic field density reaches the remanence intensity of the permanent magnet, the distance from the permanent magnet to the nozzle is about 1.71 cm, and according to the above formula, it can be seen that the magnetic nanofluid will be subjected to the traction force of the magnetic field when it is sprayed from the nozzle. However, the traction force is small at the beginning because the relationship between the magnetic induction intensity of the permanent magnet and the distance is about the inverse ratio of the third power. Therefore, when the magnetic nanofluid gradually enters deeply into the magnetic field area, the traction force it receives will gradually increase and reach the maximum value.

Magnetic nanofluid can be regarded as composed of numerous single micro-droplets, and a single micro-droplet contains numerous magnetic nanoparticles, so the magnetic nanoparticles should be analyzed first.

If the magnetic field effect is mainly considered, according to the force balance equation, the motion equation of the magnetic nanoparticles is written in vector form as:

$$\overrightarrow{F_m} + \overrightarrow{F_d} + \overrightarrow{F_g} = m\vec{a} \tag{4}$$

Wherein, $\overrightarrow{F_m}$ represents the magnetic force applied to the magnetic nanoparticles, $\overrightarrow{F_d}$ represents the viscous force applied to the magnetic nanoparticles, $\overrightarrow{F_g}$ represents the gravity applied to the magnetic nanoparticles, and $m\vec{a}$ represents the inertial force. Since the magnetic nanoparticles are small in size, the gravity applied to the magnetic nanoparticles can be neglected compared with the magnetic force and the viscous force.

The magnetic force to which the magnetic nanoparticles are subjected due to the external magnetic field gradient is:

$$\overrightarrow{F_m} = \mu_0 V_p \overrightarrow{M_p} \cdot \nabla \vec{H} \tag{5}$$

Wherein, μ_0 is the vacuum magnetic permeability ($4\pi\times10^{-7}$T·m/A), V_p is the particle volume, $\overrightarrow{M_p}$ is the particle magnetization strength, and $\nabla\vec{H}$ is the magnetic field strength vector.

Since magnetic nanoparticles can be seen as spherical, the volume of magnetic nanoparticles can be calculated according to the equation of $V_p=4\pi R_p^3/3$, wherein R_p is the radius of magnetic nanoparticles (10 nm = 1×10^{-6} cm), which is about 4.2×10^{-18} cm^3.

Because in magnetic fluid, the magnetic nanoparticles will move along the direction of the magnetic induction line, so that the magnetic nanoparticles are easy to realize $\overrightarrow{M_p}$ is parallel to $\vec{H}$, the magnetization of the magnetic body can be approximately expressed as $\overrightarrow{M_p} = X\vec{H}$, wherein X is the difference between the magnetic susceptibility of the magnetic nanoparticles and the fluid, $X=X_p-X_f$. Generally, compared with the magnetic susceptibility X_p of the magnetic nanoparticles, the magnetic susceptibility X_f of the fluid is very small, and X_f is often neglected, so the difference of the magnetic susceptibility can be approximately expressed as $X=X_p$, if using $\overrightarrow{M_p}$ and a_p represent the magnetization and demagnetization factor of ferromagnetic particles, respectively, then the $\overrightarrow{M_p}$, α_p and magnetic susceptibility X_p of the ferromagnetic particles are calculated respectively by the following equations:

$$\overrightarrow{M_p} = 3\alpha_p\vec{H}, \alpha_p = min(\frac{X_{p,0}}{3+X_{p,0}}, \frac{M_{p,s}}{3H}),\ X_p = 3\frac{\alpha_p}{1-\alpha_p} \tag{6}$$

Wherein $X_{p,0}$ and $M_{p,s}$ represent the initial susceptibility and saturation magnetization of magnetic nanoparticles respectively, H is the magnitude of the external background magnetic field at the magnetic nanoparticles, taking $X_{p,0}$=100SI, $M_{p,s}$=79.8 emu/g, α_p=0.97, according to the formula $H=B/\mu$, $\mu=\mu_0(1+X)$, then obtaining that μ is about 0.012 T·cm/A, H at the nozzle is about 0.247 A/cm, and the magnetization of magnetic nanoparticles is about 0.72A/cm.

And because there is no conduction current inside the magnetic fluid, so $\nabla\times\vec{H}=0$, and $\nabla(\vec{H}\cdot\vec{H})=\nabla H^2=2\mathrm{H}\nabla\mathrm{H}$, plus the demagnetization effect of the ferromagnetic particles of the kau rate, the magnetic force to which the magnetic nanoparticles are subjected will be simplified as:

$$\overrightarrow{F_m} = \frac{1}{2}\mu_0 V_p X_p (1-\alpha_p)\nabla(\overrightarrow{H^2}) \tag{7}$$

If only the magnetic field strength in the horizontal direction is considered, the magnetic force per unit magnetic nanoparticle at the nozzle is about 2.32×10^{-19} N, and the magnetic force per unit magnetic nanoparticle at the workpiece surface is about 1.09×10^{-18} N.

Under the action of the magnetic field, the magnetic nanoparticles are pulled by the magnetic field, and the macroscopic performance is that the magnetic fluid moves under the action of the magnetic field, which is equivalent to the effect that the magnetic nanoparticles will give the magnetic fluid a body force, and the body force is mainly composed of the magnetic body force $\overrightarrow{F_{mv}}$ applied by the external magnetic field, which is as:

$$\overrightarrow{F_{mv}} = n\overrightarrow{F_m} = \frac{1}{2}n\mu_0 V_p X_p (1-\alpha_p)\nabla(\overrightarrow{H^2}) \tag{8}$$

wherein n represents the total content of magnetic nanoparticles in the magnetic nanofluid, if it is assumed that the magnetic fluid in unit time is calculated, then the flow rate in unit time is about 0.67 ml, the magnetic nanoparticles contained therein are about 0.1 cm^3 according to the density and volume content thereof, and the mass is about 0.367 g, and the content of the magnetic nanoparticles in the magnetic nanofluid in unit time calculated according to the mass fraction and molar mass thereof is about 9.61×10^{20}, then the magnetic field force applied to the magnetic fluid in unit time at the nozzle is about 222.86 N, and the magnetic field force applied to the magnetic fluid in unit time at the surface of the workpiece is about 1047.49 N.

It can be seen from the formula of magnetic field force that the magnetic force produced by the external magnetic field on the magnetic nanoparticles is related to many parameters such as particle volume, magnetic properties of the material itself, magnetic field intensity and magnetic field gradient. The magnetic force is zero in the absence of a magnetic field or in a uniform magnetic field. The magnetic force increases with the increase of magnetic field intensity and gradient, but when the magnetic field intensity reaches a certain value and the magnetic particles are saturated, the increase of magnetic field intensity has no effect on the electromagnetic force, but the magnetic force is still proportional to the magnetic field gradient. At the same time, it can be seen that the distribution characteristics of magnetic force can be effectively changed by adjusting the magnitude and direction of magnetic field intensity and gradient through the design of external magnetic field.

In the present example, magnetic field assisted grinding has the characteristics of deep grinding and the processing mode of forward and backward grinding. Because of this characteristic and processing mode, a height difference is formed on both sides of the workpiece surface, and the existence of this height difference makes the magnetic nanofluid unable to accurately act on the unprocessed and processed regions. The analysis of the height difference and the formed cutting depth is as follows:

The topography of the grinding surface of the workpiece V-4 is a collection of track curves of several single abrasive grains on the grinding wheel, so the grinding effect of single abrasive grain on workpiece is studied firstly. When studying the mechanism of grinding workpiece with a single abrasive grain, it is necessary to study the motion relationship between the abrasive grain and the workpiece. Here, the workpiece is selected to be 40 mm long, 30 mm wide and 20 mm high.

v_w is moving speed of the workpiece and is of 0.05 m/s, and r_s is the radius of the grinding wheel and is of 150 mm. Because θ is very small, $\sin\theta \approx \theta$, $\theta_a^2 = 2(1-\cos\theta_a)$, then the rotation angle of the grinding wheel can be expressed as $\theta_a = \omega \cdot t$, wherein the angular velocity of the grinding wheel is $\omega = v_s / r_s$, wherein v_s is the rotational speed of the grinding wheel and is of 30 m/s, ω is 200 rad/s, and θ_a is about 120. Combining the above equation, obtaining $t = d_s\theta_a / 2v_s$, wherein d_s is the wheel diameter and is of 300 mm and t is about 0.6 s.

In the process of single abrasive cutting workpiece, the grinding parameters of single abrasive and the motion parameters of workpiece will affect the geometric relationship between single abrasive and workpiece. The moving contact arc length of single abrasive l_k is mainly related to the moving speed of workpiece, the cutting depth of single abrasive, the diameter of grinding wheel and the rotating speed of grinding wheel.

According to the relative motion relationship between the grinding wheel and the workpiece, the process can be regarded as the workpiece is stationary and the trajectory *AC* formed by the motion action of the abrasive grain is a pendulum, and the equation of the pendulum is:

$$\begin{cases} x = r_s\, sin\,\psi \pm v_\psi \\ y = r_s(1 - cos\,\psi) \end{cases} \tag{9}$$

Wherein, Ψ is the angular displacement of the abrasive grain, v_Ψ is the horizontal movement distance of the grinding wheel, because the value of Ψ is small, therefore $\sin\Psi = \Psi$, then:

$$\psi = \sqrt{1 - cos^2\,\psi} = \left[1 - \left(\frac{r_s - a_p}{r_s}\right)^2\right]^{\frac{1}{2}} = 2\left[\frac{a_p}{d_s} - \frac{a_p^2}{d_s^2}\right]^{\frac{1}{2}} \approx 2\sqrt{\frac{a_p}{d_s}} \tag{10}$$

$$v_\psi = \frac{\psi \cdot v_0}{2\pi} \tag{11}$$

Wherein, v_0 is the horizontal travel distance of the workpiece corresponding to each revolution of the grinding wheel, so:

$$v_\psi = \frac{\psi \cdot v_0}{2\pi} = \frac{v_w}{60n_s \times 2\pi}\psi = \frac{v_w}{60n_s \times 2\pi}\frac{r_s}{r_s}\psi = \frac{r_s v_w}{60 v_s}\psi \tag{12}$$

And by the calculation, v_ψ is about 4.79×10^{-4} mm, and x is about 1.72×10^{-2} mm.

In the grinding process, due to the properties of the workpiece material, the workpiece will also produce elastic deformation during grinding. In addition, the residual stress after grinding leads to the deformation of the workpiece surface, which makes the actual grinding track generated on the workpiece surface higher than the theoretical track. Therefore, the actual curve of the workpiece surface should be the superposition of the theoretical curve and the elastic recovery of the workpiece.

In cutting process, take the machined workpiece surface as a datum plane, the equation of the cutting depth of abrasive particles can be express as:

$$h = a_p - \frac{x^2}{d_s \cdot \left(\frac{v_w}{v_s + 1}\right)^2} \tag{13}$$

Because the magnetic field-assisted grinding is a kind of deep grinding method, the single grinding depth is much larger than that of ordinary grinding, so if the single grinding depth a_p is 5mm, the calculated h is about 4.62 mm.

4.2.2. Device Design

The present example provides a device for supplying and recovering minimal quantity lubricant in a magnetic field-assisted abrasive grinding, as shown in Figure 5, comprising a controllable magnetic field assembly I, a controllable nozzle assembly II, a recovering and filtering device III, a controlling and monitoring assembly IV and a grinding wheel guard assembly V; wherein, the controllable magnetic field assembly I may adjust and change a position of a magnetic field according to the needs of machining, the controllable nozzle assembly II may adjust a length of the nozzle by stretching according to the needs of machining, and the recovering and filtering device III may separate magnetic nanoparticles from impurity iron chips after machining and further filter for recycling.

As shown in Figure 6, the grinding wheel guard assembly V comprises a wind deflector V-1, a grinding wheel V-2, a grinding wheel guard V-3, a magnetic worktable V-5, and a magnetic fixture V-6, wherein the magnetic worktable V-5 is mounted on a lower side of the grinding wheel guard V-3, and the magnetic fixture V-6 for clamping the workpiece V-4 is mounted on the magnetic worktable V-5. The controllable magnetic field assembly I comprises a permanent magnet I-1 and a first guide rail mechanism I-2, wherein the first guide rail mechanism I-2 is mounted in the groove V-1-1, and the permanent magnet I-1 is connected to the first guide rail mechanism I-2. Specifically, as shown in Figures 6 and 7, the first guide rail mechanism I-2 comprises a rack guide rail I-2-1, a first gear I-2-2, a sliding plate I-2-4, a servo motor I-2-3, a roller I-2-6, etc., the rack guide rail I-2-1 is in an arc shape adapted to the groove V-1-1 and is connected to the groove V-1-1 through bolts, and the rack guide rail I-2-1 is convex towards the outside.

The controllable magnetic field assembly I comprises a permanent magnet I-1 and a first guide rail mechanism I-2, wherein the first guide rail mechanism I-2 is mounted in the groove V-1-1, and the permanent magnet I-1 is connected to the first guide rail mechanism I-2. Specifically, as shown in Figure 7 and 8, the first guide rail mechanism I-2 comprises a rack guide rail I-2-1, a first gear I-2-2, a sliding plate I-2-4, a servo motor I-2-3, a roller I-2-6, etc., the rack guide rail I-2-1 is in an arc shape adapted to the groove V-1-1 and is connected to the groove V-1-1 through bolts, and the rack guide rail I-2-1 is convex towards the outside.

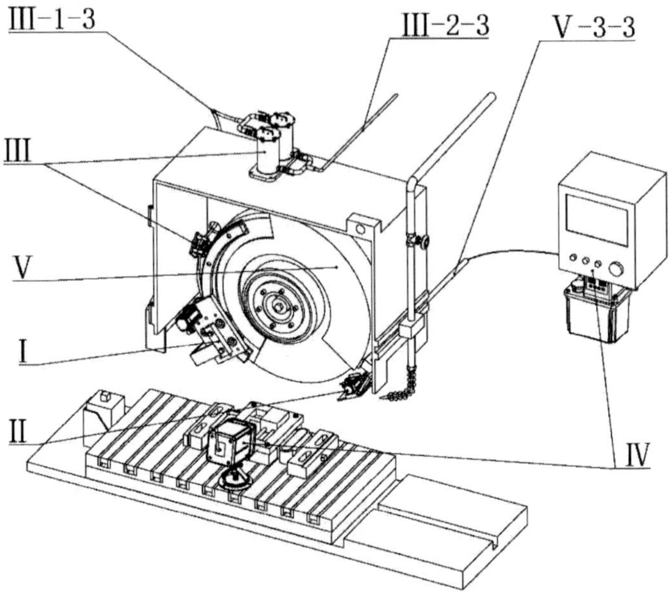

(a)

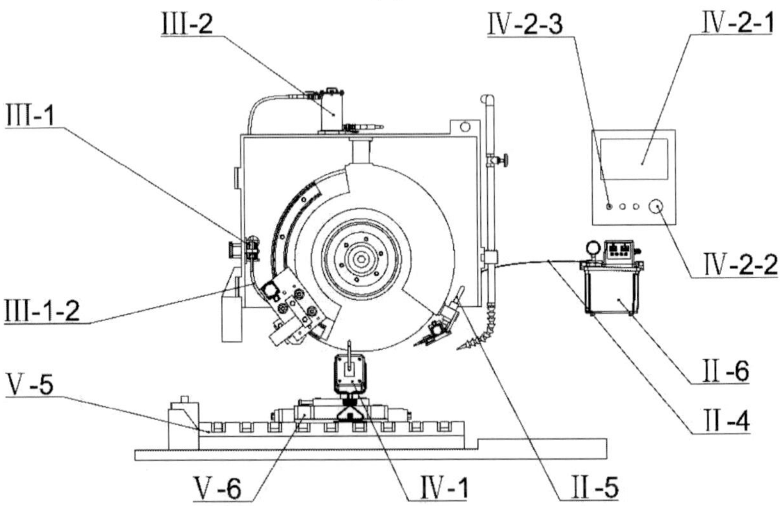

(b)

Figure 5. (a) Axonometric view of device, (b) main view of device.

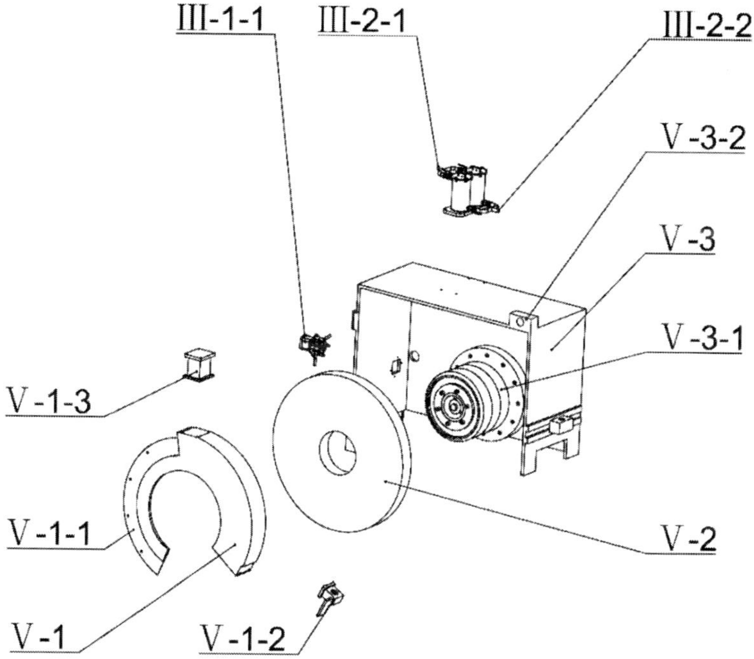

Figure 6. Exploded view of an internal system of a grinding wheel guard.

In the present example, an outer side of the rack guide rail I-2-1 is provided with teeth, so the rack guide rail I-2-1 may engage with the first gear I-2-2. The first gear I-2-2 is mounted on the sliding plate I-2-4 and is located on a first side of the sliding plate I-2-4. A servo motor I-2-3 is mounted on a second side of the sliding plate I-2-4, and the servo motor I-2-3 is connected to the first gear I-2-2. In the present example, the sliding plate I-2-4 is a rectangular plate.

The surface of the first side of the sliding plate I-2-4 on which the first gear I-2-2 is mounted is taken as a back surface, and the back surface of the sliding plate I-2-4 is further provided with two groups of rollers I-2-6, and the rollers I-2-6 are rotationally connected with the sliding plate I-2-4. Wherein, a plurality of rollers I-2-6 in each the group are arranged at intervals along an extending direction of the rack guide rail I-2-1, and the rollers I-2-6 contact a side wall of the rack guide rail I-2-1 to play a guiding role; under a driving action of the servo motor I-2-3, the first gear I-2-2 rotates, and the sliding plate I-2-4 may make a circular motion of a certain angle along the arc-shaped rack guide rail I-2-1 under the action of the rollers I-2-6. In the present example, two rollers are provided in each the group of the rollers I-2-6.

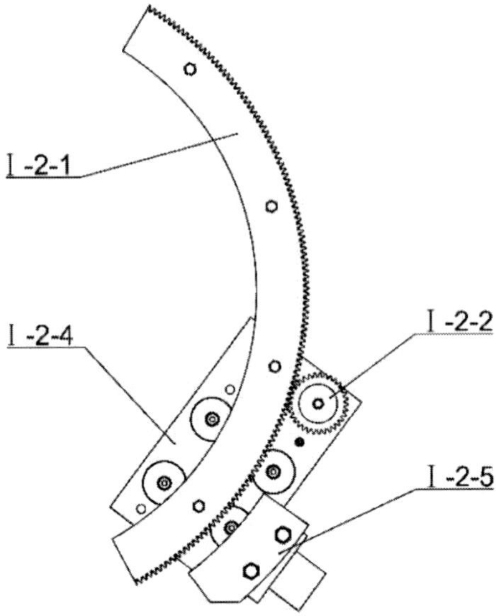

(a)

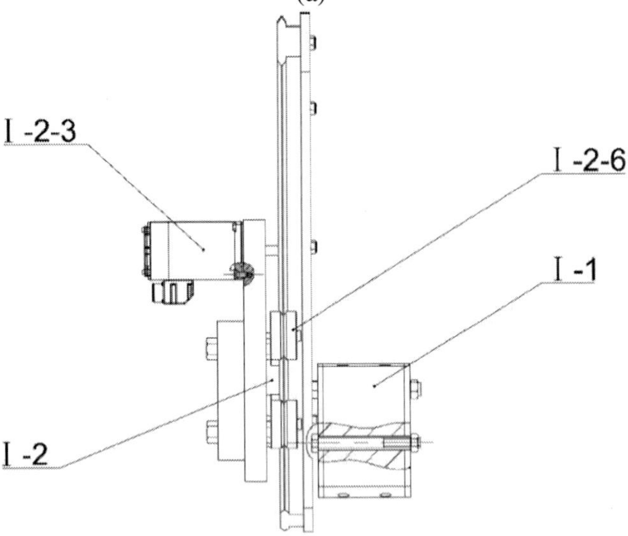

(b)

Figure 7. (a) rear view of a first guide rail mechanism, (b) side cross-sectional view of the first guide rail mechanism.

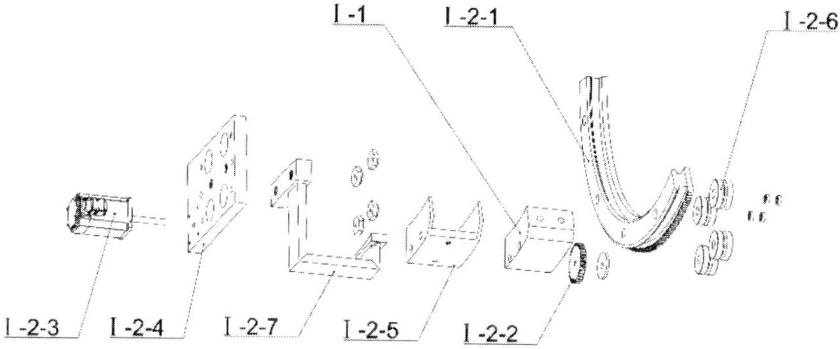

Figure 8. Exploded view of the first guide rail mechanism.

As shown in Figure 9, the sliding plate I-2-4 is connected with a clamping plate I-2-5 through the pillar I-2-7, a first end of the pillar I-2-7 is connected to a front surface of the sliding plate I-2-4 through bolts, and a second end of the pillar I-2-7 is connected to the clamping plate I-2-5. The pillar I-2-7 is of a bending structure. In the present example, the pillar I-2-7 comprises a first horizontal section, a vertical section and a second horizontal section which are connected in sequence, wherein a length of the second horizontal section is less than that of the first horizontal section, and the clamping plate I-2-5 is provided on the second horizontal section. The clamping plate I-2-5 is of a U-shaped structure.

The permanent magnet I-1 is connected in the clamping plate I-2-5 through bolts, in the present example, the permanent magnet I-1 is a cylinder with a certain radian, and is coaxial with the grinding wheel V-2, and the permanent magnet I-1 can perform a circular arc motion at a certain angle under the driving of the sliding plate I-2-4, so as to adjust the position of the permanent magnet I-1, thereby facilitating an instant adjustment of a height difference between two sides of a workpiece generated by forward and backward grinding during the magnetic field-assisted abrasive grinding, and facilitating the magnetic nanofluid to better act on the surface of the workpiece.

As shown in Figure 10(a) and 10(b), the permanent magnet I-1 is internally provided with a hole I-1-3 for connecting a third tube III-1-2 of a peristaltic pump III-1, an end surface of the third tube III-1-2 is flush with an end surface I-1-1 of the permanent magnet I-1, and a first side of the end surface I-1-1 is provided with an inclined place-being-cut I-1-2 for preventing collision with the workpiece when rotating and adjusting the angle of the permanent magnet I-1. A volume of the permanent magnet I-1 can be calculated according to the cylinder formula, and the parameters of permanent magnet I-1 can be selected according to actual requirements.

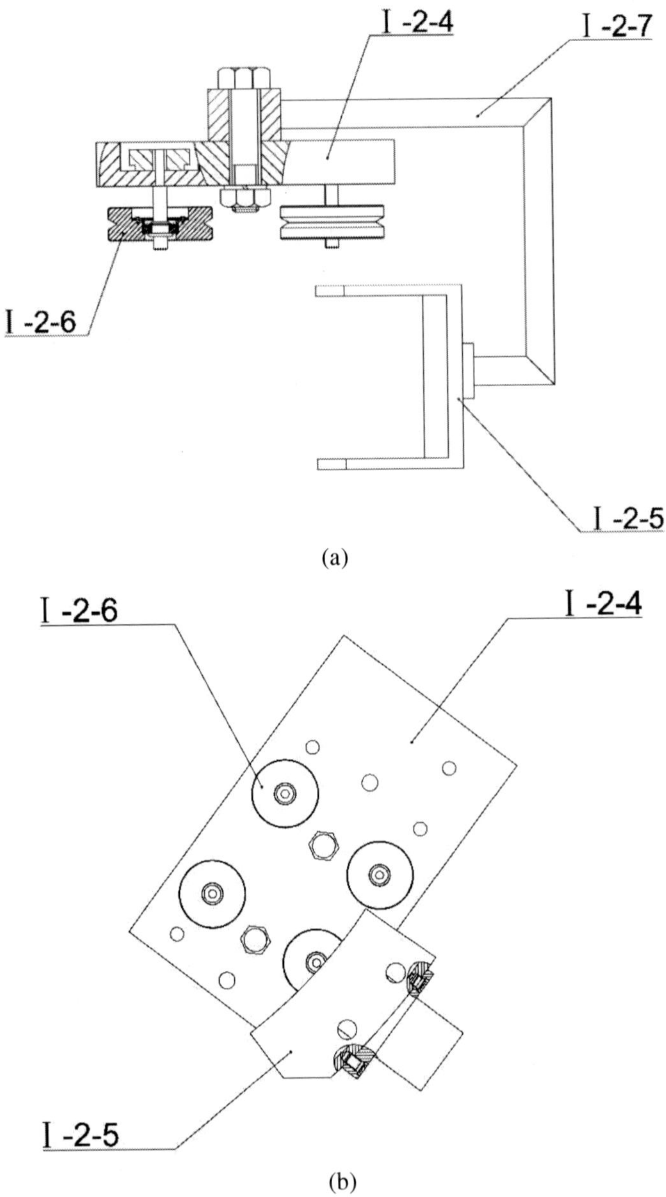

Figure 9. Different view of the clamping plate and the sliding plate. (a) side cross-sectional view, (b) front view.

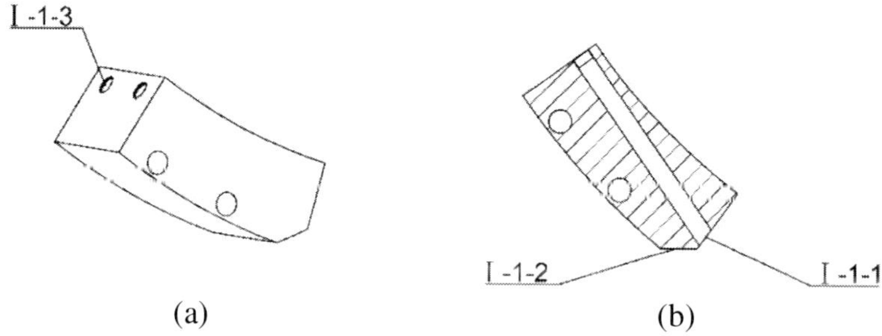

Figure 10. Different view of permanent magnets. (a) axonometric view, (b) side cross-sectional view.

During magnetic field-assisted abrasive grinding process, the permanent magnet I-1 will continuously attract magnetic substances, the magnetic nanoparticles therein and iron filings generated during machining will continuously accumulate on the end face I-1-1 of the permanent magnet I-1 with the accumulation of machining time, and if it is not cleaned in time, it will affect the machining to a certain extent. Therefore, it is necessary to provide the recovering and filtering device III.

As shown in Figure 4, the recovering and filtering device III comprises the peristaltic pump III-1 and a filter assembly III-2, wherein the peristaltic pump III-1 is mounted on a top of the grinding wheel guard V-3 and is connected to a first end of a peristaltic pump motor III-1-1 through a fourth tube III-1-3, a second end of the peristaltic pump motor III-1-1 is connected to a third tube III-1-2 which passes through the hole I-1-3 of the permanent magnet I-1.

As shown in Figure 6, the peristaltic pump III-1 is provided with an inlet III-2-1 and an outlet III-2-2, wherein the inlet III-2-1 is connected to the fourth tube III-1-3, and the outlet III-2-2 is connected to the second tube III-2-3. The peristaltic pump III-1, driven by the peristaltic pump motor III-1-1, absorbs the magnetic nanoparticles and iron filings retained on the end face I-1-1 of the permanent magnet I-1 through the third tube III-1-2, and conveys them to the filter assembly III-2. Since the diameter difference between the magnetic nanoparticles and the iron filings is large, the magnetic nanoparticles can be well separated from the iron filings, and then the magnetic nanoparticles are conveyed to the inside of the box structure of the grinding wheel guard V-3 through the second tube III-2-3 for post-treatment, to facilitate the recycling of the magnetic nanoparticles.

As shown in Figure 11 and 12, the controllable nozzle assembly II comprises a nozzle II-1, a second guide rail mechanism II-2, a movable plate

II-3, a magnetic nanofluid delivery tube II-4, etc., wherein the nozzle II-1 and the second guide rail mechanism II-2 are mounted on the movable plate II-3, and the nozzle II-1 is connected to a thread groove in the first table with cylinder V-1-2 through threads.

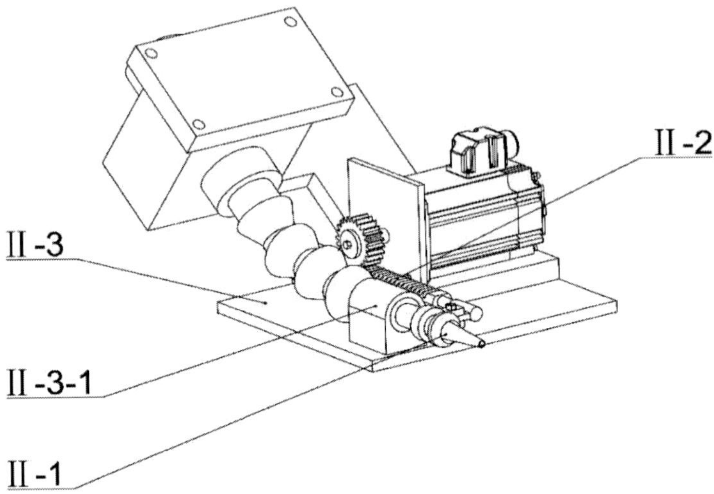

(a)

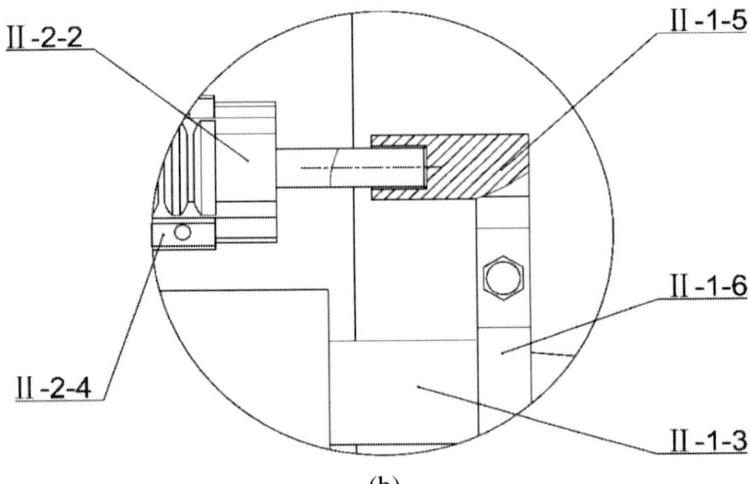

(b)

Figure 11. (a) axonometric view of a second guide rail mechanism, (b) cross-sectional view of a linear guide connected to a threaded sleeve.

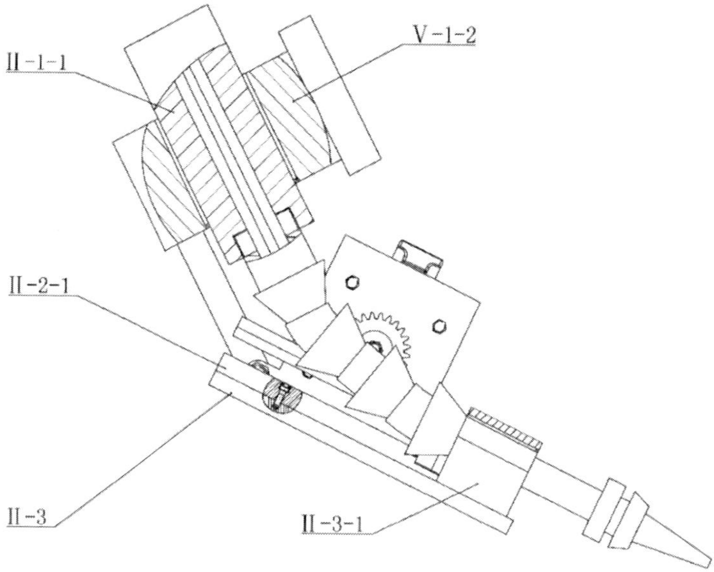

Figure 12. Side cross-sectional view of the second guide rail mechanism.

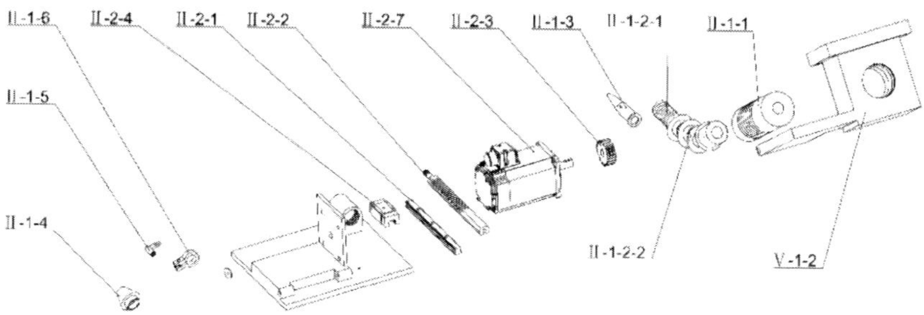

Figure 13. Exploded view of the second guide rail mechanism.

As shown in Figures 13, 14, and 15, the nozzle II-1 comprises a telescopic front section II-1-3, a middle section II-1-2 and a rear section II-1-1 which are arranged in sequence, an outer side of the telescopic front section II-1-3 is provided with a nested-ring II-1-4, the middle section II-1-2 comprises a piston sleeve II-1-2-1 and a universal bamboo joint tube II-1-2-2 connected to the piston sleeve II-1-2-1, and the telescopic front section II-1-3 is matched with the piston sleeve II-1-2-1 and can extend and contract along an inside of the piston sleeve II-1-2-1. As shown in Figure 14 (a), an end of the piston sleeve II-1-2-1 is provided with a limiting bulge to prevent the telescopic front section II-1-3 from falling out. The universal bamboo joint tube II-1-2-2 can adjust the angle of the nozzle before machining, and the piston sleeve II-1-2-1 is used for the telescopic front section II-1-3 to perform telescopic movement inside it.

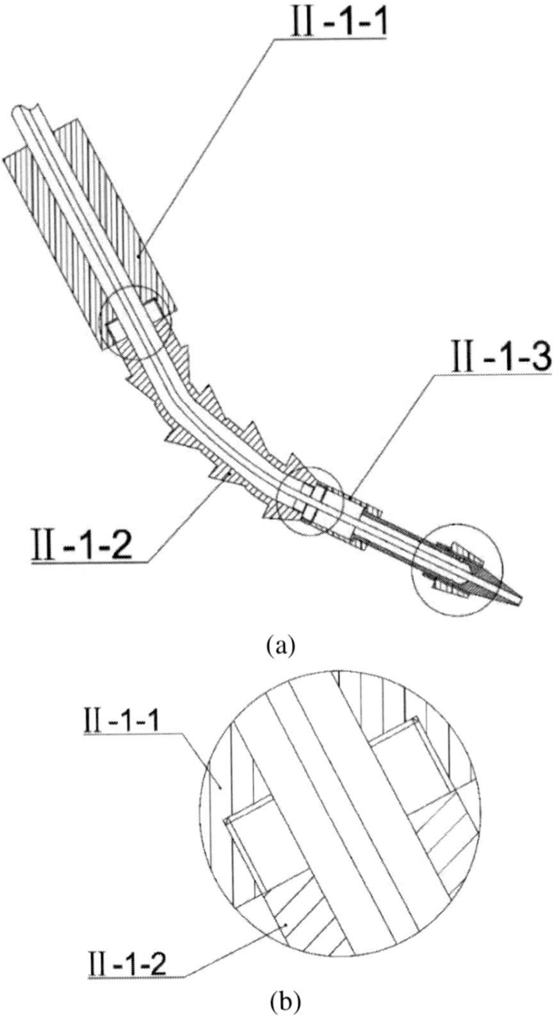

Figure 14. (Continued).

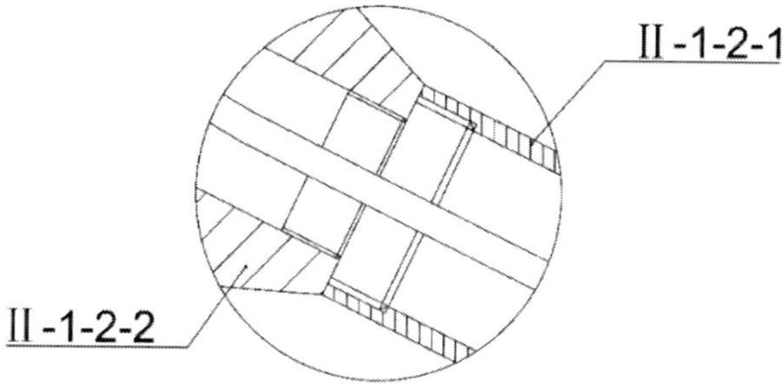

(c)

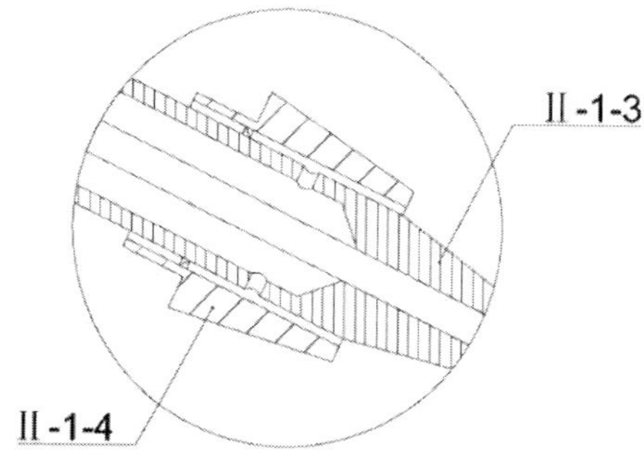

(d)

Figure 14. (a) side cross-sectional view of a nozzle (b) cross-sectional view of the connection between a middle portion of the nozzle and a rear portion of the nozzle (c) cross-sectional view of the connection between a high-pressure gas delivery tube and the middle portion of the nozzle (d) cross-sectional view of the connection between the telescopic front portion and a nested-ring.

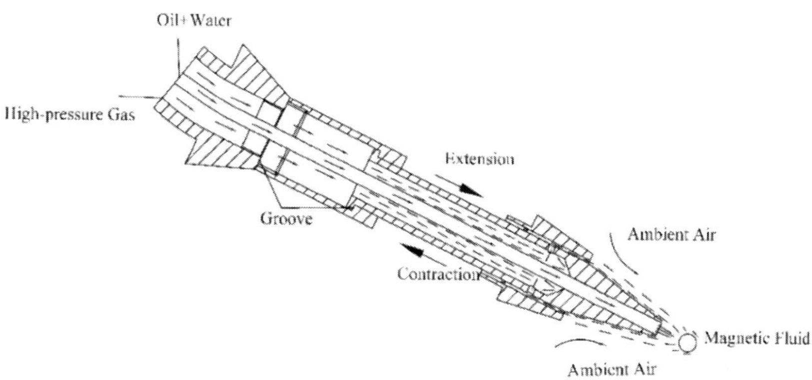

Figure 15. Schematic view of a telescoping motion of the nozzle and an oil-water-gas three-phase flow motion.

As shown in Figures 11(b), 13 and 16, the second guide rail mechanism II-2 comprises a second gear II-2-3, a rack II-2-2, a linear guide rail II-2-1 and a slider II-2-4, wherein the linear guide rail II-2-1 is fixed on an upper surface of the movable plate II-3, the slider II-2-4 is matched with the linear guide rail II-2-1, and the rack II-2-2 is fixed on a top of the slider II-2-4 and is parallel to the linear guide rail II-2-1. The linear guide II-2-1 and the slider II-2-4 constitute the guide assembly. The rack II-2-2 is engaged with the second gear II-2-3 and the second gear II-2-3 is connected to the servo motor II-2-5.

The telescopic front section II-1-3 is connected to a threaded sleeve II-1-5 at an end part of the rack II-2-2 through the nested-ring II-1-6, the second gear II-2-3 is rotated through the rotation of the servo motor II-2-5 to drive the rack II-2-2 to move, further to change telescopic and spraying range of the nozzle II-1, so that the height difference between two sides of a workpiece generated by forward and backward grinding during magnetic field assisted abrasive grinding is conveniently adjusted in real time, and the magnetic nanofluid is favorably acted on the surface of the workpiece.

As shown in Figure 13, the piston sleeve II-1-2 -1 is provided with external threads, a sleeve II-3-1 mounted on the movable plate II-3 is provided with internal threads, and the piston sleeve II-1-2-1 is matched with the sleeve II-3-1 through threads. The servo motor II-2-5 is fixed on the movable plate II-3 through screws, and the back side of the movable plate II-3 is rotatably connected with the second table with cylinder V-1-3. The movable plate II-3 is favorable for changing the position of the servo motor II-2-5 when the angle of the nozzle II-1 is adjusted before machining, so that the servo motor II-2-5 is kept relatively static to the nozzle II-1 to ensure the normal expansion and contraction of the nozzle II-1.

As shown in Figure 5(b), a first end of the magnetic nanofluid delivery tube II-4 is connected to an oil delivery port of the minimal quantity lubricant pump box II-6, and a second end of the magnetic nanofluid delivery tube II-4 enters an interior of the high-pressure gas delivery tube II-5 from an rear end of the box structure and is clamped and fixed with an interior of the telescopic front section II-1-3, and a diameter of the magnetic nanofluid delivery tube II-4 is much smaller than a diameter of the high-pressure gas delivery tube II-5.

A first end of the high-pressure gas delivery tube II-5 is connected to the box structure through an interior of the first tube V-3-6 so as to facilitate an input of high-pressure gas at the rear end, and a second end of the high-pressure gas delivery tube II-5 enters an interior of the nozzle through firstly the hole of the grinding wheel guard V-3 and then the rear section II-1-1 of the nozzle, and is connected to the universal bamboo joint tube II-1-2-2 through threads. The first tube V-3-6 is welded and fixed with the grinding wheel guard V-3.

As shown in Figure 5, the controlling and monitoring system IV comprises a vision camera IV-1 and a system control box IV-2, wherein the system control box IV-2 comprises a display screen IV-2-1 and a control center IV-2-2, and a bottom of the vision camera IV-1 is adsorbed by the magnetic worktable V-5 before machining, and a function of the vision camera IV-1 is to monitor and detect the lubrication properties and cooling performance of the magnetic nanofluid on the workpiece V-4 during machining.

The vision camera IV-1 captures a movement of the magnetic nanofluid mixed with the fluorescent agent during machining, collects the images of the temperature and the wear condition of the machining surface of the workpiece, and displays the collected images on the display screen IV-2-1 through a wireless transmission device IV-2-3.

As shown in Figure 16, during machining, inputting machining parameters (e.g., grinding wheel speed V_c or grinding depth a_p, worktable reciprocating speed V_s and spindle feed speed V_f) and size data of the workpiece V-4; comparing, by the control center IV-2-2, the input data with the parameter values set by the database, wherein these parameter values are set by the shape and size of the workpiece and the way of forward and backward grinding, there is an optimal machining parameter solution; then, analyzing and obtaining, by the control center IV-2-2, the optimal pulse signal amount required by the motor; and then transmitting the pulse signal to a drive board of the motor of each component through an expansion board of the motor, and then converting, by the drive board of the motor, the pulse signal into the rotation angle of the motor of each component, so as to control the position change of the permanent magnet I-1 and the nozzle II-1 and the operation of the recovering and filtering device III.

When the corresponding motor completes the corresponding angle rotation, the control center IV-2-2 will feed back information to the display screen IV-2-1 through a serial communication to display the adjustment of the position. When the machining is completed, the control center IV-2-2 receives the completion instruction and resets the permanent magnet I-1 and nozzle II-1 according to the motor rotation angle. Meanwhile, the situation of the reset will also be fed back to the display screen IV-2-1 through the means of serial communication.

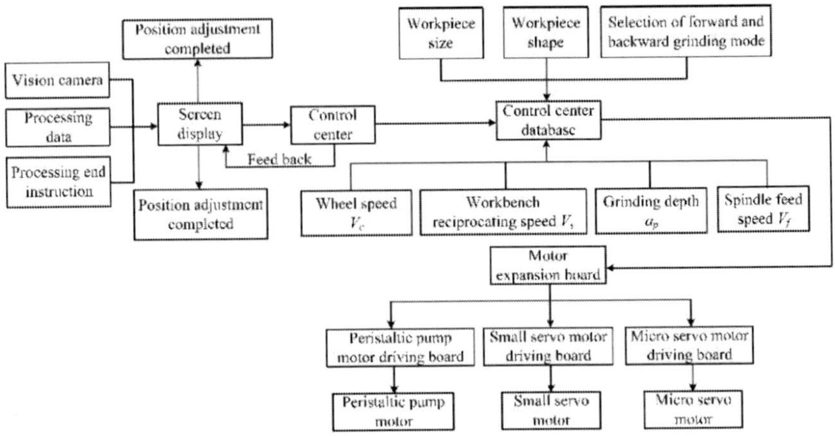

Figure 16. Flow chart of a principle of a control system.

The working process of the present example is as follows:

Step 1: adsorbing, by the magnetic worktable V-5, the magnetic fixture V-6 and the bottom of the vision camera IV-1; fixing, by the magnetic fixture V-6, the workpiece V-4, positioning the grinding wheel V-2 in the machining position above the workpiece V-4, and placing the visual camera IV-1 at the angle of view facing the machining area of the grinding surface of the workpiece.

Step 2: closing the grinding wheel guard V-3, and sucking the grinding wheel guard V-3 tight by using the magnetic switch V-3-2.

Step 3: starting the grinding, and controlling, by the control system, the controllable telescopic nozzle II-1 to spray a multiphase flow spray formed by magnetic nanofluid and compressed air into the grinding area between the workpiece and the grinding wheel.

Step 4: dispersing, by using the wind deflector V-1, the air barrier layer produced by the high-speed rotation of the grinding wheel V-2, to make the multiphase flow spray from the nozzle II-1 spray to the grinding area better, and also to facilitate the traction of the magnetic nanofluid by the permanent magnet I-1.

Step 5: along with the grinding, monitoring, by the visual camera IV-1, the lubricating properties and cooling performance of the magnetic nanofluid on the workpiece V-4. Because of the characteristics of magnetic field-assisted deep grinding and the difference of forward and backward grinding, the horizontal height of the left and right sides of the surface of the workpiece V-4 will be different. The control system controls and adjusts the angular positions of the permanent magnet I-1 and the nozzle II-1, on one hand, the real-time angle adjustment is beneficial to the permanent magnet to better draw the sprayed magnetic nanofluid, and on the other hand, the permanent magnet and the surface of a workpiece form a constantly changing magnetic field area, so that the magnetic nanofluid can better form an oil film on the surface of the workpiece with height difference, and the excellent lubricating and cooling characteristics of the magnetic nanofluid are exerted.

Step 6: with the progress of grinding and the accumulation of time, continuously attracting, by the permanent magnet I-1, magnetic substances, and the magnetic nanoparticles and the iron filings generated in the machining in the magnetic substances will continuously accumulate on the end face of the permanent magnet with the accumulation of machining time. Absorbing, by the peristaltic pump III-1 driven by the peristaltic pump motor III-1-1, the magnetic nanoparticles and the iron filings retained on the end face of the permanent magnet I-1 through the recovering tube and conveying them to the recovering and filtering device III; because the diameter difference between the magnetic nanoparticles and the iron filings is large, the magnetic nanoparticles can be well separated from the iron filings, and then are conveyed to the interior of the box structure through the tube on the outlet at the other end of the recovering and filtering device III for post-treatment, so that the recycling of the magnetic nanoparticles is facilitated.

Step 7: After the machining is completed, shutting the control system down and demagnetizing the magnetic worktable V-5.

Further, in grinding, the energy consumed for removing a unit volume of material is much greater than other cutting methods, and a large amount of heat is generated in the grinding zone. The excessive grinding zone temperature will not only affect the quality of the machined surface and the service life of the grinding wheel, but also affect the performance of the lubricating fluid. When the temperature rises, the viscosity of grinding fluid will decrease, which will affect the forming ability of grinding fluid on the machined surface, and reduce the thickness and bearing capacity of lubricating oil film. As the viscosity of the grinding fluid is reduced and the fluidity is enhanced, when the grinding wheel contacts with the surface of the workpiece, it is easy to cause damage to the oil film. After the oil film is damaged, the grinding wheel will form direct contact friction with the surface of the workpiece, so that the temperature of the grinding zone rises sharply, which is very unfavorable to the grinding process, and will form a vicious circle of high temperature – viscosity of the grinding fluid is reduced – the temperature is further raised – the viscosity of the grinding fluid is further reduced.

To solve the above problems of nano minimal quantity lubrication, the present example adopts the technical solution of forming the microscopic magnetic fluid by adding magnetic nanoparticles (e.g., the nanoparticles that can conduct magnetism and show magnetism under the action of external magnetic field) into the grinding fluid, and forming an oil film with good lubrication and heat dissipation performance on the machined surface under the traction action of the superimposed magnetic field area formed by the permanent magnet I-1 and the working surface.

The magnetic nanofluid (a mixed solution of magnetic nanoparticles and grinding base fluid according to a certain proportion) enters the nozzle II-1 by flowing through the liquid path, while the compressed gas enters the nozzle II-1 by flowing through the gas path. The magnetic nanofluid and the compressed air are mixed and accelerated in the nozzle II-1 and then sprayed out.

In the present example, the angle α of the nozzle II-1 is set to be 15° – 45°, the injection flow rate of the nozzle II-1 is 2.5 – 5.5 ml/min, and the pressure of the compressed air is 4.0 – 10 bar. The particle size of the nanoparticles is ≤ 100 nm, and the volume content of the nano particles is 1% – 30 vol%. The magnetic nanofluid is prepared by selecting Fe_3O_4 magnetic nanoparticles with the weight fraction of 0.5 wt.%, the density is of 3.67 g/cm^3 and the average particle size is of 20 nm; the lubricating base oil is soybean oil (vegetable oil) and is mixed with graphene powder, a fluorescent agent and a dispersant. The magnetic Fe_3O_4 nanoparticles in the prepared grinding fluid can be adsorbed on the surface of graphene to form magnetic lubricating mixed particles.

4.4. Preparation and Physical Properties of Nanolubricant

4.4.1. Nanomaterials

Carbon nanolubricant has been one of research hotspots in the tribological field [3, 4]. Fullerene, CNT and GR are typical representatives of carbon nano materials [5-7]. In particular, GR is a two-dimensional (2D) cellular crystal structure formed by tight packing of monolayer carbon atoms [8]. GR is a known type of new 2D nano material with the lowest thickness, highest strength and best electric and thermal conductivity in the world [9]. It is valuable that the heat conductivity coefficient of GR is 1.7 times that of CNTs and the shear force among monolayer GR is very small. GR is an excellent kind of solid lubrication material [10-12]. During grinding, GR shows excellent cooling lubrication performances. The carbon nanolubricant needs to be magnetized in order to be adsorbed by the magnetized grinding wheel. The common magnetofluid is prepared by Fe_3O_4 which doesn't have good cooling and lubrication performances (Figure 17).

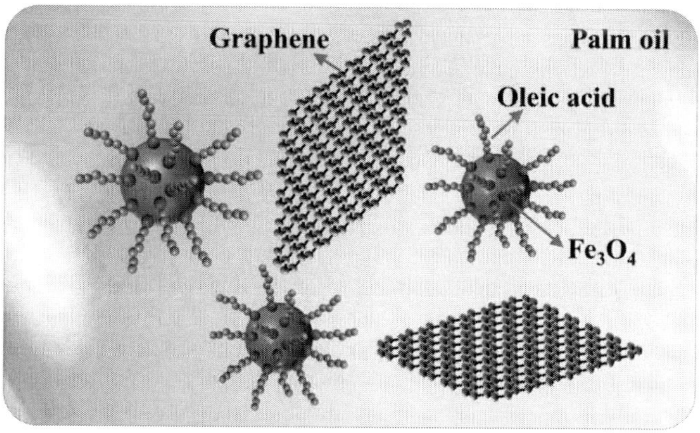

Figure 17. Diagram of magnetic nanofluid.

Nanoparticles used in the experiment include graphene, Fe_3O_4, Fe_3O_4@GR composite. SEM images are shown in Figure 18 and the physical properties are listed in Table 1. All nanoparticles were bought from Suzhou Tanfeng GR Technology Co., Ltd. The Fe_3O_4@GR composite nanoparticles form chemical modification of Fe_3O_4 onto GR surface, thus realizing magnetic functionalization of GR nanoparticles. The steps are introduced as follows: Fe_3O_4 and GR nanoparticles are mixed in the absolute ethyl alcohol solution for ultrasonic vibration to mix evenly. Next, a modification experiment with specific solution in a constant-temperature water bath is carried out. Finally, the composite nanoparticles can be gained through filtering and drying.

Table 1. Physical properties of nanoparticles

Nanolubricant	Thermal conductivit	Density	Modulus of elasticity	Micromorphology
GR	5300 W/m·K	0.7 g/cm^3	1.1 TPa	Layered
Fe_3O_4	157 W/m·K	5.18 g/cm^3	--	Spheroidal
Fe_3O_4@GR	--	--	--	Layered

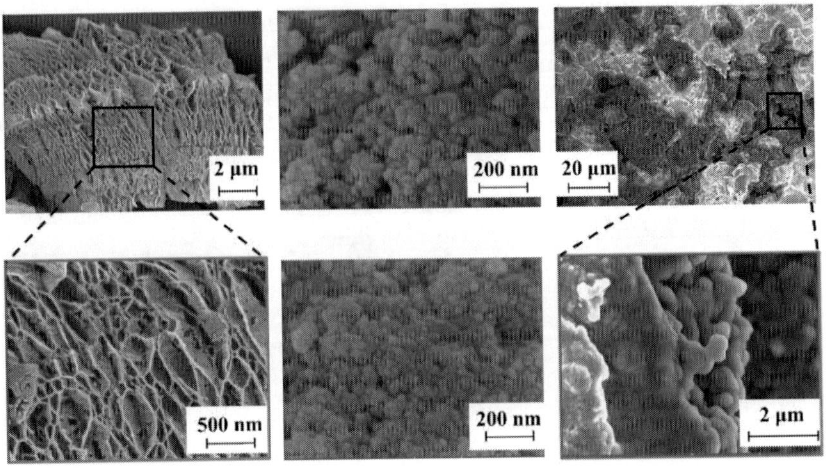

(a) (b) (c)

Figure 18. The SEM of nanoparticles. (a) Fewer layers of GR, (b) Fe_3O_4, (c) Fe_3O_4@GR.

4.4.2. Preparation Method

Since dispersibility of nanoparticles in nanofluid is a key factor that influences lubrication performance. Hence, this study uses the most common two-step method with the best effect to prepare nanofluids (Figure 19). During preparation, nanoparticles and base oil are mixed, to which Alkylphenol polyoxyethylene ether-10(APE-10) is added in as the dispersing agent. Furthermore, the prepared nanofluids are on the CNC oscillator, assisted with 20 min of ultrasonic vibration to improve dispersibility. According to a study of Gao et al. [13], APE-10 has better dispersion performances for carbon nanoparticles than other dispersing agent. Hence, APE-10 is applied as the dispersing agent to prepare nanofluids.

In previous studies, Guo et al. [14] found through their study that palm oil shows excellent lubrication performances in NMQL due to the high viscosity, long modular chain and polar groups. Additionally, oleic acid is an excellent kind of dispersing agent of Fe_3O_4 and oleic acid accounts for about 42% in palm oil, higher than those in other types of vegetable oils. Therefore, palm oil is chosen as the base oil of nanofluid. This assures not only dispersibility of Fe_3O_4, but also has excellent lubrication performance. The major physical properties of palm oil are listed in Table 2.

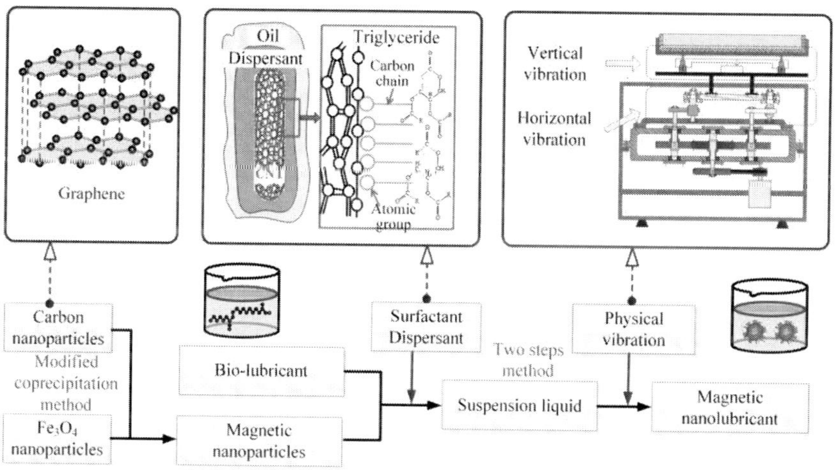

Figure 19. Preparation of magnetic nanofluid.

Table 2. Physical properties of palm oil

Physical property	Value
Colour	Pale yellow
Density (20 °C)	0.93 g/ml^3
Viscosity (20 °C)	66.1 cp
Flash point	165°C
Acid number	0.16 mg/g
Saponification number	195.07 mg/g
Types and contents of main fatty acids	Oleic acid 42.02%, Palmitic acid 40.24%, Linoleic acid 12.32%, Stearic acid4.27%

4.4.3. Viscosity and Density

Viscosity is the variable change and adhesive force generated by intermolecular movement and it is one of factors influencing cooling and lubrication performances of getable lubricant. The viscosity of lubricant mainly influences its infiltration performances, lubrication performances and heat exchange performances during grinding. ① Effects of viscosity on infiltration performances. After entering into the grinding zone at a certain speed and angle, the mobility of lubricant with high viscosity deteriorates due to holding by the viscous force, and it cannot infiltrate into the grinding wheel-workpiece interface fully. ② Effects of viscosity on cooling performances. The lubricant entering into the grinding zone flows at a certain speed. During relative movement of lubricating oil film on the grinding wheel-workpiece interface, there's relative movement among liquid layers. Hence, the heat transfer process of lubricant microdroplets conforms to the heat convection theory of liquid and the heat transfer coefficient is related with viscosity.

According to existing studies, it has been proved that adding nanoparticles into lubricating oil may cause changes of viscosity of fluid. Since different nanoparticles are added in, viscosity changes differently. Therefore, viscosity of different nanofluids used in the experiment (palm oil, Fe_3O_4 nanofluids, GR nanofluids, Fe_3O_4@GR nanofluids and Fe_3O_4/GR composite nanofluids) was measured firstl in this study. During measurement, the Brookfield DV-2T viscometer was applied, getting viscosity values of different types of nanofluids (Figure 20).

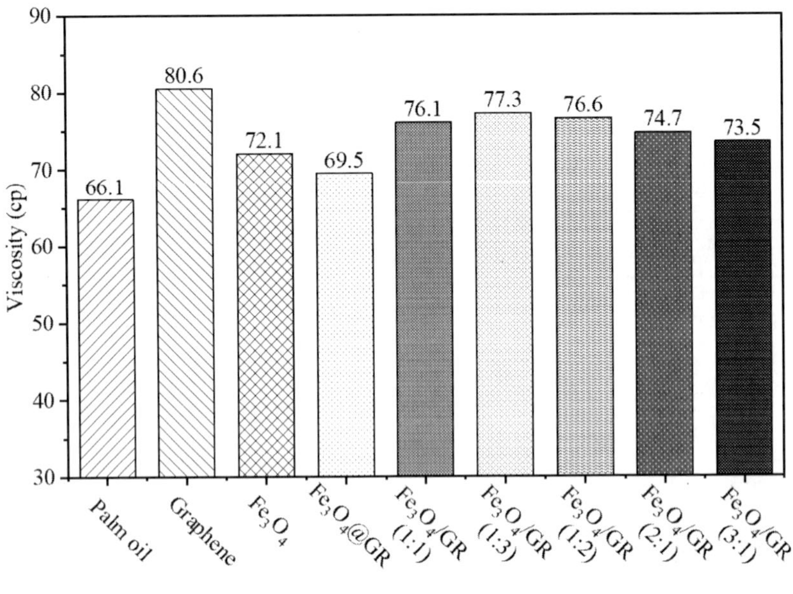

Figure 20. Viscosity of different lubricants.

After palm oil is added into nanoparticles, viscosity of nanofluids is increased to some extent. Specifically, viscosity of GR nanofluids (80.6 cp) is the highest, which is 21.9% higher than palm oil (66.1 cp). This is because frictional force among nanoparticle molecules increases viscosity of the base oil. The frictional force produced by lamellar GR nanoparticles in oil is higher than that of spherical Fe_3O_4, thus increasing viscosity of GR nanofluids. According to further observation, it is found that the viscosity values of

Fe_3O_4/GR composites with different mixing ratios are between those of GR nanofluids and Fe_3O_4 nanofluids. With the increase of Fe_3O_4 nanoparticle content, viscosity presents a decreasing trend. This is also caused by the spherical structure of Fe_3O_4. Since Fe_3O_4 is modified in graphene, the viscosity value of the composite nanofluid is lower than that of nanofluids prepared by physical mixing method.

The magnetic nanolubricant prepared in this study is essentially ferrofluid. According to the basic theory of hydrodynamics of ferrofluid, its viscosity is related with volume concentration of Fe_3O_4. It can be described by the Einstein formula when the volume concentration is lower than 2% and it can be described by the Vand formula when the volume concentration is higher than 2%. The viscosity of magnetofluid increases to some extent when it is in the magnetic field because ferromagnetic particles form oriented arrangement in base liquid as response to effect of the magnetic field, the frictional force among molecules increases, thus increasing viscosity. Since magnetic nanofluid is applied to magnetic field, the magnetic field will influence movement and arrangement of Fe_3O_4 nanoparticles in oil, thus influencing viscosity of nanofluids. Hence, key attention is paid to viscosity measurement of three types of magnetic nanofluids (Fe_3O_4 nanofluids, Fe_3O_4@GR nanofluids and Fe_3O_4/GR composite nanofluids) under the influence of magnetic field. Moreover, influences of physical and chemical properties of magnetic nanolubricant on impacting and adsorption behaviors are discussed.

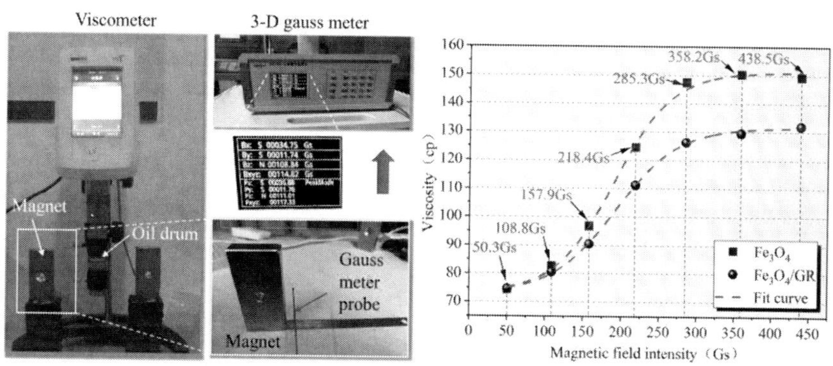

(a) (b)

Figure 21. Variation of lubricant viscosity under magnetic field. (a) Viscosity test method under magnetic field (b) variation laws of viscosity of nanofluids under different magnetic field intensity.

The viscosity measurement device under the action of magnetic field is shown in Figure 21(a). Permanent magnets were put at two sides of the oil cylinder of the viscometer and viscosity values of nanofluids at several typical distances were measured by changing the distance from the permanent magnet to the oil cylinder. Moreover, the magnetic field strength at the above distances was measured by using a 3 Gauss meter (DX-350). As a result, variation laws of viscosity of nanofluids with magnetic field strength were disclosed (Figure 21(b)).

When the permanent magnet is far away from the oil cylinder, viscosity values of three types of magnetic nanofluids range from 71 cp to 74 cp when the magnetic field strength is small (50 Gs). When the magnetic field strength reaches 285 GS, viscosity values of Fe_3O_4 nanofluids, Fe_3O_4@GR nanofluids and Fe_3O_4/GR composite nanofluids are 147.5 cp, 82.4 cp and 126.3 cp, respectively. With the increase of magnetic field strength, viscosity curves of Fe_3O_4 nanofluids, Fe_3O_4@GR nanofluids and Fe_3O_4/GR composite nanofluids all present S-shaped rising trend. When the magnetic field strength is higher than 285 Gs, viscosity of Fe_3O_4 nanofluids, Fe_3O_4@GR nanofluids and Fe_3O_4/GR composite nanofluids presents a stable fluctuation state. The viscosity curves of Fe_3O_4 nanofluids and Fe_3O_4/GR composite nanofluids increase quickly, while the viscosity curve of Fe_3O_4@GR nanofluids fluctuates slightly. The viscosity curve of Fe_3O_4 nanofluids fluctuates more obviously than that of composite nanofluid. This is because Fe_3O_4 content in the composite nanofluid is lower than that in Fe_3O_4 nanofluids, and it is influenced less by the magnetic field. In the composite nanofluid, there are fewer Fe_3O_4 combined with GR molecule. Therefore, magnetic field influences viscosity of composite nanofluid slowly. To sum up, nanofluids with different magnetism have different response degrees to magnetic field. Furthermore, S-shaped curve was applied for fitting of variation trend of viscosity. The fitted curve is shown in the following figure. Fitting results are shown in Eq. (14) and Table 3. In this study, Fe_3O_4 nanolubricant was applied as the representative sample.

$$\mu = \frac{K}{(1+a\cdot e^{-bH})} + C \tag{14}$$

Magnetic nanolubricant is the solid-liquid mixture. Solid particles include Fe_3O_4 nanoparticles, GR nanoparticles and Fe_3O_4@GR nanoparticles. According to the law of conservation of mass, the density expression of magnetic lubricant can be inferred [15]:

$$\rho_m = (1-\varphi)\rho_{Nf} + \varphi\rho_{Np} \tag{15}$$

Where φ is the volume fraction of solid particles. ρ_{Nf} is the liquid density (kg/m^3). ρ_{Np} is the real density of solid particles (kg/m^3).

Table 3. Viscosity fitting results of magnetic nanolubricant

Nanolubricant	*K*	*a*	*b*	*c*	Adjusted R-square
Fe_3O_4	74.3	200.1	0.02766	74.3	0.997
Fe_3O_4@GR	11.76	87.35	0.02405	71.1	0.994
Fe_3O_4/GR	59.54	73.11	0.02247	72.35	0.999

4.4.4. Saturation Magnetization and Relative Permeability

Magnetic nanofluid is a kind of superparamagnetism material and its saturated magnetization intensity presents a linear growth relationship with volume fraction of Fe_3O_4. The viscosity of ferromagnetic fluid increases with the increase of magnetic field intensity. When the applied magnetic field reaches the saturated magnetization intensity of ferromagnetic fluid, the viscosity tends to be stable and doesn't change any more. This is because all ferromagnetic particles are in ordered arrangement along the external magnetic field under the magnetization intensity. According to test results of viscosity in the above section, viscosity of the Fe_3O_4/GR lubricant tends to be stable after the magnetic field strength exceeds 85.3 Gs(22.7 kA/m). Hence, it can be inferred that the saturated magnetization intensity of Fe_3O_4/GR lubricant is $M_s \approx 22.7$ kA/m. Additionally, the relative magnetic permeability of ferromagnetic fluid is$\mu_r \approx 1$. According to H-M curve test of lubricant by Radionov et al. [16], M presents a power function with the increase of magnetic field strength (H). Hence, the following equation can be established by comparing magnetic fluids in this study:

$$\begin{cases} M = 0 & (H = 0) \\ M = -50.84 \cdot H^{-0.3512} + 29.36 & (H > 0) \end{cases} \text{(kA/m)} \tag{16}$$

Conclusion

The new magnetic field assisted NMQL technology changes the existing viscous force dragging infiltration of the grinding wheel into energy field adsorption infiltration. The infiltration mechanism of lubricant on grains-workpiece interface changes fundamentally, which can improve infiltration film forming performances of the nanolubricant significantly.

In this chapter, two magnetic field assisted NMQL devices are proposed: the attractive traction device of magnetic field and the adsorptive traction device of magnetic field. The former one forms a magnetic field by setting magnets in front of the infiltration to provide power for infiltration of magnetic nanolubricant. The later one applies magnetic fields at two sides of the grinding wheel and magnetic nanolubricant adsorbed onto the grinding wheel is carried into the cutting zone by the grinding wheel, thus improving infiltration ability of the lubricant. Besides, magnetic nanolubricant samples were prepared, and their physical properties and magnetic fluid parameters were tested, getting magnetic nanolubricant with excellent performances.

References

[1] Wang J, Guo XC, He Y, et al. (2018). Tribological characteristics of graphene as grease additive under different contact forms. *Tribol. Int.,* 127, 457-469. 10.1016/j.triboint.2018.06.026.

[2] Li R, Lu TY. (2015). The influence of carbon nanotubes and graphene as additives in lubricant oil on friction and wear. *China Sciencepaper,* 10(10), 1123-1126.

[3] Gong K, Wu X, Zhao G, et al. (2017). Nanosized MoS2 deposited on graphene as lubricant additive in polyalkylene glycol for steel/steel contact at elevated temperature. *Tribol. Int.,* 110, 1-7.

[4] Shang WJ, Cai T, Zhang YX, et al. (2018). Facile one pot pyrolysis synthesis of carbon quantum dots and graphene oxide nanomaterials: All carbon hybrids as eco-environmental lubricants for low friction and remarkable wear-resistance. *Tribol. Int.,* 118, 373-380. 10.1016/j.triboint.2017.09.029.

[5] Ku BC, Han YC, Lee JE, et al. (2010). Tribological effects of fullerene (C 60) nanoparticles added in mineral lubricants according to its viscosity. *International Journal of Precision Engineering Manufacturing,* 11, 607-611.

[6] Wu P, Chen XC, Zhang CH, et al. (2019). Synergistic tribological behaviors of graphene oxide and nanodiamond as lubricating additives in water. *Tribol. Int.,* 132, 177-184. 10.1016/j.triboint.2018.12.021.

[7] Liu L, Chen S. (2009). Research progress on fullerene and carbon nanotubes in lubricating systems. *Materials Reports,* 23(7), 13-16.

[8] Li QY, Zhang S, Qi YZ, et al. (2017). Friction of two-dimensional materials at the nanoscale: Behavior and Mechanisms. *Chinese Journal of Solid Mechanics,* 38(3), 189-214.

[9] Pu JB, Wang LP, Xue QJ. (2014). Progress of tribology of graphene and graphene-based composite lubricating materials. *Tribology*, 34(1), 93-112.

[10] Cheng JX, Xie F, Li BC. (2016). Application of nano-carbon lubricating oil additives. *Contemporary Chemical Industry,* 45(4), 843-846.

[11] Zhang ST, Wang K, Hu ZX. (2018). Effect of graphene additive on tribological properties of lube base oil. *Petroleum Processing and Petrochemicals,* 49(9), 91-95.

[12] Li ZJ, Zhang YZ, He Q, et al. (2017). Research progress on the application of grapheme as additives in lubrication. *Lubrication Engineering,* 42(10), 133-140.

[13] Gao T, Li CH, Zhang YB, et al. (2019). Dispersing mechanism and tribological performance of vegetable oil-based CNT nanofluids with different surfactants. *Tribol. Int.,* 131, 51-63. 10.1016/j.triboint.2018.10.025.

[14] Guo SM, Li CH, Zhang YB, et al. (2017). Experimental evaluation of the lubrication performance of mixtures of castor oil with other vegetable oils in MQL grinding of nickel-based alloy. *J. Clean Prod.,* 140, 1060-1076. 10.1016/j.jclepro.2016.10.073.

[15] Yang MD, The study on factors of pressure resistance of magnetic fluid seals based on MATLAB and COMSOL co-simulation, Beijing Jiaotong University, Beijing, 2021.

[16] Radionov A, Podoltsev A, Zahorulko AJPE. (2012). Finite-element analysis of magnetic field and the flow of magnetic fluid in the core of magnetic-fluid seal for rotational shaft. 39, 327-338.

Chapter 5

The Distribution and Evolution of the Magnetic Field in Magnetic Traction Nanolubricant Grinding

Abstract

The magnetic nanolubricant will be attractive by magnetic force in a gradient magnetic field. Therefore, the formation and distribution of the magnetic field is an important boundary condition to characterize the magnetic force in the grinding zone. In this chapter, a magnetic field auxiliary device grinding process equipment is developed, which employs the permanent magnet and the ferromagnetic wheel to apply a gradient magnetic field in the grinding area. The magnetization mechanism of the wheel in the magnetic field and its influence on the distribution of the magnetic inductance line are revealed, and the changing trend of magnetic field intensity and gradient is explored. The parameters setting range of the processing unit is obtained based on the force and transport of the lubricant droplets.

Keywords: magnetic field, distribution rule, grinding wheel, physical model, finite element simulation

5.1. Introduction

Magnetic field assisted processing is a method of machining the workpiece with the action of magnetic field [1-5]. The precondition for the new technology to take effect is to apply magnetic field onto the grinding device limited within a fixed space and make magnetic fluid adsorbed onto the grinding wheel surface through the magnetic field force, thus improving infiltration performances of the grinding zone. Different from magnetorheological finishing (MRF) device [6, 7], the line direction and action zone of the magnetic induction of the magnetic field generator cannot be locked within the lubricant adsorption zone directly, which makes it difficult to apply the magnetic field. As a result, this study used permanent magnets to magnetize the grinding wheel substrate, produce a magnetic field through the grinding wheel, and guide the distribution of magnetic induction lines. On this basis, the application mode of magnetic field and magnetic field assisted device are designed by combining with geometric conditions of the grinding device. Nevertheless, the positional relationship between permanent magnet and grinding wheel has different influences on distribution of magnetic induction lines, thus changing the magnetic field strength. Therefore, the influencing laws of arrangement of permanent magnets and geological positions of magnetic field assisted device on magnetic field strength and distribution were analyzed through theoretical analysis and numerical simulation, getting physical boundary conditions for nanolubricant infiltration and antifriction performances.

5.2. Model of the Grinding Zone Magnetic Field

The research method on variation laws of magnetic field based on the magnetic field measurement system is characterized by low efficiency and high cost [8-10]. At present, automatic measuring instruments can be applied to measure magnetic field of small-sized permanent magnet, but it can only measure surface magnetic field [11, 12]. Moreover, the 3D characteristics of magnetic field distribution can only be reflected by multiple 3D reconstructions of one-dimensional, two-dimensional measurements and complicated data. On this basis, this part simulates the 3D magnetic field quickly and accurately by using the finite element analysis software COMSOL Multiphysics (COMSOL).

5.2.1. Theoretical Modelling of Magnetic Field Distribution

The basic theory based on electromagnetic field which is proposed by three mathematical models (magnetic doublet model, equivalent magnetic charge model and equivalent current model) to describe magnetic field is of equivalence to the Maxwell equation set. The Maxwell equation set describes the relations of electric field and magnetic field with charge density and current density:

$$\nabla \times \boldsymbol{E} = -\frac{\partial \boldsymbol{B}}{\partial t} \tag{1}$$

$$\nabla \times \boldsymbol{H} = \boldsymbol{J} + \frac{\partial \boldsymbol{D}}{\partial t} \tag{2}$$

$$\nabla \cdot \boldsymbol{D} = \rho \tag{3}$$

$$\nabla \cdot \boldsymbol{B} = 0 \tag{4}$$

where $\boldsymbol{D}$ is the electric displacement vector (C/m^2), $\boldsymbol{E}$ is the electric field intensity (V/m), $\boldsymbol{J}$ is the current density (A/m^2), $\boldsymbol{H}$ is magnetic field strength (A/m), $\boldsymbol{B}$ is the magnetic induction intensity (T), and ρ is charge density (C/m^3).

In the equation set, the Eq. (1) describes Faraday's law of electromagnetic induction. Eq. (3) describes the Gauss law of electrostatic field. Eq. (2) is the differential form of law of total current. In this study, permanent magnets were used as the magnetic field generator. Although metal matrix of the grinding wheel will produce weak currents when cutting the magnetic induction lines, the order of magnitudes of the magnetic field strength produced by the current is far lower than that of the permanent magnet and it can be even ignored. Additionally, the grinding wheel base is uniform and the magnets are installed symmetrically at two sides of the grinding wheel. Rotation of the grinding wheel won't change the magnetic field distribution. Therefore, this study chose Eq. (4), or what is known as the differential form of principle of continuity of magnetic flux, to describe current-free static magnetic field problems.

Moreover, the magnetic scalar potential ($V_m(A)$) can be defined through the following relations when the Eq. (2) is 0 in the current-free zone:

$$\boldsymbol{H} = -\nabla V_m \tag{5}$$

For rare earth permanent magnet (hard magnetic material) in static magnetic field, it can be described by the magnetized constitutive relation:

$$\boldsymbol{B} = \mu_0(\boldsymbol{H} + \boldsymbol{M}) \tag{6}$$

Where μ_0 is the permeability of vacuum ($4\pi \times 10^{-7}$ N/A^2) and M refers to magnetization intensity (A/m).

The grinding wheel material (soft magnetic material) in static magnetic field can be described by the constitutive relation of relative magnetic permeability:

$$\boldsymbol{B} = \mu_0 \mu_r \boldsymbol{H} \tag{7}$$

Where μ_r is the relative magnetic permeability of substances.

5.2.2. Physical Modelling of Equipment

(i) Geometric model and boundary setting. A 3D geometric modeling technique was applied for numerical simulation in order to explore distribution laws of grinding wheel and 2 pieces of magnets in the space magnetic field (Figure 1). The geometric model is designed according to structural characteristics of the precise plane numerical control grinder K-P36 and it is composed of 4 domains: ① build a grinding wheel model: the grinding wheel size is 300 mm×76.2 mm×20 mm; ② two pieces of permanent magnets at two sides of the grinding wheel shall be installed at the exposed part of the grinding wheel. Hence, the permanent magnet model size is 100 mm×50 mm×20 mm. ③ using the grinding wheel as the axis and the midpoint of thickness as the origin to build a spherical air model with a diameter of 1,000 mm. Magnetic insulation boundary conditions on the spherical surface are: $n \cdot B=0$.

The model applies triangular meshing. This study focuses on magnetic field distribution laws in the grinding wheel, and the distribution laws of magnetic field surrounding the grinding wheel will be influenced by grid size. Hence, the meshing schemes shall be refined.

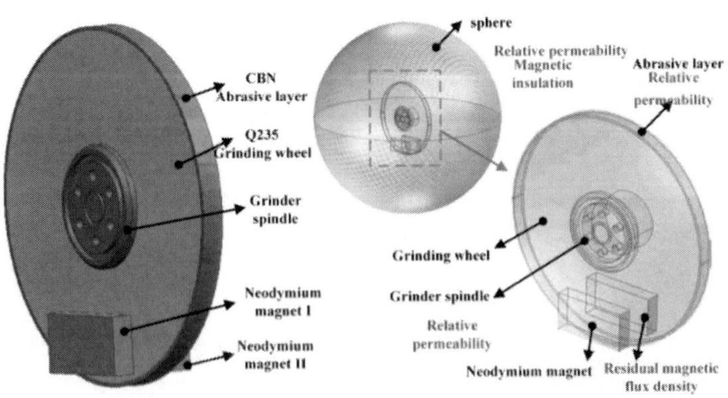

Figure 1. Construction of numerical simulation geometric model.

(ii) Setting of model input parameters. The sphere in the model allows gas input from the platform and its relative magnetic permeability is set at 1. Furthermore, neodymium magnet (NdFeB) is chosen as the permanent magnet material and the models are defined N54~N35 in national standards according to residual magnetic flux density from high to low. For example, the residual magnetic flux density of N54 is 1.47 T and the residual magnetic flux density of N35 is 1.21 T. The relative magnetic permeability of NdFeB is the set as the same with that of air (1). The base materials of grinding wheel usually use low-carbon steel and medium-carbon steel. In this study, Q235 is chosen as the research object. The magnetic characteristic parameters of materials are shown in Table 1. The magnetization curve of Q235 is shown in Figure 2 [13].

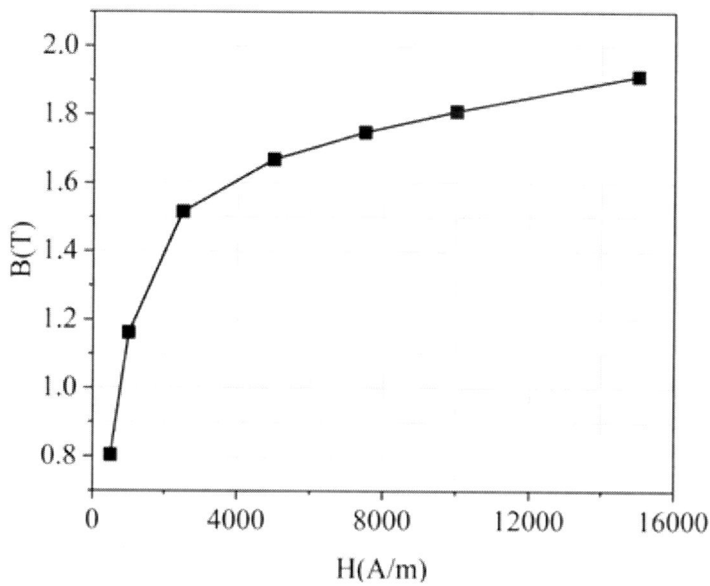

Figure 2. Magnetization curve of Q235.

Table 1. Magnetic characteristic parameters of Q235 grinding wheel material

Coercive force H_c (A/m)	Coercive force B_r (T)	Maximum relative permeability μ_{rm} (H/m)	Maximum permeability corresponding to the magnetic field strength $H_{\mu m}$ (A/m)
130~210	--	400	--

5.3. The Effect of Permanent Magnet Properties on Magnetic Field Distribution

5.3.1. Magnetic Pole Arrangement

Distribution of the permanent magnet has significant influences on magnetization degree and magnetic field distribution of grinding wheel. Therefore, four pole distribution modes of permanent magnets were designed according to positional relations between permanent magnets and the grinding wheel in the magnetic field assisted device, namely, repulsion between the left and right N-S poles (type I), attraction between the left and right N-S poles (type II), homodromous between upper and lower N-S poles (type III) and opposites between upper and lower N-S poles (type IV), as shown in Figure 3(a). To analyze magnetic field strength and distributions produced by four distribution modes, key attention is paid to the magnetic field on three directions: horizontal and radial directions below the grinding wheel and circumferential direction of the grinding wheel. The schematic diagram of directions of grinding wheel surface is shown in Figure 3(b).

Furthermore, a simulation calculation of magnetic field size and distribution surrounding the grinding wheel under four distribution modes was carried out, thus creating the magnetic field distribution and variation laws of magnetic field strength. The variation curves of horizontal and radial magnetic field strength below the grinding wheel under different distribution modes are shown in Figure 4. The cloud charts of magnetic flux density distribution and magnetic field distribution produced by permanent magnets on grinding wheel surface under four distribution modes are shown from Figure 5 to Figure 8.

For type I distribution, it can be seen from the cloud chart in Figure 5(a) that the magnetized shape below the grinding wheel looks like a “door” and a strong magnetic field is produced at the grinding wheel position corresponding to edges of permanent magnet, that is, the dynamic magnetic region of the grinding wheel. It can be known from the thermodynamic diagram of magnetic field distribution below the grinding wheel that the magnetic field strength beneath the grinding wheel is the highest and it decreases gradually along circumferential direction of the grinding wheel. With the increase of radial distance beneath the grinding wheel, the magnetic field strength decreases gradually. As shown in Figure 4(b), the magnetic field strength under type I presents a decreasing trend with the increase of distance, thus generating a gradient magnetic field. In Figure 4(a), the variation curve of horizontal magnetic field strength below the grinding wheel is high at two sides and low in the middle, but the variation rate is relatively low. For example, the magnetic field strength in the middle is 84,036 A/m, which is 26,145 A/m lower than that at the right side (110,181 A/m).

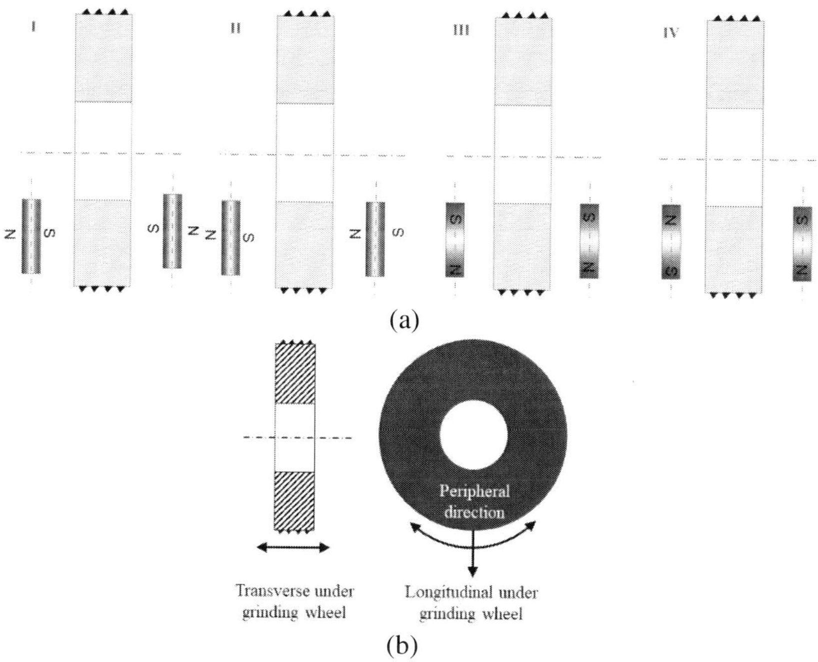

Figure 3. Distribution of permanent magnet pole and different directions of grinding wheel. (a) Distribution of permanent magnet pole, (b) Different directions of the grinding wheel surface.

In Figure 5(a), the magnetic induction lines of permanent magnet start from the N pole. Since the relative magnetic permeability of grinding wheel material is higher than the air magnetic permeability, the magnetic induction lines will spread along the grinding wheel after reaching the side surface and then return to the S pole, producing a vertical magnetic field. When the same magnetic poles contact mutually, the produced magnetic induction lines repel mutually and radiate outward. Hence, it can be approximately believed that the produced magnetic fields have consistent directions and two magnetic fields superposed. The produced magnetic field strength is twice that of a single magnet [14]. Therefore, when the same magnetic poles of the permanent magnet under type I are in the same two sides of grinding wheel, the magnetic induction lines in the grinding wheel are opposite in directions and the produced magnetic field strength is twice that produced by a single permanent magnet on the grinding wheel.

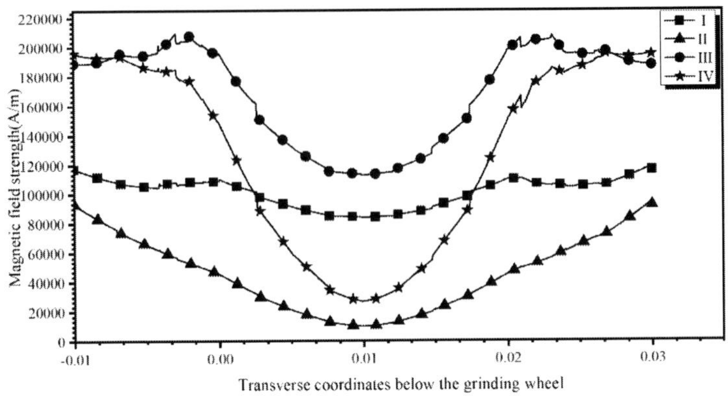

(a)

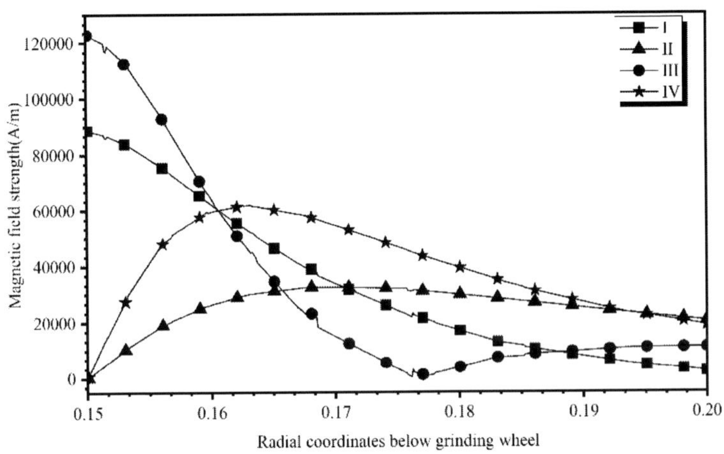

(b)

Figure 4. Magnetic field intensity distribution in different directions on grinding wheel surface. (a) Transverse direction, (b) Radial direction.

For type II distribution, it can be seen from the cloud chart in Figure 6(a) that a relatively strong magnetic field is produced at the grinding wheel position corresponding to the middle part of permanent magnets after magnetization below the grinding wheel. It can be known from the thermodynamic diagram of magnetic field distribution below the grinding wheel that with the increase of radial distance, the magnetic field strength increases firstly and then decreases. However, the variation rate of magnetic field strength is small. This agrees with the variation trend of strength at type II distribution in Figure 4(a). The variation curve of horizontal magnetic field strength below the grinding wheel under type II in Figure 4(a) is the same with that under type I. The magnetic field strength at the middle position is 10,109 A/m, which decreases by 35,628 A/m compared to that on the right side (45,737 A/m).

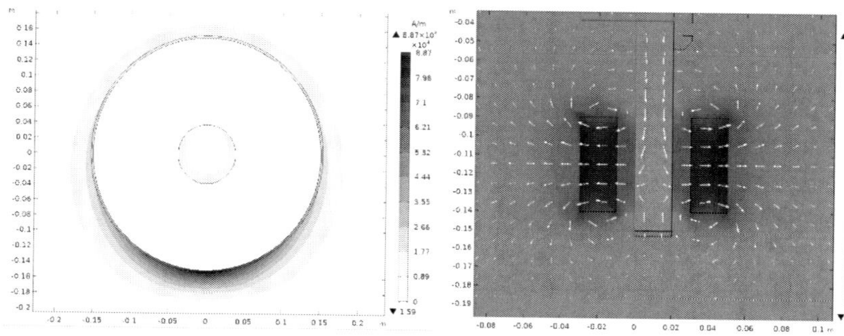

(a) (b)

Figure 5. Magnetic field distribution of type I magnetic pole arrangement. (a) Radial section, (b) Axial section.

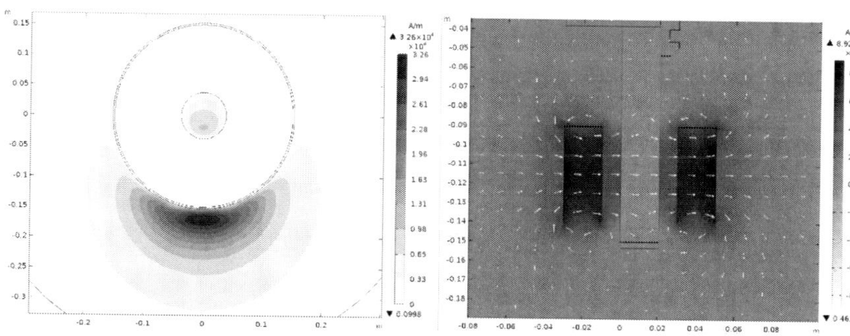

(a) (b)

Figure 6. Magnetic field distribution of type II magnetic pole arrangement. (a) Radial section, (b) Axial section.

Similarly, it can be seen from Figure 6(a) that the magnetic induction lines of permanent magnet start from the N pole and return to the S pole on the other side after running through the magnetically conductive grinding wheel, forming the magnetic induction line distribution pattern in the figure. It produces a horizontal magnetic field. hence, the magnetic induction lines in

Figure 6(a) are in disordered distribution perpendicular to thickness of the grinding wheel. The contact of heteronymous magnetic poles shows no difference from magnetic induction lines of normal single piece of permanent magnet.

For type III distribution, it can be seen from the cloud chart in Figure 7(a) that the position with a higher magnetic field strength in the grinding wheel after magnetization below the grinding wheel is the same as that under type II. It can be seen from the thermodynamic diagram of magnetic field distribution below the grinding wheel that the magnetic field strength below the grinding wheel is the maximum and higher than that under type I, decreasing gradually toward two sides of the grinding wheel along the circumferential direction of the grinding wheel. Below the grinding wheel, the magnetic field strength declines gradually with the increase of distance to the grinding wheel surface. This agrees with the variation curve under type III in Figure 7(b). With respect to the horizontal distribution laws below the grinding wheel in Figure 4(a), the magnetic field strength in the middle position under type III is 112,923 A/m, which decreases by 84,956 A/m compared to that on the right side (197,879 A/m). It can be seen from Figure 7(b) that same with that under type I, the magnetic induction lines start from the N pole, spread along the grinding wheel after reaching the grinding wheel side surface, and finally return to S pole, producing a vertical magnetic field.

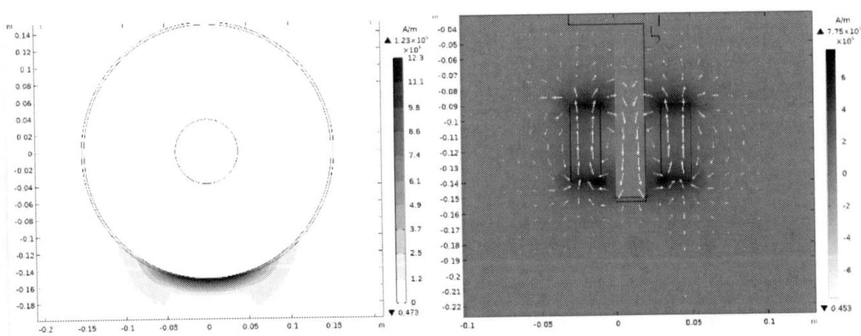

(a) (b)

Figure 7. Magnetic field distribution of type III magnetic pole arrangement. (a) Radial section, (b) Axial section.

The type IV distribution is shown in Figure 8. According to the cloud chart in Figure 8(a), the internal magnetic field strength after grinding wheel magnetization is relatively high at the bottom of grinding wheel. It can be seen from the thermodynamic chart that the magnetic field strength increases quickly in the beginning and then decreases gradually with the increase of radial distance below the grinding wheel. This is consistent with the variation curve of type IV in Figure 4(a).

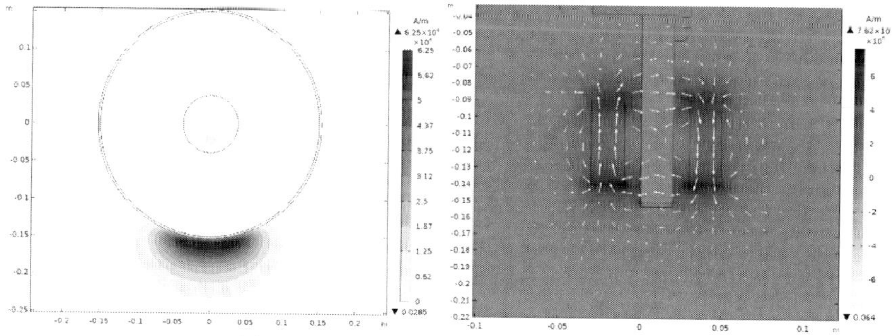

(a) (b)

Figure 8. Magnetic field distribution of type IV magnetic pole arrangement. (a) Radial section, (b) Axial section.

The magnetic field strength decreases gradually from the middle to two sides along the circumferential direction. For the horizontal direction, the IV type distribution has great influences on horizontal magnetic field distribution. The magnetic field strength in the middle position is 27,036 A/m, which decreases by 123,930 A/m compared to that on the right side (150,966 A/m).

According to distribution of magnetic induction lines in Figure 8(b), four magnetic poles of the permanent magnets are at positions with some intervals and there are intersections of two magnetic induction lines between heteronymous magnetic poles, thus forming magnetic induction lines in irregular distribution in the grinding wheel. The magnetic induction lines start from the right N pole will return to two S poles along the grinding wheel. Not all magnetic induction lines run through the middle position below the grinding wheel, thus decreasing the magnetic field strength in the middle obviously. At two sides close to the permanent magnets, there are more magnetic induction lines than those in the middle, forming a higher magnetic field strength.

To sum up, the magnetic field strength decreases firstly and then increases from left to right below the grinding wheel under the permanent magnetic distribution modes of type III and type IV, reaching the minimum in the middle position. Since magnetic field strengths at two sides are too high, a lot of magnetic nanolubricant will be adsorbed onto two edges when it is sprayed onto the grinding wheel surface, but there's a very low content on the surface near the middle position. For type I and type II, the magnetic nanolubricant in the uniformly distributed magnetic field is adsorbed uniformly by the grinding wheel surface (Figure 4). According to comparison of magnetic field strengths, it can get magnetic field with a higher magnetic field strength under type I distribution.

5.3.2. Magnetic Strength

According to conclusions in above study, the magnetic field strength under type I distribution of permanent magnets is relatively higher and the magnetic field distribution is uniform. Hence, permanent magnets are in type I distribution in the following text. For permanent magnets with different intensities of magnetism, the magnetic field strengths produced by magnetization of the grinding wheel vary. As a result, influencing laws of intensity of magnetism on magnetic field in the grinding wheel were analyzed through simulation calculation in this section.

The variation laws of horizontal magnetic field strength below the grinding wheel under different levels of intensity of magnetism are shown in Figure 9. Obviously, the magnetic field strength reaches the maximum at two sides of the grinding wheel and the minimum below the grinding wheel, showing a "concave" curve. With the increase of strength level, the magnetic field strength increases gradually. Similarly, the magnetic field strength on the right of the grinding wheel was taken for example for analysis. When the strength level is N54, the magnetic field strength on the right side is 110,181 A/m and it is 84,051 A/m in the middle of the grinding wheel. At N35, the magnetic field strength on the right is 90,693 A/m and it is 69,185 A/m in the middle, which is 17.7% lower compared to those at N54. Along the circumferential direction of the grinding wheel, the action zone of nanolubricant during cutting with the grinding wheel concentrates within 240°~300°. In this zone, the magnetic field strength presents the following variation laws with the increase of angle: increase-decrease-increase-decrease-increase-decrease.

The radial magnetic field strength below the grinding wheel under different strength levels is shown in Figure 10. With the increase of radial distance, the magnetic field strength decreases gradually. Take x=0.16 m for example. The minimum and maximum magnetic field strengths (50,913 A/m and 61,853 A/m) are achieved at N35 and N54, respectively. The magnetic field strength of N35 is 17.7% lower than that of N54.

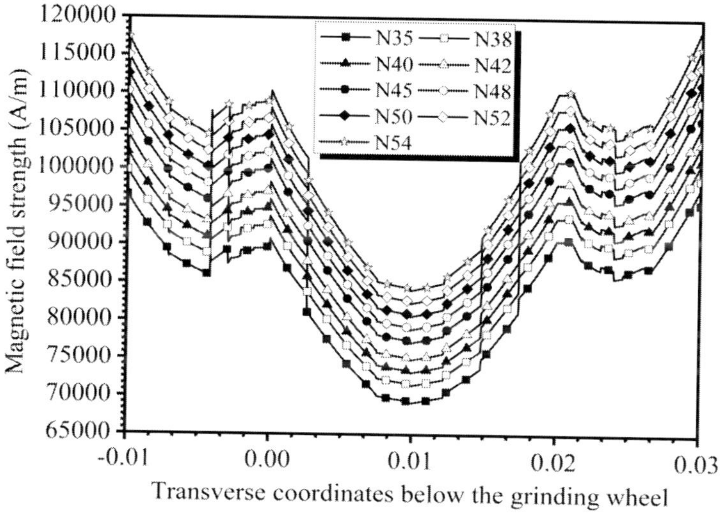

Figure 9. Variation of magnetic field intensity on transverse under grinding wheel with different magnetic strength levels.

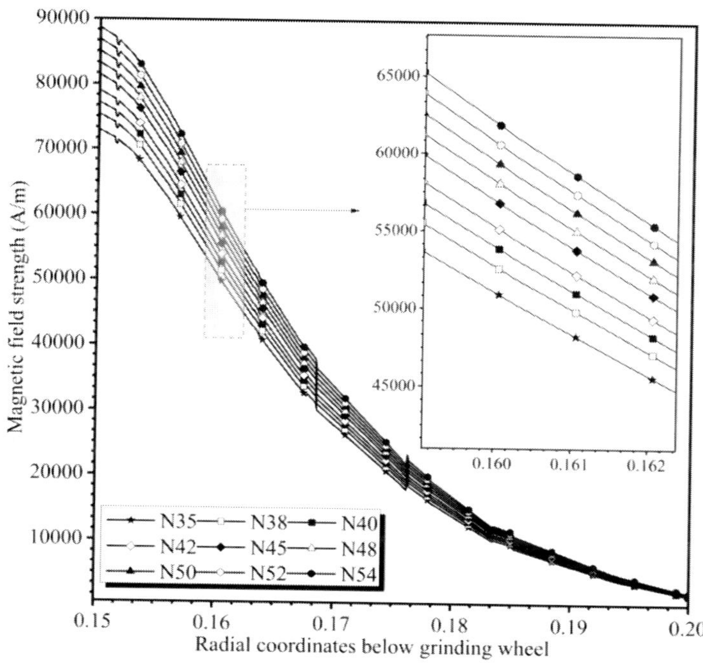

Figure 10. Variation of magnetic field intensity on radial under grinding wheel with different magnetic strength levels.

In a word, given the fixed shape and size of the permanent magnets, the magnetic field strength and magnetic field distribution surrounding the grinding wheel change slightly when the intensity of magnetism increases. Therefore, it is not necessary to choose permanent magnets with excessive magnetism after comprehensive considerations to working environment and cost.

5.4. The Effect of Device Parameters on Magnetic Field Distribution

The magnetic field strength at different magnetic induction line positions of the permanent magnets varies. The positional relationship between permanent magnets and the grinding wheel will influence distribution of the magnetic induction lines. Since the position of the permanent magnets will affect the magnetic field strength surrounding the grinding wheel, factors that may influence magnetic field distribution and magnetic field strength surrounding the grinding wheel are analyzed in this section. It can adjust three parameters in the magnetic field assisted device, which are distance between the permanent magnet and the grinding wheel side (L), the distance from the lower surface of the permanent magnets to the lower surface of the grinding wheel (H), and the included angle between the permanent magnets and grinding wheel (β). Additionally, influencing laws of the grinding wheel thickness (B) on the magnetic field strength are analyzed. Results can provide some references to the application of magnetic fields to different processing technologies.

5.4.1. Distance Between the Magnet and the Grinding Wheel (L)

The influencing laws of L on horizontal magnetic field strength below the grinding wheel are shown in Figure 11. When L=2mm, the magnetic field strength produced on the grinding wheel surface reaches the maximum and the magnetic field distribution on the surface shows the maximum variation rate. With the increase of distance, magnetic field strength declines. The magnetic field distribution along the horizontal direction becomes gentle gradually and the variation rate decreases gradually. When L=2 mm, the magnetic field strength on the right side of the grinding wheel is 218,151 A/m, which is increased by 81,945 A/m compared to that in the middle (136,206

A/m). When L=20 mm, the magnetic field strength on the right side of the grinding wheel is 62,898 A/m, which is increased by 23, 608 A/m compared to that in the middle (39,290 A/m). Moreover, when L=2 mm and 4 mm, there are irregular sudden changes of magnetic field distribution on the grinding wheel surface. This is because permanent magnets are too close to the sides of grinding wheel. Under such conditions, it is very disadvantageous for nanolubricant adsorption by the grinding wheel. Hence, L=2 mm and 4 mm shall be excluded. The magnetic field strength presents increase-decrease cycles within 240°~300° when the L is relatively high along the circumferential direction of the grinding wheel. The magnetic field distribution becomes increasingly uniform with the reduction of L. Therefore, on the one hand, we hope that the higher L, the better, so that the magnetic fields can be distributed uniformly in the grinding wheel surface and magnetic nanolubricant can be adsorbed uniformly on the grinding wheel surface. On the other hand, the higher L brings the lower magnetic field strength. As a result, the value of L shall be considered comprehensively and chosen appropriately according to practical processing conditions.

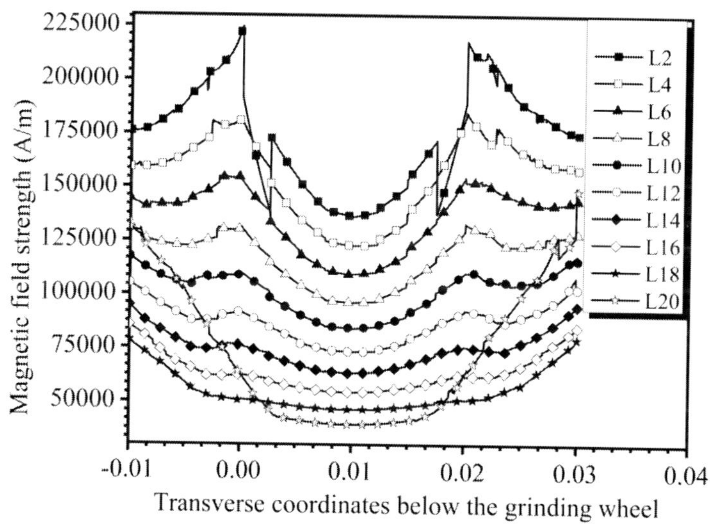

Figure 11. Variation of magnetic field intensity on transverse under grinding wheel with different L.

The influences of L on radial magnetic field strength below the grinding wheel are shown in Figure 12. With the increase of radial distance, the magnetic field strength decreases gradually. The curves begin to cross at

x=0.175 m. Later, the magnetic field strength shows the opposite variation trend with the previous variation trend. For example, when x=0.16 m, the maximum and minimum magnetic field strengths are achieved at L=2 mm and L=20 mm, which are 88,364 A/m and 35,505 A/m, respectively. The magnetic field strength at L=20 mm is 59.8% lower than that at L=2 mm. In a word, the value of L shall be chosen within the range of 6~16 mm.

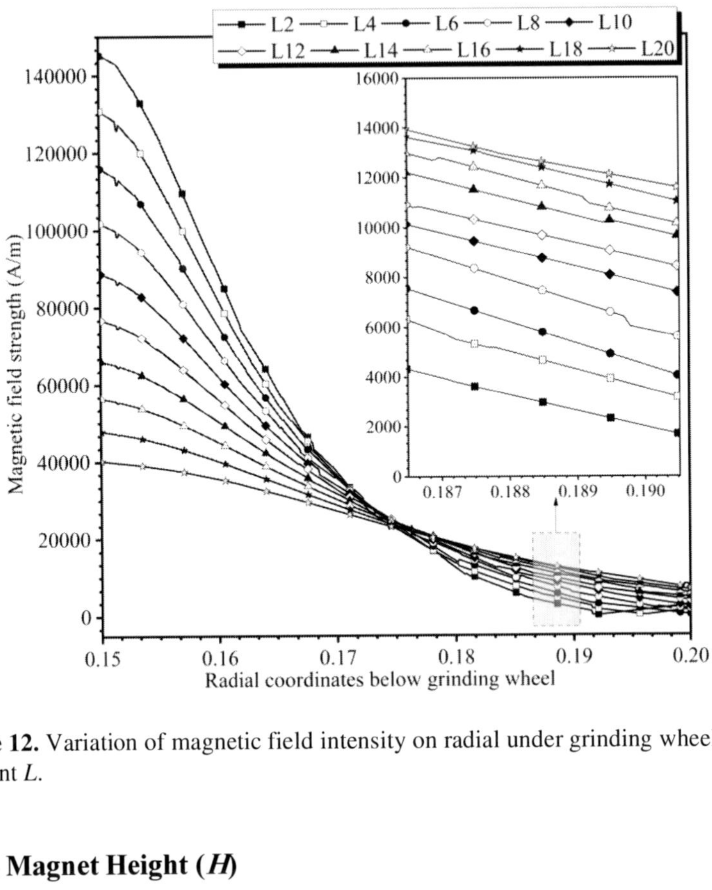

Figure 12. Variation of magnetic field intensity on radial under grinding wheel with different L.

5.4.2. Magnet Height (H)

Changes of height of permanent magnets will affect the positions of grinding wheel and workpiece. During grinding, especially during molding process, restrictions of workpiece, clamp and grinding parameters shall be taken into account. With comprehensive considerations to the above factors, the range of H is determined 10~30 mm to guarantee normal working of the grinding wheel.

The variation laws of horizontal and circumferential magnetic field strength below the grinding wheel under different *H* values are shown in Figure 13. It can be seen from Figure 13, the horizontal magnetic field strength decreases gradually with the increase of *H*, but its distribution is not influenced obviously. When *H*=10 mm, the magnetic field strength on the right side of the grinding wheel is 110,181 A/m, which is increased by 26,130 A/m compared to that in the middle position (84,051 A/m). When *H*=30 mm, the magnetic field strength on the right side of the grinding wheel is 83,365 A/m, which is increased by 21,760 A/m compared to that in the middle position (61,605 A/m). On the circumferential magnetic field strength becomes increasingly uniform with the increase of *H* and the increase-decrease cycles decrease to some extent within 240°~300°.

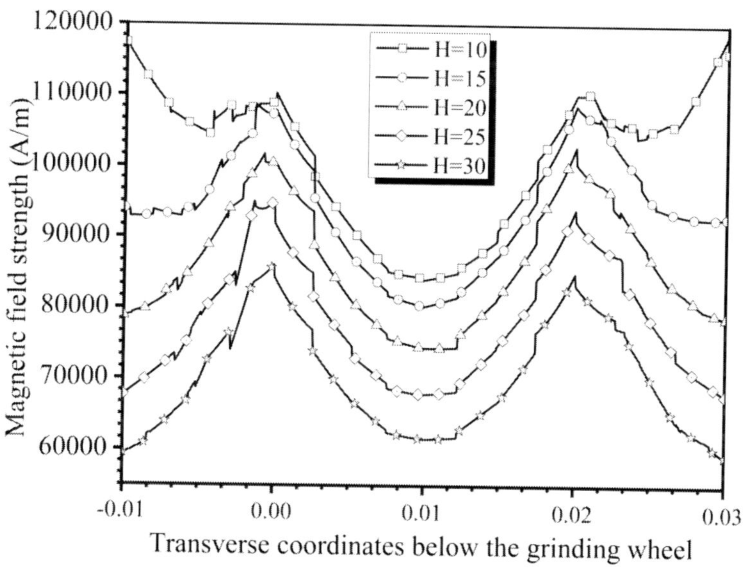

Figure 13. Variation of magnetic field intensity on transverse under grinding wheel with different *H*.

It further can be seen from Figure 14 that with the increase of radial distance below the grinding wheel, the magnetic field strength decreases gradually. The maximum magnetic field strength (61,853 A/m) is achieved at *x*=0.16 m and *H*=10 mm, and it decreases with the increase of *H*. The minimum magnetic field strength (43,515 A/m) at *H*=30 mm is 29.6% lower than that at *H*=10 mm. Therefore, H shall be the higher the better if processing conditions are allowed.

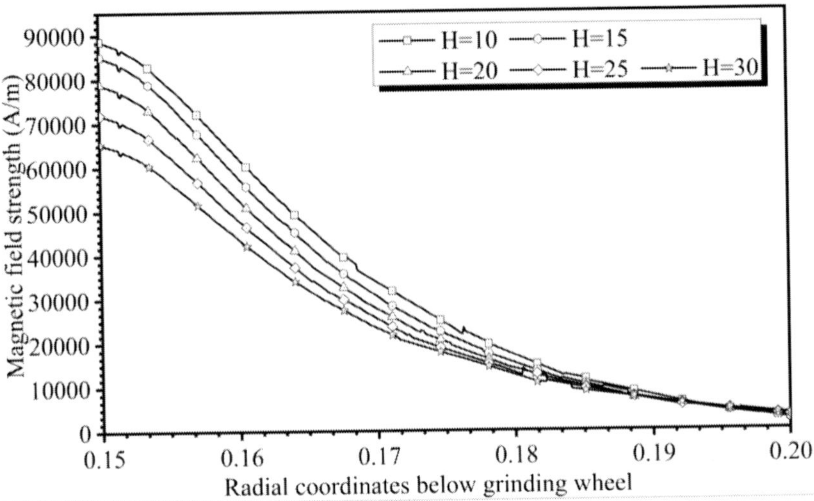

Figure 14. Variation of magnetic field intensity on radial under grinding wheel with different *H*.

5.4.3. Angle Between the Magnet and the Grinding Wheel (β)

The influencing laws of β on magnetic field strength are shown in Figure 15. In Figure 15, the horizontal magnetic fields below the grinding wheel are the most uneven whenβ=0°. The magnetic field strength on the right side of the grinding wheel reaches 108,556 A/m, but it is only 80,306 A/m in the middle, which is 26.02% lower than that on the right side. With the increase of β, the magnetic field distribution becomes gradually uniform. The magnetic field strength on the right side and middle position values 39,722 A/m and 32,781 A/m when β=40°. On the circumferential magnetic field distribution is very uniform when β=0° and it increases firstly and then decreases within 240°~300°, reaching the maximum beneath the grinding wheel. The magnetic field becomes increasingly uneven as β increases, showing several increase-decrease cycles. This demonstrates that it may not form stronger and more uniform magnetic fields on the grinding wheel by superposing more and more magnetic fields below two permanent magnets. Instead, magnetization of the grinding wheel is influenced by the complicated magnetic field of the permanent magnets, thus resulting in irregular magnetic field distributions on the grinding wheel. As a result, the lubricant distribution on the grinding wheel surface is not uniform.

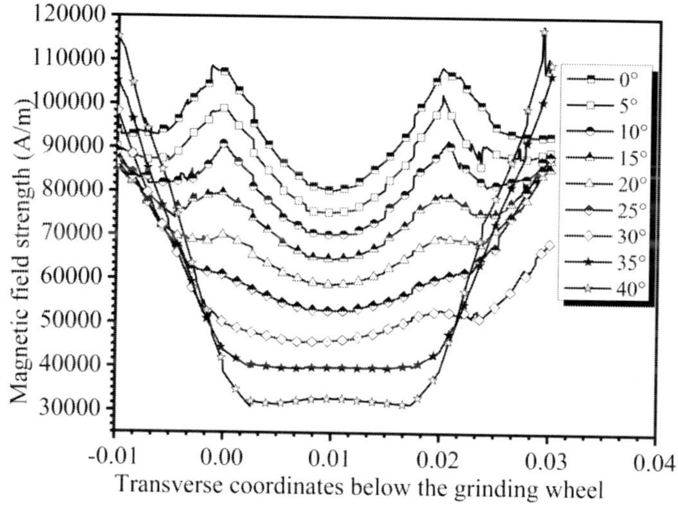

Figure 15. Variation of magnetic field intensity on transverse under grinding wheel with different β.

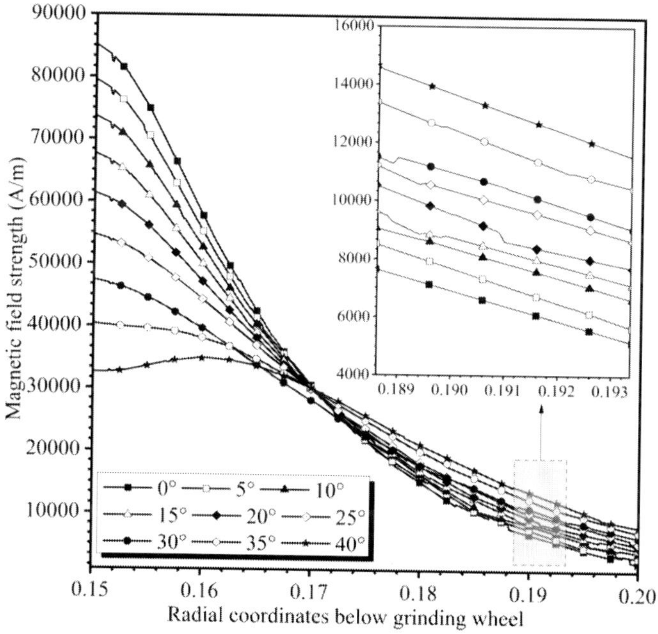

Figure 16 Variation of magnetic field intensity on radial under grinding wheel with different β.

The variation laws of radial magnetic field strength below the grinding wheel with β are shown in Figure 16. The magnetic field strength is negatively related with β. A crossing point appears when the radial distance reaches 0.17 mm, accompanied with an opposite variation trend of the strength. Given x=0.16 m, the magnetic field strength at β=40° decreases by 39.1% compared to that at β=0° (57,377 A/m). This also proves that it may not increase the magnetic field strength on the grinding wheel even though many magnetic fields are superposed. Besides, the variation gradient of curves is negatively related with β. When β=35° and β=40°, the gradient of magnetic field strength approaches 0 and even may become negative. Since the magnetic field force is proportional to the gradient of magnetic field strength, the adsorption capacity of nanolubricant declines if the gradient is too small. Hence, β=35° and β=40° shall be eliminated.

5.4.4. Grinding Wheel Thickness (*B*)

Generally, increasing thickness of the grinding wheel will change direction and distribution of magnetic induction lines, thus influencing the magnetic field strength. The variation laws of horizontal and circumferential magnetic field strength of the grinding wheel under different widths are shown in Figure 17. On the horizontal direction (Figure 17), the magnetic field distribution of grinding wheel surface is very uniform and the difference between two sides and the middle position can be ignored. In addition, the magnetic field strength on the grinding wheel decreases gradually with the increase of B. The magnetic field strength at B=50 mm decreases by 39.9% than that at B=20 mm. The grinding wheel width can significantly influence magnetic field distribution within 240°~300° on the circumferential direction. With the increase of grinding wheel width, the magnetic field distribution becomes increasingly uniform, but the magnetic field strength declines gradually.

Furthermore, the radial magnetic field strength below the grinding wheel is analyzed (Figure 18). Clearly, there's a negative correlation between magnetic field strength and B. When the radial distance is 0.19 mm, a crossing point appears, and the strength presents the opposite trend. Given x=0.16 m, the magnetic field strength is 35,506 A/m when B=20 mm and it decreases by 35.1% when B=50 mm.

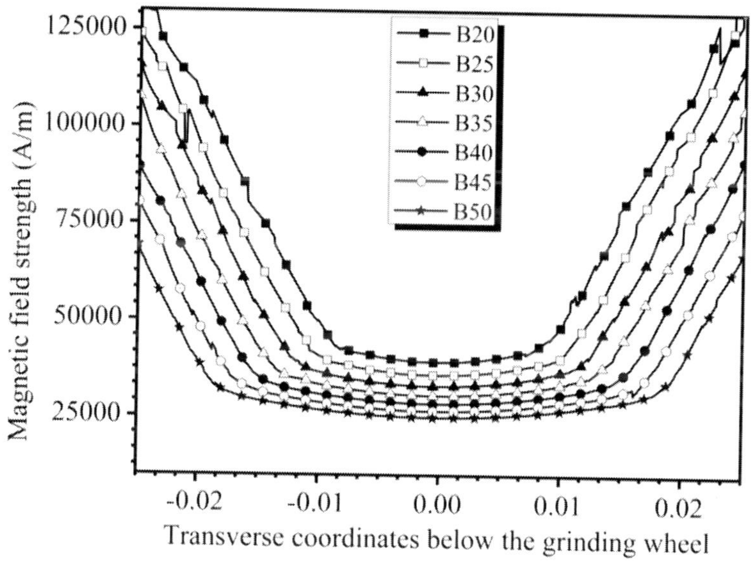

Figure 17. Variation of magnetic field intensity on transverse under grinding wheel with different *B*.

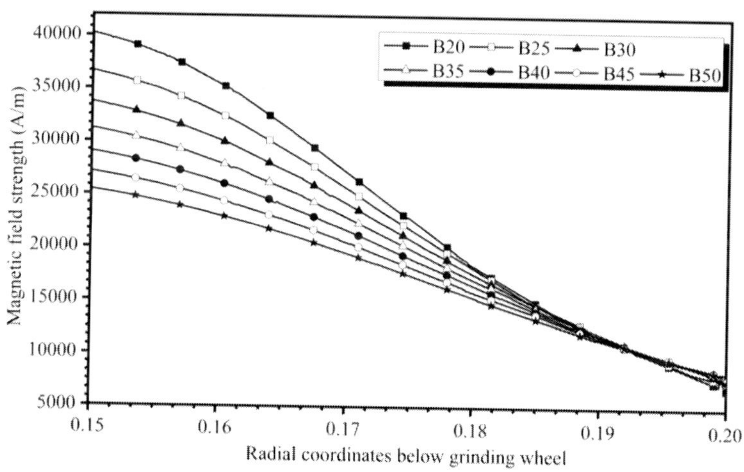

Figure 18. Variation of magnetic field intensity on radial under grinding wheel with different *B*.

5.5. Testing and Verififying the Magnetic Field

The "three-dimensional Gauss meter" is one of the most reliable methods to test magnetic field strength. However, it can only measure magnetic field distribution in open spaces due to limitations of the geometric shape of its head, but it cannot test magnetic field distribution in narrow space or in objects. Hence, it is necessary to apply numerical simulation to study the magnetic field distribution. To further verify reliability of numerical simulation results, the magnetic fields below the grinding wheel are tested and compared with simulation results in this part.

As shown in Figure 19, the variation laws of magnetic field strength measured under H=25 mm, L=12 mm and β=0° when the 20 mm CBN grinding wheel is used and magnets are in the repulsive left-right distribution are consistent with variation laws gained in simulation. Although the simulation values have some errors due to influences of the machine body on magnetic field, the order of magnitudes of such error is relatively low and it hardly influences the new technology.

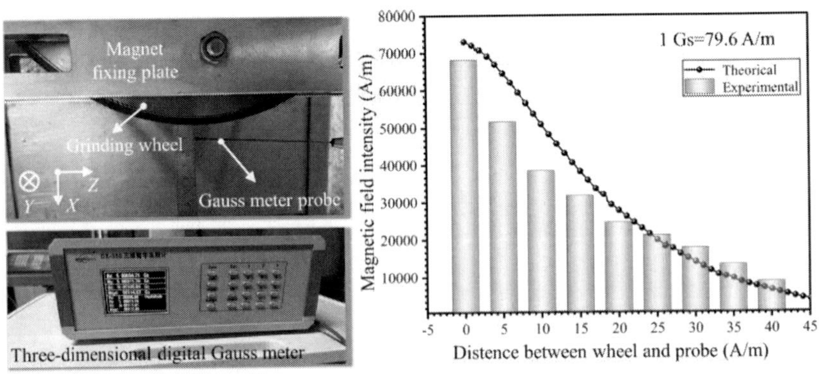

Figure 19. Magnetic field test and verification under typical working conditions.

Conclusion

In this chapter, numerical simulation on magnetic field distribution in the new technology is carried out by changing magnetic field distribution, geometric positions of the permanent magnets and parameters of permanent magnets and grinding wheel in the grinding zone. The variation laws of magnetic field strength in the grinding zone under different magnetic field distributions and parameters are analyzed. Some conclusions could be gained:

i. Different distribution modes of N-S poles will cause different distributions of magnetic induction lines and thereby producing different magnetic field strength. The magnetic field distribution and strength surrounding the grinding wheel under four pole distribution modes of permanent magnets (type I, II, III and IV) are analyzed. Results show that relatively strong uniform magnetic fields can be produced under type I, which can bring uniform distribution of magnetofluid.

ii. According to comparison of magnetic field distribution and size under different intensity of magnetism of permanent magnets, the magnetic field strength surrounding the grinding wheel changes slightly with the increase of intensity of magnetism when the permanent magnet shape and size are fixed. With the increase of grinding wheel thickness, the magnetic field strength decreases gradually, but the magnetic field distribution is influenced slightly. Hence, it is not necessary to choose permanent magnet with excessive intensity of magnetism after comprehensive considerations to working environment and cost. The width of grinding wheel has to be chosen according to comprehensive analysis of processing conditions and influences of the grinding wheel thickness are adjusted by other geometric parameters.

iii. According to studies on magnetic field strength and distribution caused by changes in geometric positions of permanent magnets surrounding the grinding wheel, it finds that: with the increase of L, the horizontal magnetic field strength distribution of the grinding wheel becomes increasingly uniform, while the circumferential distribution becomes increasingly nonuniform and the magnetic field strength decreases gradually. With the increase of β, the horizontal magnetic field distribution of grinding wheel changes from uneven to relatively uniform, and then becomes uneven, while the circumferential distribution becomes increasingly uneven. The magnetic field strength declines gradually. As H increases, the horizontal magnetic field distribution of the grinding wheel will not be influenced obviously. The circumferential distribution becomes uniform, and the magnetic field strength increases gradually. Based on above analysis, it shall set L within 6~16 mm, β within 0°~30° and increase *H* appropriately during grinding. Under these conditions, magnetic fields in uniform distribution and high strength can be formed in the grinding zone.

References

[1] Ming WY, Shen F, Zhang Z, et al. (2020). A comparative investigation on magnetic field-assisted EDM of magnetic and non-magnetic materials. *Int. J. Adv. Manuf. Technol.*, *109*(3-4), 1103-1116. 10.1007/s00170-020-05653-8.

[2] Zhang Z, Zhang Y, Ming WY, et al. (2021). A review on magnetic field assisted electrical discharge machining. *Journal of Manufacturing Processes*, *64*, 694-722. 10.1016/j.jmapro.2021.01.054.

[3] Jain VK. (2009). Magnetic field assisted abrasive based micro-/nano-finishing. *Journal of Materials Processing Technology*, *209*(20), 6022-6038.

[4] Cheng CP, Wu KL, Mai CC, et al. (2010). Magnetic field-assisted electrochemical discharge machining. *J. Micromech. Microeng.*, *20*(7), 7. 10.1088/0960-1317/20/7/075019.

[5] Bains PS, Sidhu SS, Payal HS. (2018). Investigation of magnetic field-assisted EDM of composites. *Mater. Manuf. Process.*, *33*(6), 670-675. 10.1080/10426914.2017.1364857.

[6] Huang YS, Zhao BS, Huang SQ, et al. (2014). Pulse electromagnetic field-assisted chemical mechanical polishing utilizing magnetic composite abrasives slurry and Its polishing performance. *China Mechanical Engineering*, *25*(9), 1175-1238.

[7] Xu XF, Guo Q, Huang YS, et al. (2011). Chemical mechanical polishing using magnetic composite abrasives slurry and experimental study on polishing performance. *China Mechanical Engineering*, *47*(21), 186-192.

[8] Hirose E, Tanaka KH, Takahashi T, et al. (2004). A new 3-axis magnetic field measurement system based on Hall elements. *IEEE transactions on applied superconductivity*, *14*(2), 1814-1817.

[9] Chen JZ, Zhang YM. (2003). The design for auto-instrument of magnet field measurement. *Modern Manufacturing Engineering*, *2003*(11), 58-59.

[10] Chen DX, Pan MC, Luo FL, et al. (2003). The air gap magnetic fields measurement of high speed magnetic levitation vehicle. *Electrotechnical Application*, *2003*(12), 69-70.

[11] Wang XM, Wang Y, Wang GY. (2005). The systam for measuring and analyzing superficial magnetic field of pemanent magnet. *Small & Special Electrical Machines*, *33*(10), 9-11.

[12] Wang C, Yang WY, Zhai GF, et al. (2009). Research on 3D measuring system of magnetic fiekd of small permanent magnet. *Low Voltage Apparatus*, (9), 28-31.

[13] Huang DY, Zhang T, Han B, et al. (2009). Influence of applied stress and residual stress on coercive force of Q235 mild steel. *Magnetic Materials and Devices*, *40*(6), 48-50.

[14] Zhong H, Bai WS, Hou ZJ, et al. (2014). Research of increasing the magnetic fi eld strength by connecting the same pole together. *China Modern Educational Equipment*, (3), 67-69.

Chapter 6

The Kinematics Mechanism of Microdroplets of Magnetic Nanolubricants in a Complex Airflow Field Around the Grinding Wheel

Abstract

There is a complex airflow field around the grinding wheel, which prevents the lubricant microdroplets from entering the grinding zone. Moreover, lubricant droplets enter the surrounding environment, posing a great threat to the health of workers. In MTN grinding, the magnetic microdroplet could be absorbed on the surface of the grinding wheel through the complex air flow field under the attraction of the grinding wheel. Based on this principle, the chapter reveals the kinematic behavior of the magnetic nano lubricant in the jet zone and the dynamic behavior of collision and adsorption on the workpiece surface of the grinding wheel. Further, the effect of the magnetic field intensity and jet velocity on microdroplet sliding, film formation, and falling is studied. The collision-adsorption behavior map of microdroplets under the action of magnetic field intensity and jet velocity is obtained, which provides a theoretical basis for process optimization.

Keywords: airflow field, microdroplet, kinematics, collision and adsorption, jet parameters, magnetic field

6.1. Introduction

Carbon nanolubricant, especially graphene (GR), show outstanding cooling and lubrication performances in grinding. To make carbon nanolubricant adsorbed by the magnetized grinding wheel, it is necessary to implement magnetic functionalization of carbon nanofluids. The relatively common magnetofluid is prepared by Fe_3O_4. However, Fe_3O_4 has no good cooling and lubrication performances. Therefore, attention is paid to explore preparation technologies of magnetic nanolubricant firstly and test its physical properties.

Furthermore, microdroplet groups are formed in NMQL through mixed atomization of high-pressure gases and lubricant, and they are sprayed onto the grinding zone at a certain speed to provide cooling and lubrication effect. Microdroplet groups have greater specific surface area than continuous supply mode of lubricant, thus realizing the goal of replacing flood supply of cutting liquid by MQL. Due to elasticity and liquidity of the air, an airflow field with a speed gradient and the same direction with the cutter surface movement is formed surrounding the cutter driven by its own high-speed rotation. Moreover, the speed gradient and movement direction of the airflow field become extremely complicated after interfering with the workpiece. Finally, an "air barrier" that hinders grinding fluid supply to the cutting zone, the "inflow" with the same direction to the cutting fluid supply and the "backflow" opposite to the cutting fluid supply direction are formed. As a result, the penetration and transport efficiency of lubricant are influenced significantly by the supply path of microdroplets [1, 2]. For plane grinding, the team led by Li et al. [3] explored airflow field distribution laws through numerical simulation and found that increasing linear speed of the grinding wheel could increase the thickness of "air barrier layer" as well as airflow pressure and speed in the grinding zone. This also brings significant resistance against entrance of the cutting fluid into the grinding zone [4].

The grinding wheel can be magnetized by applying permanent magnets at two sides of the grinding wheel and magnetic fields can be formed on the grinding wheel surface. If magnetic nanolubricant atomized jet is sprayed into the grinding zone, the lubricant is expected to be carried into the grinding zone from the grinding wheel surface due to adsorption of the magnetic field force, thus avoiding "backflow" and improve the infiltration performance of the lubricant [5-7]. Nevertheless, the impacting and adsorption mechanisms of microdroplets onto the grinding wheel surface under the action of magnetic field and airflow field still have to be disclosed. Moreover, the interaction relations between variables (e.g., jetting speed and magnetic field strength) and impacting adsorption shall be further studied. On this basis, this chapter will carry out related theoretical studies.

6.2. Dynamic Modeling of the Collision Adsorption of Magnetic Nanolubricants on the Grinding Wheel Surface

6.2.1. Governing Equations

(i) Governing equation of flow field

Since magnetic nanolubricant impacting onto the grinding wheel surface is a gas-liquid phase problem, the level set method with strong time economy is employed for numerical simulation. Some hypotheses are made:

① Since the movement speed of fluid is at the magnitude of 10 m/s and it is far lower than the speed of speed, the flowing behavior is not compressible.

② Viscosity of fluid is only related with magnetic field. The viscosity may not change after the magnetic field is determined. Hence, it is set as the Newtonian fluid and suppose viscosity and surface tension are constant.

③ The magnetic nanolubricant is atomized after spraying by the nozzle, but it is shrunk into spherical under the action of surface tension. Hence, the droplets can be set round at the initial moment of simulation.

④ Since the impacting point occurs in the jetting zone and temperature is maintained close to the room-temperature range, the heat transfer behaviors during impacting are ignored.

⑤ Even current is produced by droplets during movement, it is very weak, and the Lorentz force can be ignored.

For incompressible Newtonian fluid, it has to observe the law of conservation of mass and conservation of momentum, the effects of several forces shall be considered in governing equation of turbulence: pressure produced by interaction of phase interface, viscous force of the liquid on the solid interface, gravity that droplets bear, surface tension on solid-liquid-gas interface and magnetic field force of applied magnetic field on droplets. Hence, the governing equation is:

The mass conservation equation:

$$\nabla \cdot \boldsymbol{u} = 0 \tag{1}$$

The momentum conservation equation:

$$\rho(\boldsymbol{u} \cdot \nabla)u = \nabla \cdot [-p\boldsymbol{I} + \boldsymbol{K}] + \rho g + F_{ma} + F_{ce} + F_{st} \tag{2}$$

$$K = (\mu + \mu_T) \cdot (\nabla u + (\nabla u)^T) \quad (3)$$

Where p is the pressure (Pa). $\boldsymbol{K}$ is the viscous force. $\boldsymbol{u}$ is the speed vector (m/s). Due to the hypothesis that fluid is uncompressible, the speed gradient can be set as 0. F_{ce} is the centrifugal force, F_{ma} is the magnetic field force, and F_{st} is the interfacial intension. ρ is the density, g is the gravity vector (m/s^2), and I is the unit matrix. μ is the dynamic viscosity (N·s/m^2) and μ_T is the turbulence viscosity (N·s/m^2).

The level set method defines that Φ changes between 0 and 1. In other words, it values the constant 0 or 1 in two types of fluids. In this study, Φ values 0 for air and 1 for magnetic lubricant. Density and viscosity are function of Φ. The formula is:

$$\rho = \rho_1 + \Phi \cdot (\rho_2 - \rho_1) \quad (4)$$

$$\mu = \mu_1 + \Phi \cdot (\mu_2 - \mu_1) \quad (5)$$

Where ρ_1 and ρ_2 are density of fluid 1 (air) and fluid 2 (magnetic lubricant), respectively. μ_1 and μ_2 are dynamic viscosity of fluid 1 (air) and fluid 2 (magnetic lubricant).

The gas-liquid interface can be expressed by the following equation:

$$\frac{\partial \Phi}{\partial t} + \boldsymbol{u} \cdot \nabla \Phi = \gamma \nabla \cdot \left(\varepsilon \nabla \Phi - \Phi(1 - \Phi) \frac{\nabla \Phi}{|\nabla \Phi|} \right) \quad (6)$$

Where γ is the migration rate (m^3·s/kg) and ε is the control parameter of interface thickness (m).

Since calculation of surface curvature and surface normal vector is complicated, the surface tension is calculated by using the diffusion interface distribution force. The calculation formula is:

$$F_{st} = \sigma \delta k \boldsymbol{n}_{int} + \delta \nabla_S \sigma \quad (7)$$

$$k = -\nabla \cdot \boldsymbol{n}_{int} \quad (8)$$

$$\nabla_s = (\boldsymbol{I} - \boldsymbol{n}_{int} \boldsymbol{n}_{int}^T) \nabla \quad (9)$$

Where the fluid phase interface δ function (1/m) is used to process surface tension at the interface and there are many liquid-liquid interfaces to calibrate dispersion phase. $\boldsymbol{n}_{\text{int}}$ is the interface normal, σ is the surface tension coefficient (N/m) and k is the curvature (1/m).

When droplets impact onto the grinding wheel surface, it has to overcome centrifugal force to keep circumferential movement under the driving of grinding wheel. The formula of centrifugal force inputting as the body force is:

$$F_{ce} = \frac{\rho V_{dt}^2}{r} \tag{10}$$

Where V_{dt} is the tangential movement speed of droplets (m/s) and r is the radius of the grinding wheel (m).

(ii) Governing equation of magnetic field

In this study, the magnetic field and flow field coupling is the sole method to study physical mechanism of magnetofluid impacting. Hence, the magnetic field module has to be introduced into the model. Moreover, the current produced when droplets cut the magnetic induction lines is considered and Maxwell's equation set can be expressed as:

$$\nabla \times \boldsymbol{H} = \boldsymbol{J} \tag{11}$$

$$\boldsymbol{B} = \nabla \times \boldsymbol{A} \tag{12}$$

$$\boldsymbol{J} = \sigma_c \boldsymbol{E} \tag{13}$$

Where σ_{c} is the electric conductivity of magnetic lubricant (S/m). $\boldsymbol{H}$ refers to the magnetic field strength (A/m). $\boldsymbol{E}$ is the strength of electric field (V/m). A is the magnetic vector potential (Wb/m). $\boldsymbol{B}$ denotes the magnetic flux density (T). $\boldsymbol{J}$ shows the electric current density of fluid (A/m^2).

The magnetization of magnetic lubricant in the magnetic field also shall meet the following constitutive relation:

$$\boldsymbol{B} = \mu_0 \mu_r \boldsymbol{H} \tag{14}$$

Furthermore, the conduction and movement process of charge on two-phase interface shall meet the following constitutive relationship:

$$\boldsymbol{J}_c = \sigma' \boldsymbol{E} \tag{15}$$

$$D = \varepsilon_0 \varepsilon_r E \tag{16}$$

Where J_c is the conduction heat flux density (A/m^2). D is the electric displacement (C/m^2). ε_0 is the vacuum dielectric constant (F/m). ε_r refers to the dielectric constant of the computational domain (F/m).

(iii) Multi-physics coupling equation

The impacting process of magnetofluid microdroplets onto the grinding wheel surface under the effect of magnetic field is a multi-physics coupling problem covering the flow field and magnetic field. Hence, it has to form coupling of the flow field and magnetic field equation. In this study, magnetic field force that magnetic fluid bears shall be written into the fluid motion equation as the body force. According to deduction of R.E. Rosensweig to the formula of magnetofluid stresses, effects of temperature are ignored. The components of the magnetic field force (F_{ma}) per unit volume of magnetic lubricant on the X and Y directions are [8]:

$$F_{max} = \mu_0 \left(M_x \frac{\partial \boldsymbol{H}_x}{\partial x} + M_y \frac{\partial \boldsymbol{H}_x}{\partial y} \right) \tag{17}$$

$$F_{may} = \mu_0 \left(M_x \frac{\partial \boldsymbol{H}_y}{\partial x} + M_y \frac{\partial \boldsymbol{H}_y}{\partial y} \right) \tag{18}$$

6.2.2. Physical Modeling

(i) Construction of the geometric model

The key to numerical simulation lies in reproducing the magnetic field distribution in the grinding wheel/workpiece wedge-shaped region which is gained in the Chapter 5 and input it into the model in this chapter as the background magnetic field. As shown in Figure 1(a), a two-dimensional geometric model of grinding wheel (diameter: 300 mm) and surrounding space is built firstly in this study. The surrounding air layer and air layer of the infinite element domain are defined. The initial magnetic field is defined for the grinding wheel, thus getting the spatial magnetic fields. Furthermore, a rectangle with some size is built at the lower right part of the grinding wheel, that is, the jetting zone of the nozzle, and it is defined as the flow field (height: 2.5 mm; width: 4.5 mm; dip angle: 30°). The microdroplets (the radius is R_d, the distance between the center and grinding wheel surface is the jetting distance (L_{jet}), and the included angle between the initial speed vector and the grinding wheel surface, or the jetting angle, is θ_{jet}) are built according to practical physical significance. Through this method, the background magnetic field can be input accurately, and the calculation time can be shortened significantly.

(ii) Setting of boundary conditions

The boundary conditions are defined for the magnetic field and flow field according to practical situations in the model. For the magnetic field in Figure 1(a), the magnetic constitutive of the grinding wheel is set "magnetized" and the initial value of magnetic field is set M_1 [A/m]. To realize uniform variation distribution of magnetic field, the magnetizing mode (M) of the grinding wheel is defined as the linear formula related with M_1 and the vertical coordinate (y). The grain layer and air layer are both set as "relative magnetic permeability." Additionally, the boundary ⑤ is set as "magnetic insulation." For the flow field in Figure 1(a), since the grinding wheel surface, or the boundary ⑥, is a rough surface, it is set as a slip-free wall with some roughness (sand roughness=30 μm). Furthermore, the boundaries ④, ⑤ and ⑦ are open boundaries of flow field.

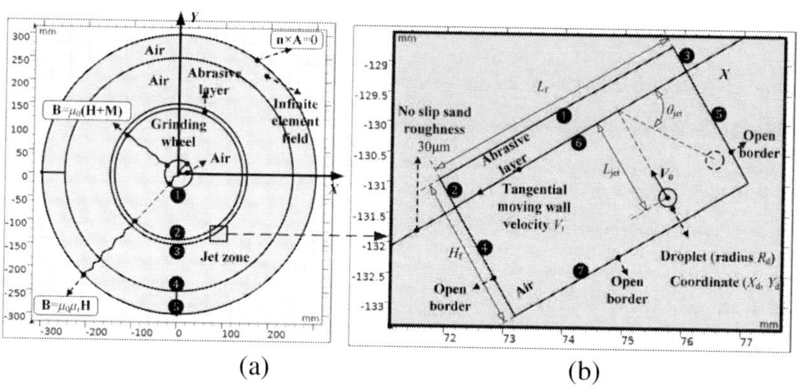

Figure 1. Geometric model and boundary conditions for collision adsorption behavior of lubricant microdroplets. (a) Magnetic field modeling in the grinding zone (b) flow field modeling in the jet zone.

(iii) Meshing

In the two-dimensional geometric model, the triangular meshing is applied. For better analysis of droplets impacting onto the grinding wheel, the meshing of grinding wheel part, droplets and air domain in the model shall be refined. It is generally defined as "extremely refined." The maximum cell size is 6 mm and the minimum cell size is 0.012 mm.

(iv) Selection of solver

In this study, it has to track kinetic and dynamic behaviors when magnetofluid impacts onto the grinding wheel. Hence, full-coupling solving is employed through phase initialization and transient solver. Specifically, the relative tolerance of the transient solver is 0.001. The calculation time range and time step length can be set according to practical needs.

6.3. The Physical Mechanism of Microdroplets: Magnetic Field Strength

6.3.1. Research Results

Based on research results in Chapter 5, magnetic fields that change uniformly can be gained surrounding the grinding wheel under the action of magnetic field. This reproduces the magnetic field distribution in practical conditions. The cloud chart of magnetic field distribution in the jetting zone is shown in Figure 2, where it also presents uniform distribution of magnetic fields. In this study, it not only has to explore influences of large-scaled magnetic field strength variation on movement behaviors of droplets, but also shall analyze action mechanism of local small-sized magnetic field changes on droplet forms. Hence, the relationship with magnetic field of grinding wheel surface is acquired by setting different boundary condition parameter M_1 (the selection of M_1 shall be multiplied) of magnetic field in the numerical simulation, thus getting magnetic field strength (M) as droplets impact onto the grinding wheel surface under different magnetic field strength. Results are shown in Table 1.

Table 1. Magnetic field intensity of wheel surface corresponding to different magnetic field intensity

Magnetic field strength M_1	Magnetic field of grinding wheel surface at impact point M
0.5×10^4 A/m	0.11×10^4 A/m
1×10^4 A/m	0.22×10^4 A/m
2×10^4 A/m	0.44×10^4 A/m
5×10^4 A/m	1.1×10^4 A/m
1×10^5 A/m	2.2×10^4 A/m
2×10^5 A/m	4.4×10^4 A/m
5×10^5 A/m	11.1×10^4 A/m

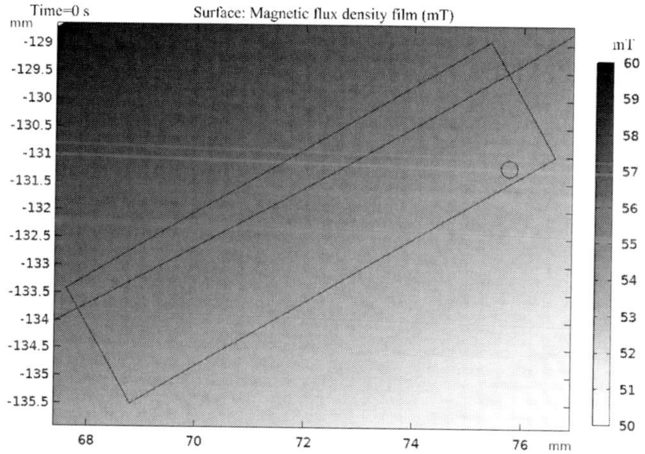

Figure 2. Magnetic field distribution in jet region.

At present, scholars have studied kinetic and dynamic behaviors when droplets impact onto the grinding wheel surface. In this study, there are four differences from previous impacting behaviors of droplets: (i) the grinding wheel in grinding has a high surface roughness and it rotates at a certain speed in the processing. Hence, there's airflow field surrounding the grinding wheel. Hence, influences of airflow field on droplet movement shall be considered when analyzing impacting behaviors of droplets. (ii) In existing studies, the liquid with a low viscosity was employed, but the vegetable oil with high viscosity was used in this study. Moreover, viscosity increases with the increase of magnetic field strength. (iii) In this study, droplets impact onto the grinding wheel surface from bottom to upward and form a certain included angle with the grinding wheel surface (incidence angle). The centrifugal force and gravity shall be considered. (iv) Influences of magnetic field on impacting behaviors of droplets shall be discussed in this study.

Through numerical simulation of droplets impacting onto the grinding wheel surface under action of magnetic field, the process that droplets impact onto the grinding wheel under different magnetic field conditions is acquired (Figure 3 to 5). The movement process of droplets when the magnetic field strength is 0 A/m (no action of magnetic field) is shown in Figure 3. Droplets begin to move toward the grinding wheel surface at the rate of 0.5 m/s since 0 ms and arrive at the grinding wheel surface at 3.6 ms. At 4 ms, their movement along the grinding wheel is dragged. As time increases, the dragging phenomenon of droplets becomes more and more obvious from 4 ms to 5 ms, and the round droplets are stretched gradually. Nevertheless, there's a trend that liquid shrinks into spheres from 5 ms to 6 ms.

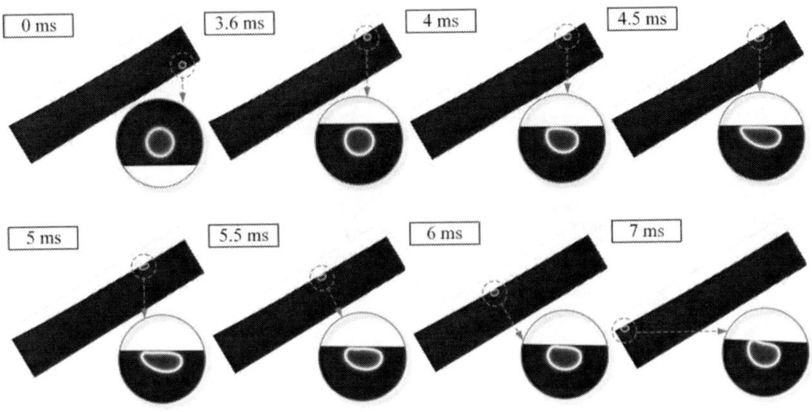

Figure 3. The process of droplet hit grinding wheel when magnetic field intensity of grinding wheel surface is 0 A/m.

Besides, the cloud chart of level set variables is also shown in Figure 3, where red represents the lubricant and blue represents the air. Clearly, there's no complete contact between droplets and grinding wheel surface. Instead, there's an "air cushion layer" between droplets and grinding wheel surface. After impacting onto the grinding wheel surface, droplets move with the grinding wheel at the rate of about 3.4 m/s. The movement speed is lower than the speed of grinding wheel. In other words, there's slippage. Such movement of droplets is named as "slippage of droplets."

The movement process of droplets when the magnetic field strength is 1×10^5 A/m is shown in Figure 4. After magnetic field is added, the time for droplets to arrive at the grinding wheel surface decreases to some extent. Droplets contact the grinding wheel surface and impact onto the grinding wheel at 3.5 ms. Similarly, droplets begin to be dragged along the movement direction of grinding wheel at 4 ms. As time increases to 5 ms, droplets are dragged quickly and contact the grinding wheel surface. At 5.2 ms, the front ends of droplets are pumped in, forming an uneven oil layer on the grinding wheel surface. As droplets move on the grinding wheel surface, the pumping phenomenon of droplets is more significant, and droplet volume decreases gradually. Since a phenomenon similar to pumping occurs during movement of droplets, it is called "pumping-induced film formation."

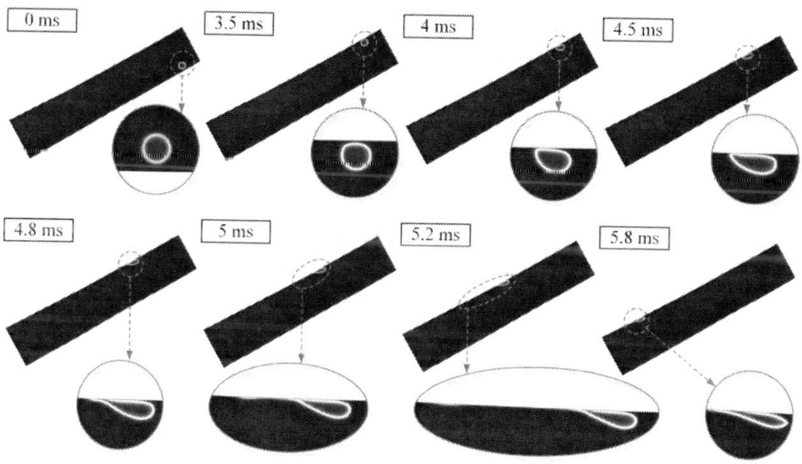

Figure 4. The process of droplet hit grinding wheel when magnetic field intensity of grinding wheel surface is 2.2×10^4 A/m.

When the magnetic field strength on the grinding wheel surface increases to 11.1×10^4 A/m, different forms appear during movement of droplets (Figure 5). Obviously, droplets arrive at grinding wheel surface at 2.1 ms and are dragged at 2.6 ms, which are earlier than those under no magnetic field and low magnetic field strength. The droplets contact with the grinding wheel surface and time is shortened obviously. The droplets begin to be pumped in at 2.8 ms and form an oil film on the grinding wheel surface. As time goes on, the pumping phenomenon becomes more and more obvious, and the droplet volume decreases gradually. at 3.8 ms, the liquid is hardly pumped out and a relatively uniform oil film is formed on the grinding wheel surface. The film forming speed of droplets is increased significantly compared to that under a low magnetic field strength.

Based on the above studies, droplets form a lubrication film on the grinding wheel surface under the action of magnetic field. Although we have improved the quality of lubricating oil film on the grinding wheel surface through assistance of magnetic field, the film formation speed is low when the magnetic field is small. Besides, it is observed that the lubricating oil film is thin and uneven. Under such circumstances, the lubrication state between grinding wheel and workpiece interface is relatively poor, which is disadvantageous for lubrication. Hence, shortening the time for film formation of lubricant is the ideal state that we want.

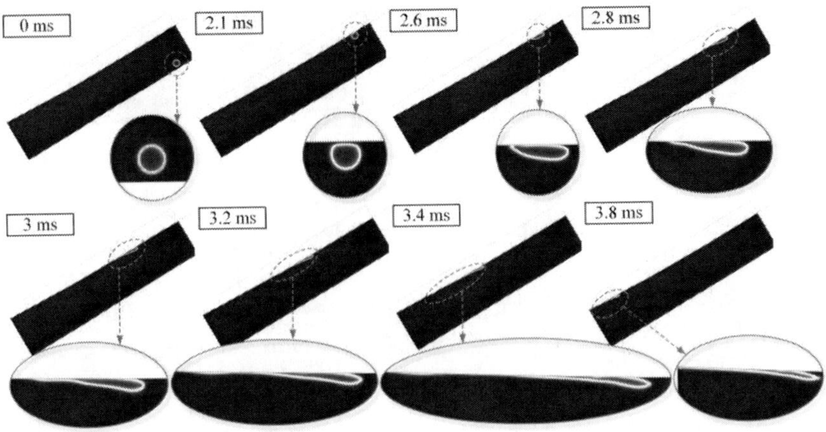

Figure 5. The process of droplet hit grinding wheel when magnetic field intensity of grinding wheel surface is 11.1×10^4 A/m.

In previous studies, magnetic field strength of grinding wheel surface has obvious influences on movement speed of droplets. For further quantitative characterization of variations of droplet speed with magnetic field strength, a transversal line from initial position of droplets to the grinding wheel surface is made. Numerical values of droplet speeds on the transversal line are extracted. On this basis, variations of droplet speed with magnetic field strength are acquired (Figure 6). It can be seen from Figure 6 that due to airflow resistance, droplet speed presents a descending trend in the beginning (initial stage). After the kinetic energy provided by magnetic field and the friction energy consumed in the droplet movement reach the balance, the speed begins to be increased or decreased at a stable rate (balance stage). With the increase of time, the speed curve changes suddenly when droplets impact onto the grinding wheel surface and droplets impact onto the grinding wheel surface at the lowest point (impacting stage).

Obviously, droplet speeds under different magnetic field strengths all decrease in the initial stage, showing small difference of curves. In the balance stage, the magnetic field strength has different influences on droplet speed. With the increase of magnetic field strength, the balance stage of droplets is shortened significantly, while the slope of speed curve changes from negative to positive, showing an increasing trend. The slope of speed curve reflects the relationship between the kinetic energy provided by magnetic field and friction energy consumed during droplet movement in the airflow field. It is 0 when magnetic field strength is 2.2×10^4 A/m. In other words, the attractive energy of such magnetic field strength to droplets offsets with energy consumed for droplet movement. With the increase of magnetic field strength, the slope is large when the magnetic field strength reaches 11.1×10^4 A/m, indicating the significant attraction of magnetic field to droplets.

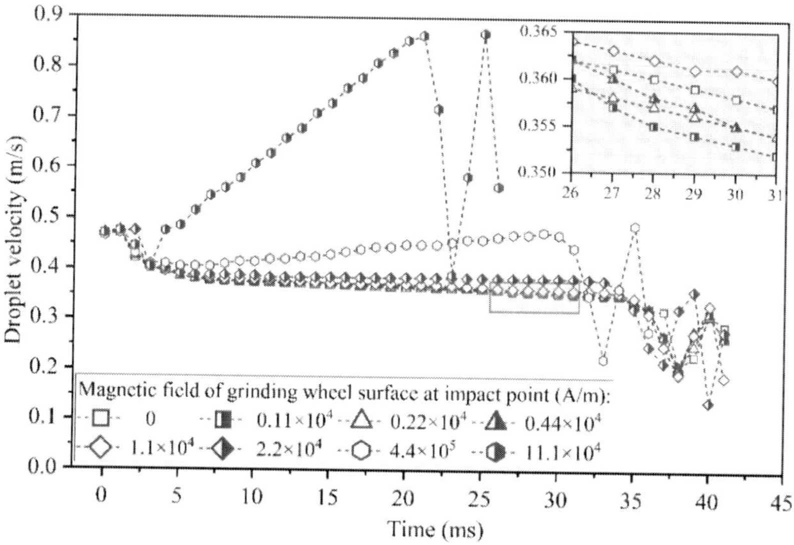

Figure 6. Droplet velocity changes under different magnetic field intensities.

6.3.2. Microdroplet Sliding Behavior

In the previous section, we found that there are slippage behaviors of droplets on the grinding wheel surface, which is related with mechanical behaviors of droplets. Hence, the pressure diagram of droplets and flow diagram of airflow field when there's no magnetic field are further observed in this section (Figure 7). Pressure distributions at different time nodes of liquid movement are shown in Figure 7(a), (b) and (c). The pressure isogram and speed flow diagram at 4.5 ms are shown in Figure 7(d) and (e). The stress analysis of droplets during movement is shown in Figure 7(f). Before contact with the grinding wheel surface, droplets bear air resistance opposite with the droplet speed and inertia force in the same direction of droplet speed except for gravity. Droplets begin to slip when they contact the grinding wheel surface. In addition to gravity and centrifugal force produced by movement of grinding wheel, viscous force of droplets and airflow field push slippage of droplets.

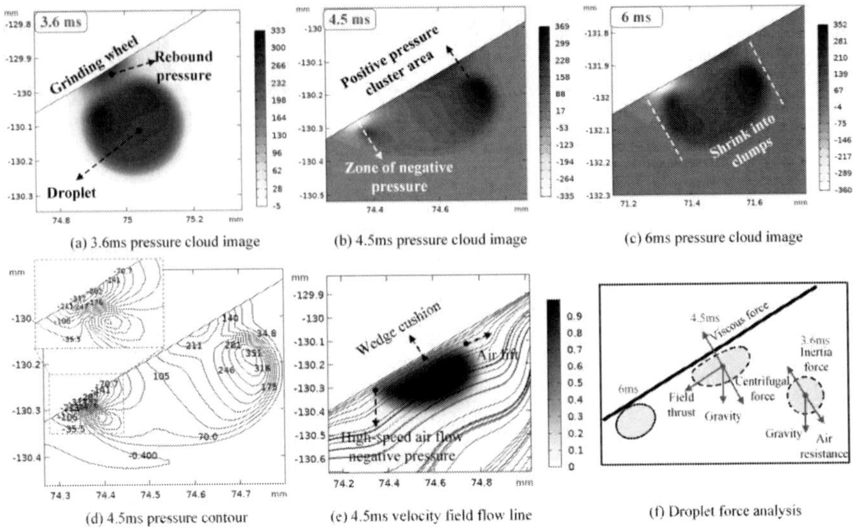

Figure 7. Distribution of force and space pressure during microdroplet slip. (a) 3.6ms pressure cloud image, (b) 4.5ms pressure cloud image, (c) 6ms pressure cloud image, (d) 4.5ms pressure contour, (e) 4.5ms velocity field flow line, (f) drop force analysis.

It can be seen from Figure 7 that although droplets do not have contact with the grinding wheel surface completely at 3.6 ms, a rebounding pressure is formed between droplets and grinding wheel surface, showing an air cushion layer. This is because eddies are formed between the rough grinding wheel surface and large airflow under the collaborative effect, which is known as the vortex layer. As a result, there's outward pressure along the normal line of grinding wheel surface (Figure 7(a)). A negative pressure zone appears in front end of droplet at 4.5 ms (Figure 7(b)) and the maximum negative pressure reaches 317 Pa (Figure 7(d)). On one hand, the grinding wheel speed is far higher than the speed of droplets. On the other hand, a wedge-shaped air cushion is formed when airflow runs through the small narrow wedge zone between the grinding wheel surface and droplets. The front ends of droplets have a great speed and the airflow at tails concentrates to lift droplets. At 6 ms (Figure 7(c)), lubricant shrinks together as a response to the surface tension. When fluid with high viscosity slips on the grinding wheel surface, an apparent slipping layer (air cushion layer) and high-viscosity liquid layer will be formed. This is called as apparent slippage. In studies on slippage mechanism, scholars have proposed the apparent slippage model. Liquid close to the grinding wheel surface changes from liquid state to the unstable gas stage during the slip process. Some studies have demonstrated that thickness of the slippage layer (air cushion layer) is related with viscosity of fluid. The apparent slippage model can be expressed as [9, 10]:

$$b = \delta\left(\frac{\eta_b}{\eta_s} - 1\right) \tag{19}$$

Where η_b and η_s are viscosity coefficients of liquid and gas, respectively. δ is the thickness of the slippage boundary layer. b is the slip length coefficient.

6.3.3. Film Formation Behavior of Microdroplet Suction

In existing studies, scholars have proved that edges of droplets will be dragged into finger-like shapes when they impact onto a dry surface. This is because viscosity, surface tension and other properties of droplets are affected upon contact with the wall surface, especially on smooth wall surface influenced by small airflow [11]. Under working conditions in this study, magnetic field has significant influences on adsorption and film forming of droplets. Moreover, the rotating grinding wheel surface has some effect on airflow field. Therefore, pressure distribution and stresses during movement of droplets under the action of magnetic field are analyzed (Figure 8).

Same with the situation without magnetic field, a rebounding pressure is formed between droplets and grinding wheel surface before full contact and an air cushion layer is developed (Figure 8(a) and (b)). At 4.8 ms (Figure 8(c)), the front ends of droplets are dragged to contact with the grinding wheel surface under the action of magnetic field, causing sucking into film. The rear ends of droplets slip with the grinding wheel. Moreover, a larger extending negative pressure zone is formed in the front end and the pressure reaches as high as 1,070 Pa (Figure 8(a)). The pumping effect of droplets is greater.

It can be seen from flow lines of speed field in Figure 8(e) that droplets which haven't had contact with the grinding wheel are driven to move by surrounding airflow at the same speed. As a result, no negative pressure is formed, thus failing to form film. Under this circumstance, droplets bear centrifugal force of grinding wheel and viscous force of droplets except for gravity. Influenced by magnetic field and airflow field, there's magnetic field force perpendicular to the grinding wheel surface and airflow field force along the movement direction of grinding wheel.

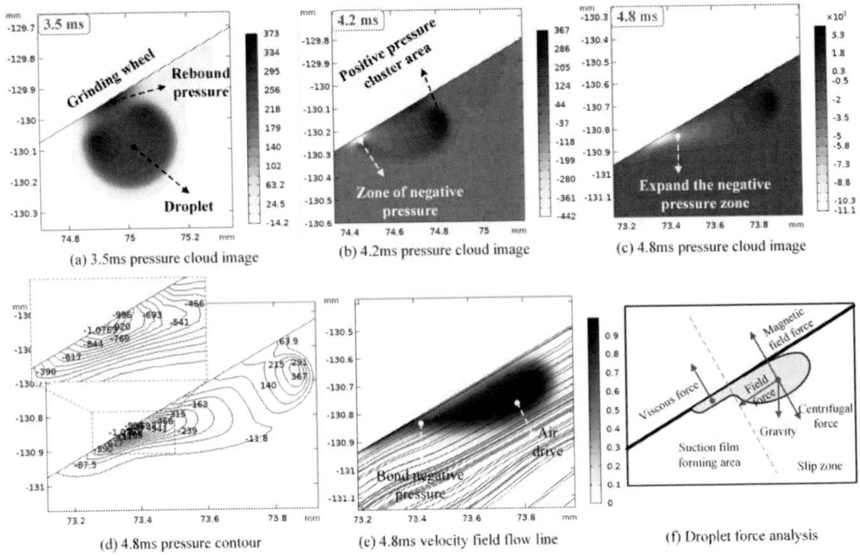

Figure 8. Droplet force and spatial pressure distribution during film formation by microdroplet suction. (a) 3.5ms pressure cloud image, (b) 4.2ms pressure cloud image, (c) 4.8ms pressure cloud image, (d) 4.8ms pressure contour, (e) 4.8ms velocity field flow line, (f) drop force analysis.

6.4. The Physical Mechanism of Microdroplets: Droplet Velocity

6.4.1. Research Results

In the previous section, the movement behaviors when droplets impact onto the grinding wheel surface under different magnetic field strength are studied. In fact, movement speed of droplets is an important factor that influences their impacting behaviors. Moreover, scholars have devoted themselves to studying behavioral laws of droplets impacting onto the wall surface. Nevertheless, the working conditions in this study are different from previous conditions for impacting of droplets. This has been introduced in the previous section and it is not introduced here anymore. In this section, influences of droplet speed (v_0=0.2, 0.3, 0.4, 0.5, 0.6, 0.7, 0.8, 0.9, 1, 3, 5, 7, 9, 11, 13 and 15 m/s) are analyzed. Moreover, the collaborative influencing laws of magnetic field strength and droplet speed are investigated.

The motion tracks of droplets when they impact onto the grinding wheel surface at the speed of v_0=0.3 m/s under different magnetic field strength are shown in Figure 9. For the convenience of analysis of droplet behaviors during movement, coordinates of droplets at different time nodes are extracted and plotted in the same diagram. The motion tracks of droplets when the magnetic field strength is 0 A/m, that is, no application of magnetic field, are displayed in Figure 9(a). Clearly, droplets move toward the grinding wheel surface since 0 ms and their tracks deviate at 6 ms. On the one hand, this is attributed to the collaborative effect of droplet gravity and airflow field during movement. On the other hand, the droplet speed is too low to provide sufficient kinetic energy. At 8 ms, droplets impact with the high-speed airflow layer of the grinding wheel surface and then move away from the grinding wheel surface due to the rebounding by the air layer. Here, the third type of droplet impacting behaviors is discovered, which is defined as "falling of droplets."

The motion tracks of droplets are changed after magnetic field is applied onto the grinding wheel surface. The motion tracks of droplets when magnetic field strength is 4.4×10^4 A/m are shown in 3-15(b). After magnetic field is applied, droplets arrive at the grinding wheel surface at 4.8 ms, which is significantly earlier than that before. Moreover, the deviation degree of motion tracks of droplets decreases obviously. As time goes on, there's no falling behaviors of droplets in (a). Under the collaborative effect of magnetic field force and viscous force, droplets begin to slip on the grinding wheel surface.

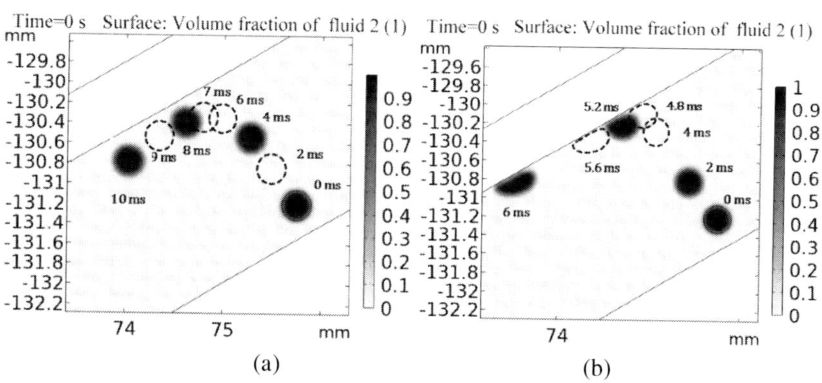

Figure 9. Trajectories of droplet hit grinding wheel surface under different magnetic field intensities (v0=0.3 m/s). (a) No magnetic field, (b) Magnetic field with 4.4×10^4 A/m.

After droplet speed increases, slippage and sucking into film appear during impacting. Since these two phenomena have been analyzed in the previous section, it is not introduced here anymore. When droplet speed exceeds 1 m/s, the liquid develops new behaviors during movement: spreading into film. In this section, the droplet speed is set 5m/s for analysis. Results are shown in Figure 10.

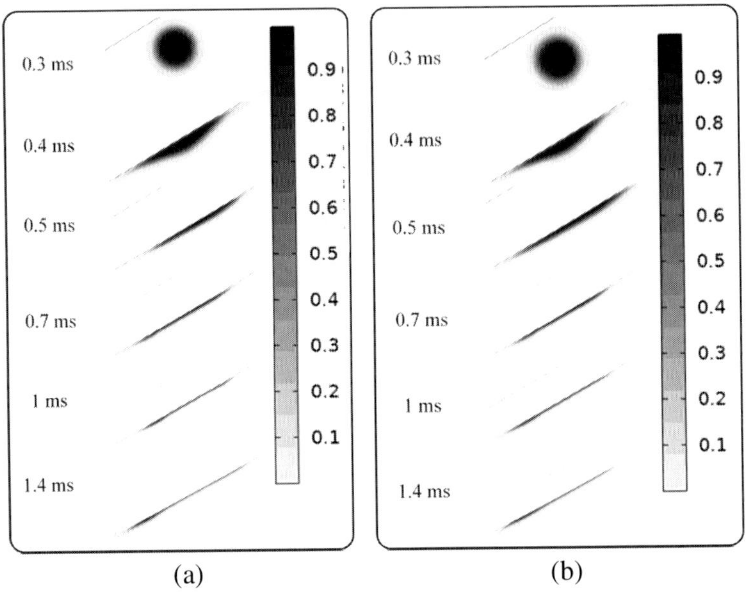

Figure 10. Droplet morphology changes under different magnetic fields (v0=5 m/s). (a) No magnetic field, (b) Magnetic field with 4.4×10^4 A/m.

It can be seen from Figure 10 that droplets impact onto the grinding wheel surface at 0.4 ms, but they spread directly on the grinding wheel surface directly, which is different from previous impacting behaviors. Moreover, there's no air layer between the grinding wheel surface and droplets. In other words, there's no slippage of droplets. Furthermore, spreading length of droplets will be extended as time goes on. In Figure 10(b), the spreading behaviors and spreading length of droplets have no obvious changes after magnetic fields are applied onto the grinding wheel surface, and there's hardly any difference with those under the absence of magnetic fields. This reflects that magnetic field has mild influences on film formation of high-speed droplets. It is important to note that the observed extending behaviors of droplets are different from phenomenon in traditional impacting process of droplets onto the wall surface. Since the high-speed rotating grinding wheel carries droplets into the grinding zone, only the spreading process of the droplets is observed. However, shrinkage and rebounding may not occur in a short time. Hence, they are not very important to working conditions in this study.

Furthermore, the variation laws of β with time under different droplet speeds are measured for quantitative characterization of spreading degree of droplets (Figure 11). Specifically, β is the ratio of diameter after droplet spreading (d_1) and the initial diameter of droplets (d).

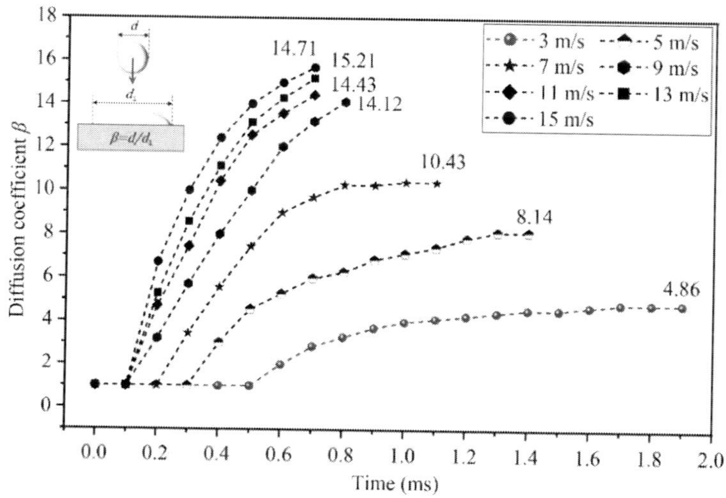

Figure 11. Diffusion coefficient at different droplet speeds.

It can be seen from Figure 11 that the variation laws of β under high speed and low speed are different and it values 1 when droplets begin to move and contact with the grinding wheel surface. When the droplet speed is relatively low (3 m/s, 5 m/s and 7 m/s), β increases firstly and then becomes stable after impacting onto the grinding wheel surface, but it increases quickly under high speeds (9 m/s, 11 m/s, 13 m/s and 15 m/s). Since the jet region in simulation hasn't extended to the entrance of grinding zone, the curve lasts for a very short period and droplets have moved out of boundaries. Additionally, β increases with the increase of droplet speed and reaches the maximum (14.71) at 15 m/s. It is 4.86 at 3 m/s, which is decreased by 66.9%.

It is important to note that β should falls within an appropriate interval. If β is too high, the lubricating oil film is too thin to provide satisfying lubrication effect. If β is too low, the oil film is too thick and it is easy to stack at entrance of the microscale grinding zone when passing through, thus decreasing utilization of lubricant.

6.4.2. Film Formation Behavior of Microdroplet Spreading

Based on the above analysis, we found the fourth type of behaviors of droplets when they impact onto the grinding wheel surface: spreading to film. To further study dynamic behaviors of droplet spreading, the cloud diagram of droplet pressure and stress variation with time when droplets spread into film are analyzed (Figure 12).

The pressure cloud charts from the beginning of contacting with grinding wheel surface to spreading into film of droplets (from 0.5 ms to 0.9 ms) are shown from Figure 12(a)-(c). When droplets begin to move, they bear air resistance and inertia force except for gravity (Figure 12(f)). When droplets contact the grinding wheel surface, the high-speed air layer on the grinding wheel surface produces a rebounding pressure (Figure 12(a)). Furthermore, the kinetic energy is relatively high at 0.6 ms due to the high droplet speed. Droplets overcome the rebounding pressure and spread on the grinding wheel surface and contact it directly (Figure 12(b)). Slippage and sucking into film of droplets are different from spreading into film. The droplet speed during slippage and sucking into film is low and kinetic energy is too small to overcome rebounding pressure of the air layer. Hence, an air cushion layer is formed between droplets and grinding wheel surface and it exists all the time. Besides, droplets extend horizontally toward two sides since they have some kinetic energy and are influenced by rebounding pressure of grinding wheel surface. At this moment, droplets bear magnetic field force along normal direction of the grinding wheel, viscous force and the opposite centrifugal force. there's viscous frictional force along the tangential direction of the grinding wheel. At 0.9 ms, pressure on droplet surface changes after droplets are extended to some extent and form film. In other words, the region occupied by droplets becomes negative pressure. It can be seen from pressure isogram at 1.1 ms that the maximum negative pressure reaches about 383.3 Pa (Figure 12(c) and (d)). The surrounding airflow is the positive pressure region. Under such circumstances, droplets are squeezed and spread on the grinding wheel surface continuously. This also can be seen in Figure 11, where the spreading coefficient increases gradually.

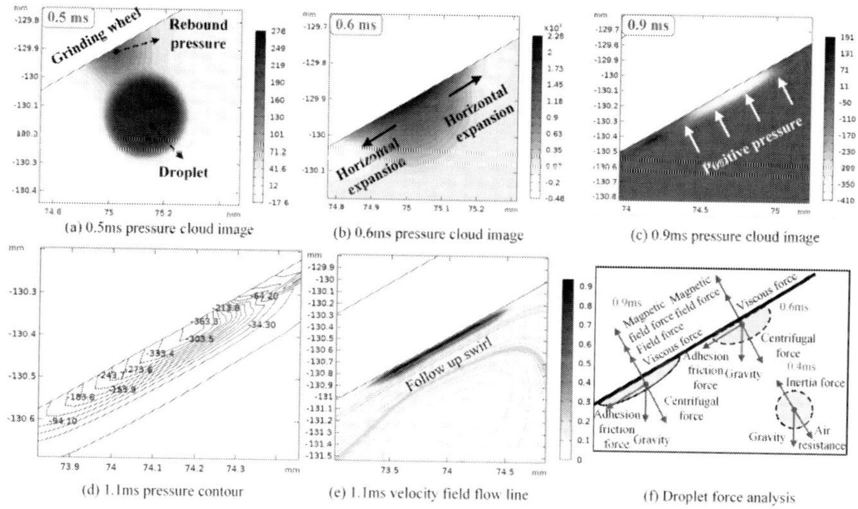

Figure 12. Droplet force and spatial pressure distribution during film spreading. (a) 0.5ms pressure cloud image, (b) 0.6ms pressure cloud image, (c) 0.9ms pressure cloud image, (d) 1.1ms pressure contour, (e) 1.1ms velocity field flow line, (f) drop force analysis.

The above phenomenon is caused by movement behaviors of droplets (Figure 12(e)). Obviously, droplets with great kinetic energy move with the grinding wheel after contact with the grinding wheel surface directly. They can be viewed as a part of the grinding wheel. On the one hand, movement of droplets may disturb the surrounding airflow field and it moves accordingly. On the other hand, the movement speed of droplets and grinding wheel is very high, which is far higher than speed of surrounding airflow field. Therefore, the region occupied by droplets forms negative pressure and droplets are squeezed onto the grinding wheel surface. At this moment, the normal direction of grinding wheel is squeezed by airflow field force, that is, airflow pressure against the liquid film. The viscous frictional force is the power source for droplet movement.

6.5. The Microdroplet Behavior Map Under the Combined Action of Magnetic Field Intensity and Droplet Velocity

In the above studies, we have analyzed adsorption behaviors when droplets impact onto the grinding wheel surface under different magnetic field strength, different initial speeds of droplets and their collaborative action. In this process, four types of behaviors are found, namely, falling of droplets, slippage of droplets, sucking into film and spreading into film. To further characterize influencing laws of magnetic field strength and droplet speed on droplet behaviors, a series of simulation studies are carried out. As shown in Figure 13, different numerical values of droplet speed and magnetic field strength of grinding wheel surface are chosen. For conversion among four types of behaviors, three critical speeds of droplets are defined: falling-to-slippage ($v_{f\to sl}$), slippage-to-sucking into film ($v_{sl\to su}$) and sucking into film-to spreading into film ($v_{su\to sp}$).

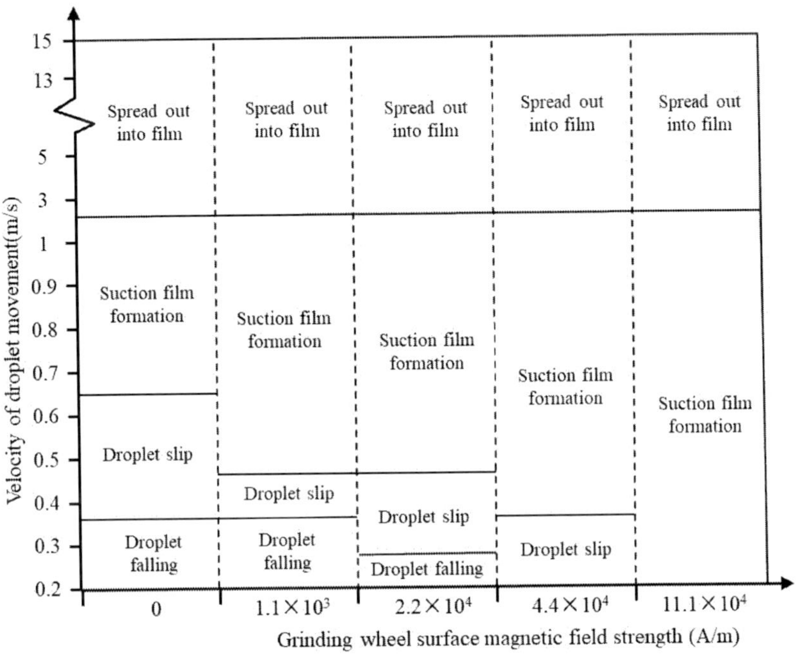

Figure 13. Droplet collision-adsorption behavior spectrum.

When magnetic field strength of grinding wheel surface is 0 A/m, that is, no magnetic field is applied, four types of behaviors are developed with the increase of droplet speeds, including falling of droplets, slippage of droplets, sucking into film and spreading into film. The $v_{f\to sl}(0)$, $v_{sl\to su}(0)$ and $v_{su\to sp}(0)$ fall within 0.3~0.4 m/s, 0.6~0.7 m/s and 1~3 m/s, respectively. Additionally, $v_{su\to sp}(0)$, (1.1×10^3 A/m), (2.2×10^4 A/m), (4.4×10^4 A/m) and (11.1×10^4 A/m) all fall within 1~3 m/s. This indicates that magnetic field plays a small role in the conversion from sucking into film to spreading into film of droplets.

After the magnetic field is applied, $v_{sl \to su}$(1.1×10^3 A/m) and (2.2×10^4 A/m) are changed and fall within 0.4~0.5 m/s. When the magnetic field strength increases to 4.4×10^4 A/m, $v_{sl \to su}$(4.4×10^4 A/m) decreases to 0.3~0.4 m/s. The droplet slippage disappears when the magnetic field strength increases to 11.1×10^4 A/m. This indicates that magnetic field has significant effects in reducing slippage behaviors of droplets. Besides, $v_{f \to sl}$(2.2×10^4 A/m) also decreases from 0.3~0.4 m/s to 0.2~0.3 m/s as a response to magnetic field and then disappears. Similarly, the application of magnetic field improves the falling behaviors of droplets.

In fact, microdroplet speed distributes within a certain range during jetting and it is usually between 0 and 20 m/s. Moreover, it decreases to some extent under the action of airflow field during movement. The low-speed droplets will be thrown out of the grinding zone by the airflow and fly in air. Hence, the existence of magnetic field allows fall, slippage and transformation into film of low-speed droplets, thus increasing utilization of lubricant significantly.

Conclusion

In this chapter, a numerical simulation of dynamic process when droplets impact onto the grinding wheel surface is carried out by combining with practical situation of droplet jets during magnetic field assisted NMQL process. Based on hypotheses that fluid is incompressible, viscosity and surface tension are constant and there's no phase changes in the impacting process, a dynamic model for magnetofluid microdroplets to impact onto the grinding wheel surface is built by using two-phase flow continuous equation and governing equations of mass conservation, momentum conservation and magnetic field. The influencing laws and adsorption mechanism of magnetic field on droplet impacting behaviors onto the grinding wheel surface are analyzed by tracking their motion tracks and behavioral states.

i. The diffusion behaviors and states between gas and liquid phases are tracked by level set method. Influencing laws of magnetic field on droplet impacting behaviors onto the grinding wheel surface are analyzed. Results show that without application of magnetic field, droplets slip with the grinding wheel after arriving at the grinding wheel surface. After magnetic fields are introduced in, droplets are pumped into film after impacting onto the grinding wheel surface, forming uneven oil film. With the increase of magnetic field strength, the time for droplets to arrive at the grinding wheel surface is shortened, and sucking into film is more obvious.

ii. It is found from analysis of droplet movement speed that with the increase of magnetic field strength, the balance stage of droplets decreases significantly, and the speed curve presents a rising trend. The slope of the speed curve changes from negative to positive. This reveals that the attraction force of magnetic field to droplets strengthens with the increase of magnetic field strength.

iii. According to simulation calculation under different jetting speeds, it is found that with the increase of speed, influences of magnetic fields on impacting tracks and forms of droplets decrease gradually. At low speeds, the application of magnetic fields changes deviation of motion tracks of droplets and eliminates falling of droplets. When speed reaches 5 m/s, droplets spread into film directly after arriving at the grinding wheel surface. After magnetic fields are applied, they will influence motion behaviors and spreading of droplets. This demonstrates the tiny effects of magnetic field on film formation of high-speed droplets.

iv. The value of β has important influences on friction reduction and wear resistance of magnetic lubricant. It varies with droplet speed. Through a comparative analysis of β under different droplet speeds, it found a positive correlation between them. Moreover, β increases quickly under a high speed, indicating that droplets impact onto the grinding wheel surface quickly and spread into film.

v. The impacting behaviors of droplets are disclosed with considerations to the collaborative effects of speed and magnetic field, strength. Results show that magnetic field can improve falling and slippage behaviors of low-speed droplets obviously and change these behaviors to sucking into film or spreading into film, thus increasing utilization of the lubricant.

References

[1] Williams JA, Tabor D. (1977). The role of lubricants in machining. *Wear*, *43*(3), 275-292.

[2] Godlevski VA, Volkov AV, Latyshev VN, et al. (1998). A description of the lubricating action of the tribo-active components of cutting fluids. *Lubrication Science*, *11*(1), 51-62.

[3] Li C, Zhang X, Zhang Q, et al. (2014). Modeling and simulation of useful fluid flow rate in grinding. *The International Journal of Advanced Manufacturing Technology*, *75*, 1587-1604.

[4] Han ZL, Li CH. (2011). Kinematics and finite element modeling and simulation of grinding processing. *Precise Manufacturing & Automation*, (4), 4-9.

[5] Zhang L, Guo XH, Zhang KD, et al. (2020). Enhancing cutting performance of uncoated cemented carbide tools by joint-use of magnetic nanofluids and micro-texture under magnetic field. *J. Mater. Process. Technol.*, *284*, 10. 10.1016/j.jmatprotec.2020.116764.

[6] Lv T, Xu XF, Weng HZ, et al. (2022). A study on lubrication and cooling performance and machining characteristics of magnetic field-assisted minimum quantity lubrication using Fe_3/O_4 nanofluid as cutting fluid. *Int. J. Adv. Manuf. Technol.*, *123*(11-12), 3857-3869. 10.1007/s00170-022-10500-z.

[7] Zhang KD, Li ZH, Wang SS, et al. (2023). Study on the cooling and lubrication mechanism of magnetic field-assisted Fe_3O_4@CNTs nanofluid in micro-textured tool cutting. *J. Manuf. Process.*, *85*, 556-568. 10.1016/j.jmapro.2022.11.081.

[8] Rosensweig RE. (1985). Ferrohydrodynamics. *The Physics of Fluids*, *7*(12), 344.

[9] Ma G, Wu C, Zhou P. (2007). Wall slip and hydrodynamics of two-dimensional journal bearing. *Tribology International*, *40*(7), 1056-1066.

[10] Choi CH, Kim CJ. (2006). Large slip of aqueous liquid flow over a nanoengineered superhydrophobic surface. *Physical review letters*, *96*(6), 066001.

[11] Sun ZC, Dong H, Zhao SW, et al. (2020). Dynamic behavior of droplet impacting on moving surface. *Journal of Engineering Thermophysics*, *41*(5), 1215-1218.

Chapter 7

The Antifriction Mechanism and Tribological Test of Magnetic Nanolubricants

Abstract

The MTN method can improve the infiltration performance of the lubricant in the grinding zone. However, due to the randomness of the grinding zone, the antifriction and antiwear properties of magnetic lubricants cannot be fully studied. Based on this, this chapter carried out innovative tribological test under simulated grinding conditions, and the tribological characteristics of magnetic nanolubricant were explored and studied. Firstly, a new type of magnetic nanolubricant Fe_3O_4/graphene was prepared. Secondly, the confirmatory test of magnetic nanolubricant was carried out. The friction coefficient curve of mixed nanolubricant is remarkably stable. The width of the wear marks decreased significantly, and the adhesion of the surface was improved. Finally, the ratio optimization of Fe_3O_4/graphene with excellent performance and strong economy was studied. Minimum mean coefficient of friction (0.32), wear area (6146μm^2) and surface roughness (1.64μm) were obtained at Fe_3O_4/graphene (2:1). The surface defects of scratches are obviously improved.

Keywords: antifriction, antiwear, tribological test, Fe_3O_4/graphene nanolubricant, coefficient of friction

7.1. Introduction

In NMQL, nanomaterial is a decisive factor of cooling and lubrication performances [1-4]. Existing nanomaterials cannot achieve high surface integrity in grinding of difficult-to-machine material Titanium (Ti) alloy. Nevertheless, there's strong randomness of workpiece surface production in grinding experiments and it is very difficult to realize full characterization of tribological properties of the interface through grinding experiment. For magnetic nanolubricant, Shahrivar et al. [5] carried out a friction-wear test study of three crickets and found that the diameter of magnetofluid scratches was 428μm before applying the magnetic field, but it decreased by 24.8% to 322μm after the application of magnetic field. Besides, it was observed from SEM images of grinding surface that the surface quality was improved after application of magnetic fields. The same conclusion was drawn in the friction-wear test of four balls [6].

Therefore, this chapter simulates the characteristics of grinding wheel-workpiece interface and carries out ball-disc friction-wear tests, aiming to disclose the friction and wear resistance mechanism of nanolubricant and provide theoretical references for NMQL processing of Ti alloy. In this chapter, the friction and wear resistance performances of Fe_3O_4@GR prepared by chemical modification method and Fe_3O_4/GR prepared by physical mixing method are verified through friction-wear tests. Furthermore, the mixing ratio of Fe_3O_4/GR with excellent performances and strong economy is optimized. The performance characterization parameters, such as coefficient of friction (CoF), wearing zone and workpiece surface roughness, are discussed and analyzed. Additionally, the friction and wearing performances of magnetic nanoparticles under the action of magnetic fields are observed.

7.2. A Novel Tribological Test: Simulating Grinding Conditions

7.2.1. An Innovative Experimental Method

In the grinding process, scholars usually characterize tribological properties of the grinding zone in the following two ways:

i. Force ratio (F_r): The CoF cannot be acquired directly in experiments due to the great randomness of grains in the grinding zone. Scholars usually replace CoF (μ) by force ratio (F_r) in analysis. However, there's great bias ($F_r > \mu$) due to material deformation, removal and adhesion in the grinding process [7, 8]. Liu et al. [9] tested in a Ti alloy grinding test that the force ratio ranges within 0.45~0.56. Xu et al. [10] carried out a friction-wear test of Ti alloy and found that the CoF could be as low as 0.38, which was significantly different from grinding force ratio. As a result, the grinding force model cannot improve the prediction accuracy since μ cannot be gained.
ii. Workpiece surface integrity: the final formation of workpiece surface in the grinding test often comes from successive cutting of multiple

grains due to the strong randomness of workpiece surface production. It cannot fully reflect the tribological characteristics of the grain-workpiece interface [11, 12]. Hence, it is difficult to achieve full characterization of tribological properties of the interface through grinding test.

An innovative friction test is proposed to study tribological properties of nanolubricant in the grinding zone. The principle of experimental apparatus is shown in Figure 1. The ball-disc friction-wear test method was applied and the grinding wheel was simulated by using balls containing SiC grains (SiC particle size is about 200 μm and it is equal to an 80# grain). The Ti-6Al-4V material is used as the disc. In the test, loads corresponding to the grinding process are employed. On this basis, the working conditions for grinding are simulated to discuss the tribological properties of interface.

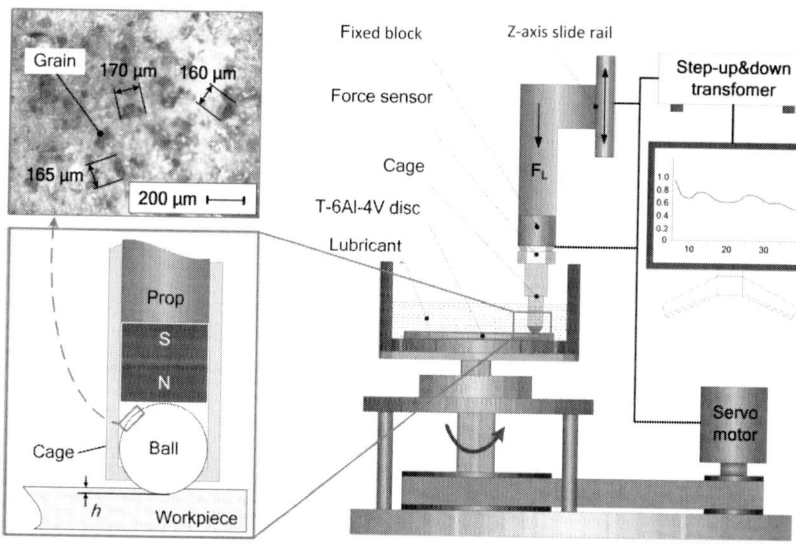

Figure 1. Experimental method of rotary ball-and-disc friction-wear test assisted by magnets.

To simulate high-pressure boundary conditions during grinding, it is particularly important to set load (F_L) in the friction-wear test. The solving process of F_L is shown in Figure 2.

According to the study of Zhang et al. [13], when grinding wheel parameters (e.g., number of grains and structure No.) are fixed, the grain probability distribution and height of grinding wheel surface can be calculated as a matrix:

$$G(z) = \frac{G(d)}{2} + G(z_g) = \begin{bmatrix} z_{11} & z_{12} & \cdots & z_{1N_y} \\ z_{21} & z_{22} & \cdots & z_{2N_y} \\ \vdots & \vdots & \ddots & \vdots \\ z_{N_x1} & z_{N_x2} & \cdots & z_{N_xN_y} \end{bmatrix} \quad (1)$$

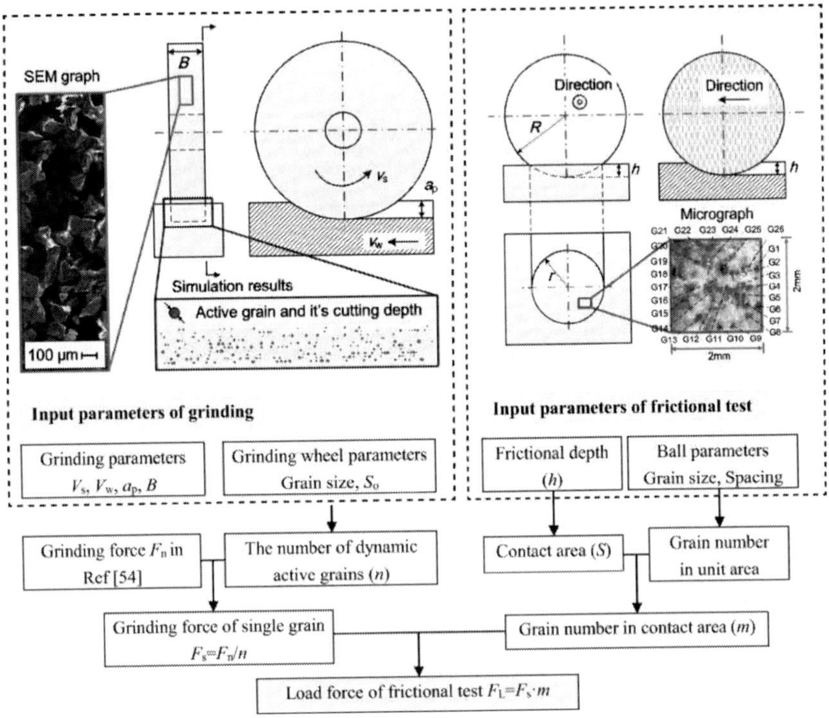

Figure 2. Setting method of loading force F_L.

In the grinding zone, the material cutting of the next grain is based on the cutting behaviors of the previous grain. Therefore, the later grain may not cut materials when the distance between two effective cutting grains (λ) is short. The grain can be dynamic effective grain to cut materials as long as λ reaches a certain value. The calculation formula of cutting depth of the dynamic effective grain is shown in Eq. (1). Thickness of the maximum undeformed debris of the nth grain is:

$$a_{g\,max(n)} = 2 \cdot \lambda_{(n\sim n-1)} \cdot \frac{V_w}{V_s} \cdot \sqrt{\frac{a_p}{D_s}} + \left(a_{p(n)} - a_{p(n-1)}\right) \tag{2}$$

Where $a_{gmax(n)}$ is the cutting depth of the nth dynamic effective grain. $\lambda_{(n\sim n-1)}$ denotes the distance between the n-1th and the nth dynamic effective grains. $a_{p(n)}$ and $a_{p(n-1)}$ are protrusion heights of the nth and the n-1th dynamic effective grains, respectively. D_s is the diameter of grinding wheel.

Therefore, the number of dynamic effective grains in the grinding zone can be calculated from Eq. (1) and Eq. (2) when the grinding parameters (V_s, V_w and a_p) are fixed. The average normal acting force of a single grain in the grinding zone can be calculated from the experimental results of grinding force: $F_s = F_n/n$.

The ball-disc contact area under some contact depth is further calculated (Figure 2). The contact area is:

$$S = \pi\left[R^2 - \left(R - a_p\right)^2\right] \tag{3}$$

According to grain size and intervals on the chosen spherical surface, the number of SiC grains in the area of S can be calculated, which is known as the number of grains (m) in the ball-disc contact zone in the friction-wear test. Hence, the load in experiment is set as:

$$F_L = \frac{F_s}{n} \tag{4}$$

For example, a grinding experiment of Ti-6Al-4V workpiece was carried out by using the SiC grinding wheel (particle size: 80 meshes; model: GC80K12V; grinding wheel size: 300×20×76.2 mm) under NMQL conditions. When the grinding parameters are V_s=24 m/s, V_w=4 m/min and a_p=10 μm, the specific normal force is F_n'=3.06 N/mm. Hence, the normal force can be calculated as $F_n = F_n' \cdot b = 3.06 \times 20 = 61.2$ N. According to Figure 2, it gets n=38 in the simulation. Hence, F_s=1.61 N. It is observed from SEM images of SiC balls that there are 13 grains per square millimeters. According to the experiment, h=20~80 μm, it can be calculated from Eq.(3) that S=0.57~2.26 mm^2. Furthermore, it can be gained from Eq.(4) that m=7~27 and F_L=11.87~43.47 N. Hence, it sets F_L=25 N in this study to simulate grinding conditions.

7.2.2. Experimental Equipment

As shown in Figure 3, the friction-west test is carried out with UMT-3 Instruments. The surface microstructure of scratches on the disc was observed and analyzed by using S-3400N scan electron microscope. The surface microstructural properties were observed, which further explained the macroscopic CoF and characterized tribological performances of different nanolubricant together. The scratch profile was observed and analyzed by using a confocal microscopy. Meanwhile, width and depth of scratches and grinding area were observed. The magnetic field strength of cylindrical permanent magnet in the friction-wear test was measured to simulate magnetic field strength surrounding the grinding wheel during grinding. In the test, the cylindrical permanent magnet was above the ball pair with a diameter of 9.2 mm. The magnetic field strength of the permanent magnet under this distance (bottom of the SiC ball) was measured about 150 Gs by using a Gauss meter.

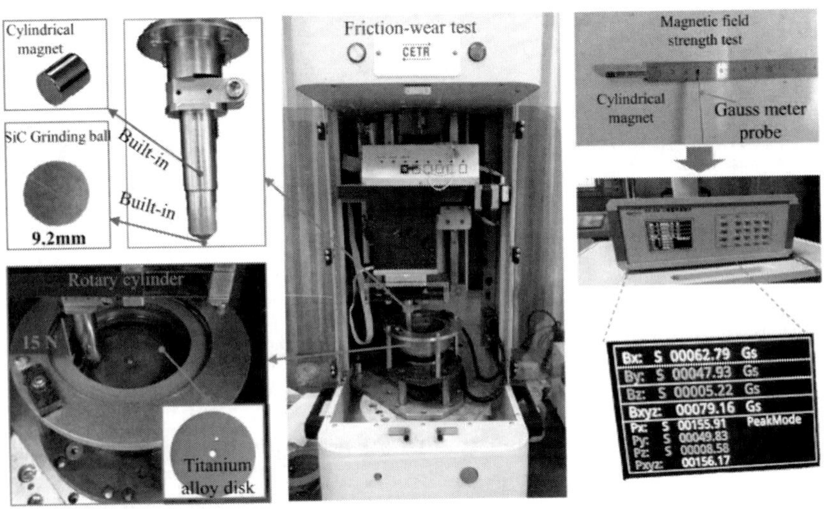

Figure 3. Equipment of rotary ball-and-disc friction-wear test assisted by magnetics.

7.2.3. Experimental Scheme

Under different nanolubricant conditions, the friction-wear test is carried out under ball-disc rotating sliding friction (UMT-3 Instruments). The normal force and tangential force in the two-pair movement process were monitored in real-time. Meanwhile, the variation curve of F_r with time was produced in the computer. The experimental instrument and principle are shown in Figure 3. To simulate grinding conditions, the friction pair ball in the experiment carries SiC grains. The grinding disc used Ti-6Al-4V material and the permanent magnet was applied as the magnetic field assisted device. The composition and performance parameters of Ti-6Al-4V are listed in Table 1 and Table 2.

Table 1. The elemental composition of Ti-6Al-4V

Material	Ti-6Al-4V
Al	6.18
V	4.19
Fe	0.3
O	0.15
N	0.05
C	0.1
H	0.015
Ti	Balance
Si	0.15

Table 2. Physical properties of Ti-6Al-4V (20°C)

Physical property	Value
Fusing point	1630～1650℃
Density	4.42 g/m^3
Modulus of elasticity	114 GPa
Specific Heat Capacity	526.3 J/(kg·℃)
Thermal conductivity	7.955 W/(m·℃)

The friction-wear test used six types of nanolubricant as well as five types of Fe_3O_4-GR nanolubricant with different mixing ratios. The control variable method was employed and the type of nanolubricant was set as the sole variable, while other conditions were set constant. The overall experimental design and experimental conditions are presented in Table 3 and Table 4. Each group of experiment was repeated for 3 times to assure accuracy of results. After finishing each group of experiment, ultrasonic cleaning of the disc and ball with petroleum ether was conducted for 30 min to eliminate residual debris, lubricant and other impurities on the disc and ball. Later, another 30min ultrasonic cleaning was performed with absolute ethyl alcohol to eliminate residual petroleum ether.

Table 3. Experimental parameters of friction-wear test

Parameter	Value
Test temperature	22±5 °C
Test time	1800 s
Loading force	25 N
Disk speed	70 r/min
SiC ball diameter	9.2 mm

Table 4. Experimental scheme design (Yes-Y, No-N)

NO.	Group	Materials	Base oil	Magnetic field
1-1	Confirmatory experimental	Fe_3O_4 (5 vol%)	Palm oil	No
1-2		Fe_3O_4 (5 vol%)	Palm oil	Yes
1-3		GR (5 vol%)	Palm oil	No
1-4		GR (5 vol%)	Palm oil	Yes
1-5		Fe_3O_4@GR (5 vol%)	Palm oil	Yes
1-6/2-1		Fe_3O_4/GR (mix ratio=1:1, 5 vol%)	Palm oil	Yes
2-2	Experimental on ratio optimization	Fe_3O_4/GR (mix ratio=1:3, 5 vol%)	Palm oil	Yes
2-3		Fe_3O_4/GR (mix ratio=1:2, 5 vol%)	Palm oil	Yes
2-4		Fe_3O_4/GR (mix ratio=2:1, 5 vol%)	Palm oil	Yes
2-5		Fe_3O_4/GR (mix ratio=3:1, 5 vol%)	Palm oil	Yes

7.3. Results of Various Magnetic Nanolubricants

7.3.1. Coefficient of Friction (CoF)

Since the grinding balls used in this study are made of SiC material and it can remove workpiece materials like a grinding wheel, the CoF curve is different from those in traditional friction-wear tests. The friction-wear curve under the use of Fe_3O_4 nanolubricant is shown in Figure 4. The traditional ball-disc friction-wear test used the metallic ball and disc. The ball and disc have a low roughness and there's frictional wear in the ball-disc pair during relative movement. In this study, the spherical surface carried SiC grains and the load setting has reached the pressure value of the grinding zone. Hence, there are two stages in the friction-wear test. (1) Material remove stage: the ball and disc interact and materials are cut. (2) Stable wearing stage: after surface materials are removed, the friction and wearing enter into the stable sliding stage. In the friction and wear studies, the frictional force at material remove contains the material removal force. Hence, the CoF is expressed as:

$$\mu = \frac{F_c + f}{F_L} \tag{5}$$

where F_c is the material removal force and μ is the CoF.

In the stable wear stage, the grain-disc contact stage varies with lubrication conditions. This is why the CoF curve is different from those in traditional friction-wear tests and presents regular fluctuations. These will be explained according to Figure 5.

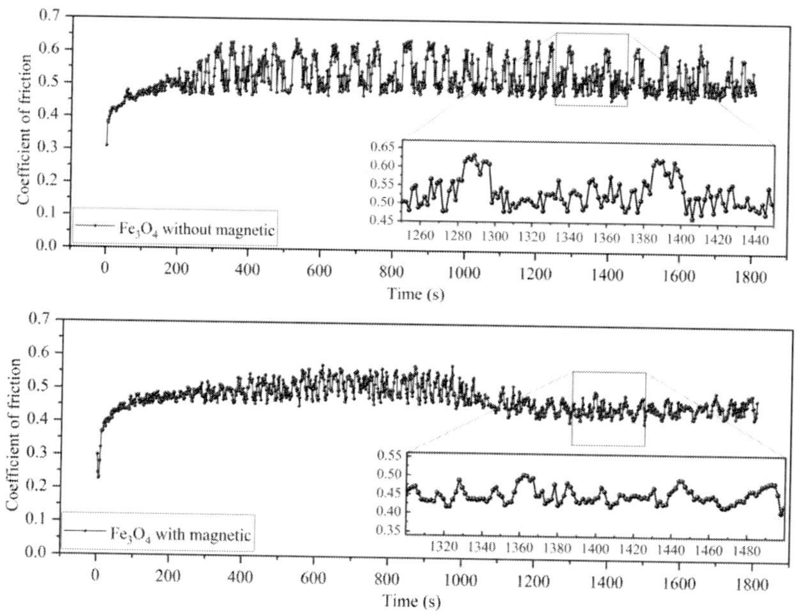

Figure 4. CoF friction curve of confirmatory experiment results (Fe_3O_4).

(1) Stable slip state (① in Figure 5): Given a good lubrication state on the grain-disc interface, there's sufficient lubricant in the interface and it is close to lubrication state of oil film: State ❶ in Figure 5. Under this circumstance, grains move slowly and stably under the support of the oil film and the material removal volume from the disc surface in unit time is small. In other words, the material removal force (F_c) is small and stable. Hence, the CoF curve is relatively stable, with mild fluctuation.

(2) Material adhesion peeling (② in Figure 5): When the lubrication condition of the grain-disc interface deteriorates, the interface is apt to boundary the lubrication state since there's little lubricant or nano materials

participating in friction and wear resistance. Microcosmically, there are solid-solid contact points on grains and disc, that is, State ❷ in Figure 5. When grains just scratch the disk, the scratches are compressed into cracks under the direct contact with grains. As grains slide the surface cracks for several times, the cracks propagate gradually and finally materials are peeled off the disc by frictional force of grains, thus resulting the adhesion peeling. Moreover, the peeled materials might be compressed onto the scratched surface by the subsequent grains, forming adhesion points. Under this circumstance, the material removal volume is low, and the produced material removal force is small. However, the CoF curve will develop some mild fluctuations (Figure 4(a)).

(3) Micro-peak breakage (③ in Figure 5): Under the above condition, materials are peeled off and valleys are formed. Meanwhile, micropeaks are formed. When grains move to the micropeaks on the disc surface, they tend to remove micropeak materials, which will produce cracks. After grains move to the micropeaks for several times, materials at the micropeaks will be finally removed. The material removal volume is relatively high. At this moment, it requires a large force to remove materials and the material removal force increases. The CoF curve fluctuates significantly. Since micropeak removal occurs after the formation of valleys, the fluctuation caused by micropeak removal is periodic to some extent. It can be seen from Figure 4(a) that the CoF curve fluctuates significantly after experiencing mild fluctuations.

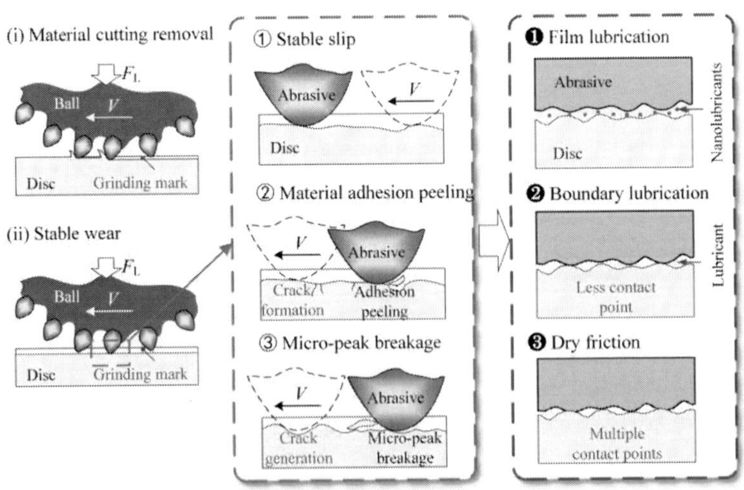

Figure 5. Material removal schematic in friction-wear test.

It is important to note that when the lubrication condition is further deteriorated and even gets close to dry friction (State ❸ in Figure 5), the number of solid-solid contact points between grains and the disc further increases, which will accelerate material adhesion peeling and micropeak removal process. Moreover, the material removal volume increases and the needed material removal force increases accordingly. Hence, the CoF curve fluctuates more obviously.

It can be seen from Figure 4(a) that before the application of magnetic field, the CoF curve under the use of Fe_3O_4 nanolubricant fluctuates the most (0.45~0.65). This is because the lubrication performances of spherical Fe_3O_4 nanolubricant between the friction pair are not stable and even may cause dry friction. The disc material removal volume by grains increases, thus causing significant fluctuation of the CoF curve. After the application of magnetic field, the CoF curve fluctuates less within the range of about 0.4~0.5 (Figure 4(b)), which is attributed to the moderate lubrication state.

CoF curves under the use of graphene, Fe_3O_4@GR and Fe_3O_4/GR mixed (mix ratio=1:1, 5 vol%) nanolubricant are shown in Figure 6.

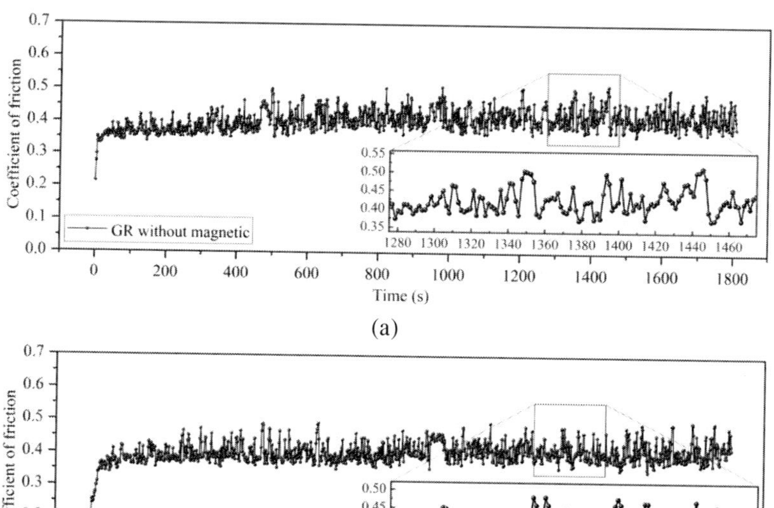

(a)

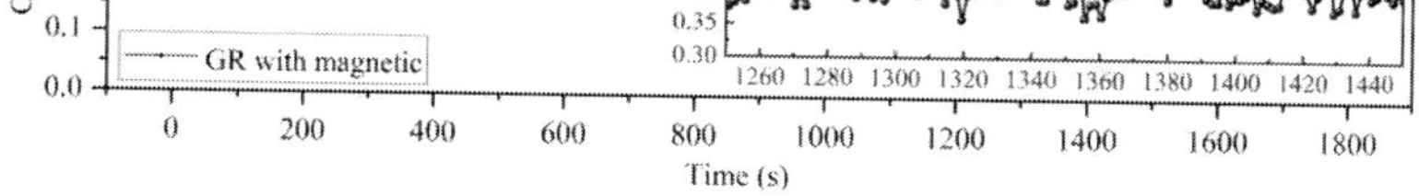

(b)

Figure 6. (Continued).

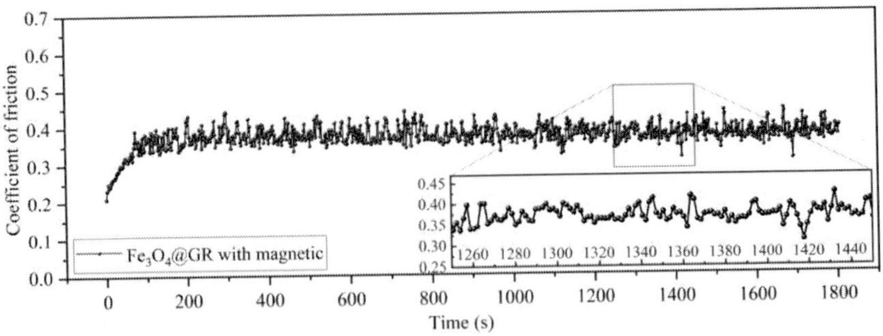

(c)

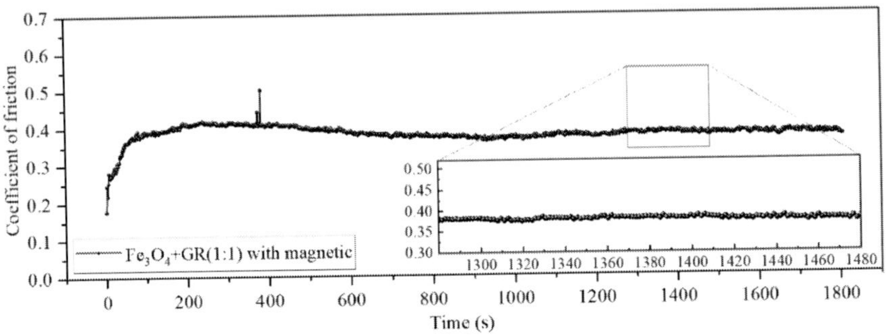

(d)

Figure 6. CoF curve of confirmatory experiment results. (a) GR without magnetic, (b) GR with magnetic, (c) Fe_3O_4@GR with magnetic, (d) Fe_3O_4+GR(1:1) with magnetic.

Under the use of GR nanolubricant, the CoF curve is not influenced significantly after the application of magnetic fields (Figure 6(a) and (b)). This is because GR nanolubricant is not magnetic. Under the interaction of friction pair, the lamellar GR forms the lubrication film on the disc surface along the shearing surface, decreasing the overall CoF. However, there are no regular forms and poses of GR between the friction pair, which will cause unstable lubrication states sometimes. Hence, the CoF curve fluctuates to some extents.

Furthermore, CoF curves of Fe_3O_4@GR and Fe_3O_4/GR (mix ratio=1:1, 5 vol%) nanolubricant under the application of magnetic fields were compared (Figure 6(c) and (d)). The fluctuation ranges of the curves decrease to some extent, indicating the good lubrication. This is because in Fe_3O_4@GR nanolubricant, Fe_3O_4 mixed with GR molecular surface is in regular arrangement under the influence of magnetic field, thus driving regular distribution of graphene. However, GR may distribute perpendicular to the disc surface, which cannot keep good and stable lubrications states well. The

CoF curve of Fe_3O_4/GR mixed nanolubricant hardly fluctuates. On one hand, Fe_3O_4 content in the Fe_3O_4/GR nanolubricant is higher than that in Fe_3O_4@GR. Spherical Fe_3O_4 can fill in spaces among grains under the action of magnetic fields, providing a relatively smooth surface for lamellar graphene. Hence, GR is easier to enter into the grain-workpiece interface, improving the infiltration capacity. On the other hand, Fe_3O_4 presents vertical linear distribution in the magnetic field. In this case, GR moves among Fe_3O_4 in regular arrangement, with stable form and pose as well as regular arrangement. Hence, the CoF curve hardly fluctuates.

A stable stage of the CoF curve was chosen and the mean CoF in this stage was calculated. Variations of the mean CoF under the different lubrication conditions were analyzed. Box plots were employed to reflect outliers and dispersion degree of CoF (Figure 7).

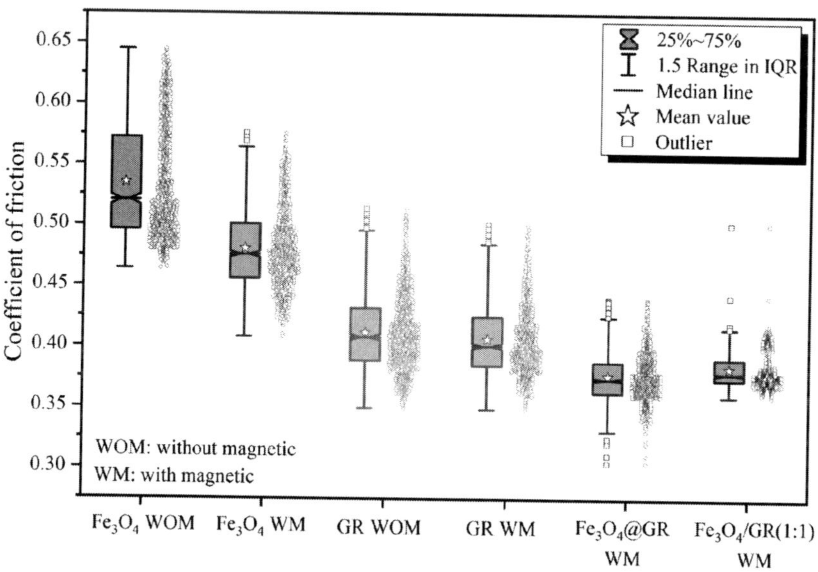

Figure 7. Box diagram of CoF in confirmatory experiment.

Obviously, the mean CoF of Fe_3O_4 nanolubricant before the application of magnetic field is the highest (0.53), but it decreases greatly by 11.4% after the application of magnetic field. Since the lamellar GR has outstanding friction and wear resistance, it still gets a relatively low CoF (0.41), which is 13.1% lower than that of Fe_3O_4 nanolubricant under the application of magnetic field. Lower CoF s are achieved under Fe_3O_4@GR and Fe_3O_4/GR (mix ratio=1:1, 5 vol%) nanolubricant, which value 0.36 and 0.37, respectively. It can be seen from the box height and curve dispersion that the CoF under the use of Fe_3O_4 nanolubricant without magnetic field is the highest. Differently, CoF s are relatively concentrated under the use of Fe_3O_4@GR and Fe_3O_4/GR (mix ratio=1:1, 5 vol%) nanolubricant. This agrees with the CoF curve.

7.3.2. Wear Contour Curve and Wear Area

Moreover, the profile curves of scratch and wearing areas under different lubrication conditions were analyzed. The wearing area solving method is shown in Figure 8. A 2D curve is chosen from the 3D map of scratch. The profile curve of scratch was plotted and the wearing area as well as disc base were marked. The wearing area (WA) was calculated to characterize wearing rate of the disc.

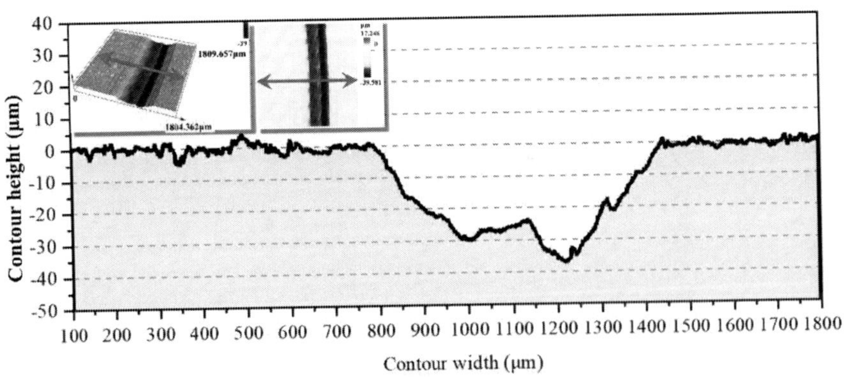

Figure 8. Typical wear curve and wear area.

In the scratch curve, wearing area represents lubrication condition. Based on above analysis, the material removal volume is small and disc surface wearing is mild under good lubrication conditions, showing a small wearing area. When lubrication condition is deteriorated, the materials are difficult to remove, thus increasing wearing area. Wearing area of scratches under different types of lubricants with and without magnetic field is shown in Figure 9.

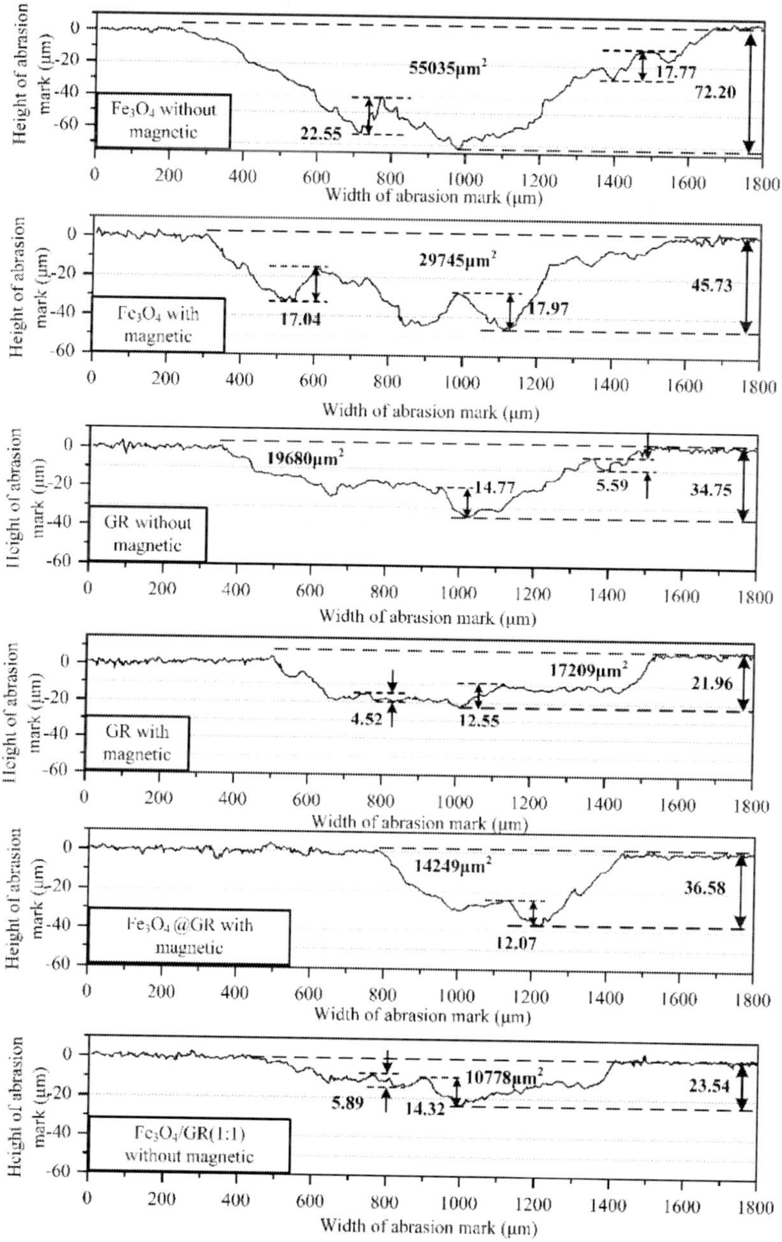

Figure 9. Wear curve under different lubricant conditions. (a) Fe_3O_4 without magnetic, (b) Fe_3O_4 with magnetic, (c) GR without magnetic, (d) GR with magnetic, (e) Fe_3O_4@GR with magnetic, (f) Fe_3O_4/GR(1:1) with magnetic.

It can be seen from Figure 9 that wearing areas under different nanolubricant conditions are significantly different. Absolutely, the wearing area and grinding depth under Fe_3O_4 nanolubricant without magnetic field are the highest, reaching 55,035 μm^2 and 72.2 μm, respectively (Figure 9(a)). After application of magnetic fields, the wearing area and grinding depth decrease by 45.9% and 36.7% compared to those without application of magnetic field g.4-9(b)). In Figure 9(c) and (d), magnetic field exert slight effects under the use of GR nanolubricant, manifested by small differences in wearing area and grinding depth. Wearing area under the use of Fe_3O_4@GR nanolubricant is relatively low and the wearing area under the use of Fe_3O_4/GR nanolubricant is the smallest (10,778 μm^2), which is 80.4% lower than the maximum (Figure 9(e) and (f)). The conclusions and laws in this section are consistent with CoF, and the mechanisms are similar.

7.3.3. The Surface Morphology of Wear Marks

The scratch morphology and surface roughness (*Sa*) under different lubrication conditions are shown in Figure 10.

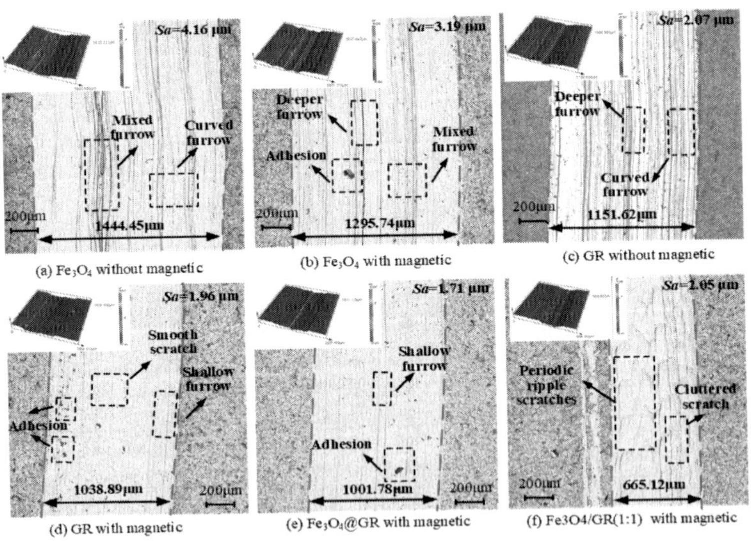

Figure 10. Scratch topography and surface roughness *Sa* in confirmatory experiment. (a) Fe_3O_4 without magnetic, (b) Fe_3O_4 with magnetic, (c) GR without magnetic, (d) GR with magnetic, (e) Fe_3O_4@GR with magnetic, (f) Fe_3O_4/GR(1:1) with magnetic.

It can be seen from Figure 10(a) and (b) that without the application of magnetic field, there are bending and disordered furrows on the scratch surface, indicating that the lubrication conditions of friction interface are unstable. After the application of magnetic fields, although there are a lot of scratches and furrows, they become relatively straight. Compared to GR nanolubricant condition with magnetic field in Figure 10(c) and (d), furrows on surface are slightly straight, but there are adhesions. Combining with CoF and wearing area, magnetic field has no influences on GR nanolubricant. For Fe_3O_4/GR nanolubricant condition in Figure 10(e) and (f), the scratch width decreases significantly and the surface adhesion is improved. However, periodic ripple scratches are developed compared to the surface under Fe_3O_4@GR nanolubricant. There are some surface defects under different lubrication conditions.

Similarly, the maximum *Sa* (4.16 μm) is achieved under Fe_3O_4 nanolubricant without magnetic field and the minimum (1.71 μm) is achieved under Fe_3O_4@GR nanolubricant, which is 58.9% lower than the maximum. The *Sa* under Fe_3O_4/GR nanolubricant condition is also relatively low.

7.3.4. Antifriction and Antiwear Mechanisms

The friction and wear resistance mechanism of different types of nanolubricant under the application of magnetic field is shown in Figure 11. As shown in Figure 11(a), spherical Fe_3O_4 nanoparticles serve for rolling antifriction in the uneven grinding zone. As time goes on, although the disordered rolling of nanoparticles provides some lubrication effects, it cannot form stable lubrication oil film. It can be known from Figure 11(e) that after magnetic fields perpendicular to the interface are applied, Fe_3O_4 nanoparticles are adsorbed onto grain holes to form stable oil film since the magnetic field strength decreases from upper to bottom. In Figure 11(f), magnetic nanoparticles will form chain structures along the magnetic field and spreading of the chain structure under the shearing effect of grains and workpiece is beneficial for Fe_3O_4 nanoparticles to spread on the workpiece interface uniformly [14, 15]. Fe_3O_4 nanoparticles have limited gain since it is not a good kind of lubrication medium although lubrication performances of nanoparticles are strengthened by magnetic field force.

In Figure 11(b), the antifriction and antiwear mechanisms of lamellar GR in the interface are different from those of spherical Fe_3O_4. It decreases friction through the weak layer-layer shearing force of nano materials. Hence, GR is a kind of outstanding lubrication medium. Moreover, the lubrication film formed by lamellar GR has strong adhesion forces on grains and workpiece surface, and the oil film is more stable than that formed by Fe_3O_4 nanoparticles. Nevertheless, the film formation efficiency is not satisfying due to the disordered arrangement of GR in the interface. The non-magnetism characteristics of GR fail to get lubrication gains even though magnetic fields are applied.

In Figure 11(c) and (g), the lamellar Fe_3O_4@GR without magnetic field brings the same antifriction and antiwear mechanism and performances with GR. After magnetic fields are applied, Fe_3O_4 nanoparticles modified onto the GR surface are in chain distribution as a response to the magnetic field, driving some GR in vertical arrangement. This is beneficial for uniform arrangement of nanoparticles in interfacial fluid, thus enhancing balance of lubrication performances. However, the vertical arrangement is disadvantageous for formation of lubricating oil film. Besides, limited by preparation technology of Fe_3O_4@GR, the Fe_3O_4 content in fluid is low under the same volume fraction and the effect of magnetic field force doesn't reach the ideal state.

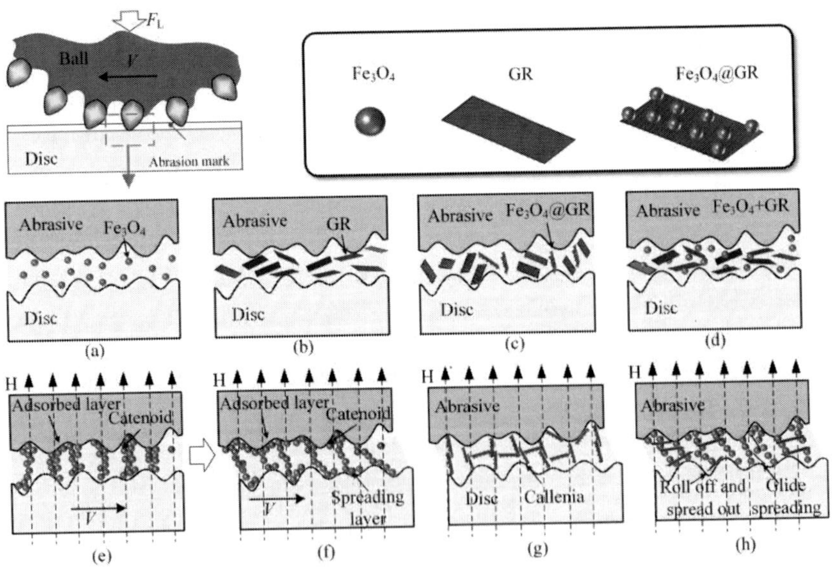

Figure 11. Schematic diagram of antifriction and antiwear mechanism under the action of magnetic field. (a) Fe_3O_4, (b) GR, (c) Fe_3O_4@GR, (d) Fe_3O_4+GR, (e) Fe_3O_4 with magnetic, (f) GR with magnetic, (g) Fe_3O_4@GR with magnetic, (h) Fe_3O_4+GR with magnetic.

In Figure 11(d) and (h), the Fe_3O_4/GR is superior to Fe_3O_4@GR. Without the application of magnetic field, spherical Fe_3O_4 scatter through the lamellar GR to form the plate-wheel structure, which is beneficial for GR to spread into film during film formation on the interface. After magnetic fields are applied, on the one hand, the volume fraction of Fe_3O_4 in Fe_3O_4/GR is higher than that in Fe_3O_4@GR given the same volume fraction, thus strengthening the response ability to magnetic field. Moreover, the balance between response ability to magnetic field and antifriction and antiwear can be adjusted easily by changing the mixing ratio of Fe_3O_4 and GR. On the other hand, Fe_3O_4 nanoparticles form a chain structure under the action of magnetic fields. GR is driven by the Fe_3O_4 chain to form perpendicular or horizontal distribution modes. Influenced by the shearing effect of interface, the perpendicular GR layer can spread to the upper peak surface at a certain inclined angle (slipping spreading), while the horizontal GR layer spread to the lower peak (rolling spreading) at a certain angle. The film formation is more uniform.

7.4. Results of Different Mixing Ratios of Fe_3O_4/Graphene Nanolubricant

Based on analysis in the previous section, Fe_3O_4/GR nanolubricant (mixing ratio=1:1) shows some advantages in term of antifriction and antiwear performances than Fe_3O_4@GR nanolubricant. The lubrication condition of Fe_3O_4@GR is unstable and nanoparticle preparation incurs a high cost. Since the mixing ratio of Fe_3O_4/GR is 1:1, whether it can beat lubrication performances better than Fe_3O_4@GR by further optimizing the mixing ratio? In the following text, an experimental study on mixing ratio optimization under the application of magnetic field is carried out. In the experiment, Fe_3O_4 nanoparticle content was increased gradually and Fe_3O_4/GR nanolubricant with mixing ratios of 1:3, 1:2, 1:1, 2:1 and 3:1 was prepared.

7.4.1. Coefficient of Friction (CoF)

The CoF curves of Fe_3O_4/GR nanolubricant with different mixing ratios under the application of magnetic field are shown in Figure 12. Clearly, the CoF curves of Fe_3O_4/GR (3:1) and Fe_3O_4/GR (1:3) (Figure 12(a) and (e)) fluctuate obviously. On the contrary, CoF curves of Fe_3O_4/GR (1:1), Fe_3O_4/GR (1:2) and Fe_3O_4/GR (2:1) are relatively stable (Figure 12(b), (c) and (d)). This reveals that the lubrication state between grains and disc is relatively poor when the mixing ratio is 3:1 and 1:3. Reasons are analyzed as follows. When there are excessive Fe_3O_4 nanoparticles, they are driven by the magnetic field to concentrate on the grain surface, causing sedimentation of nanolubricant. This influences the lubrication state of interface and the CoF curves fluctuate. Fe_3O_4/GR regulates and controls forms and poses of GR nanoparticles in the interface by oriented arrangement of magnetic Fe_3O_4 nanoparticles driven by magnetic field, realizing better antifriction and antiwear performances. However, when there are excessive GR nanoparticles, the effects of few Fe_3O_4 nanoparticles on GR nanoparticles through magnetic field deteriorate and the GR nanoparticles are in relatively disordered distribution. The CoF curves still fluctuate even under poor lubrication effects.

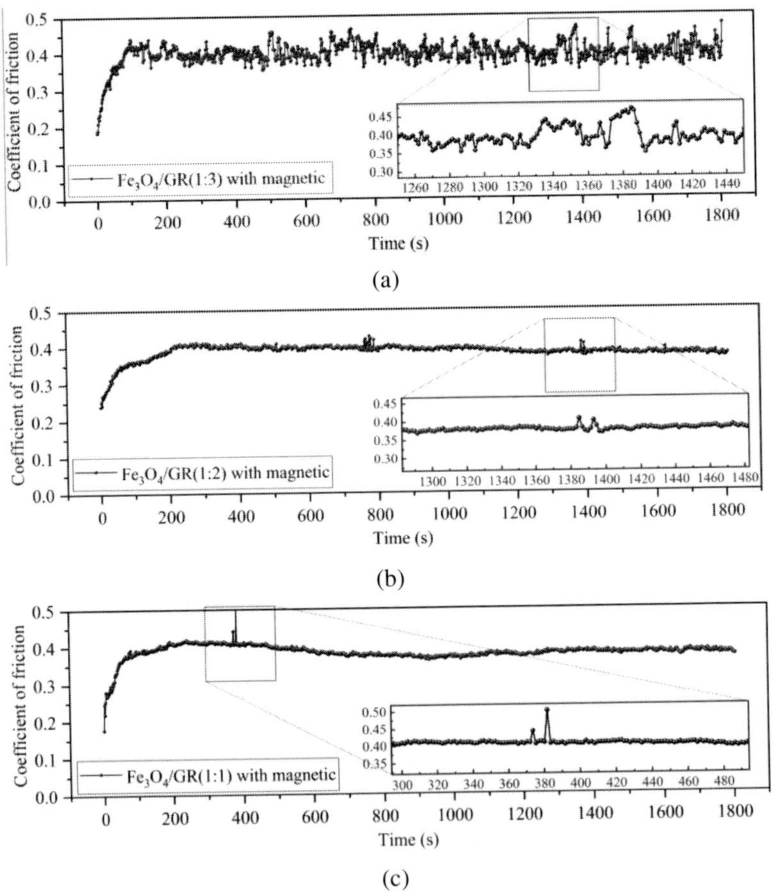

Figure 12. (Continued).

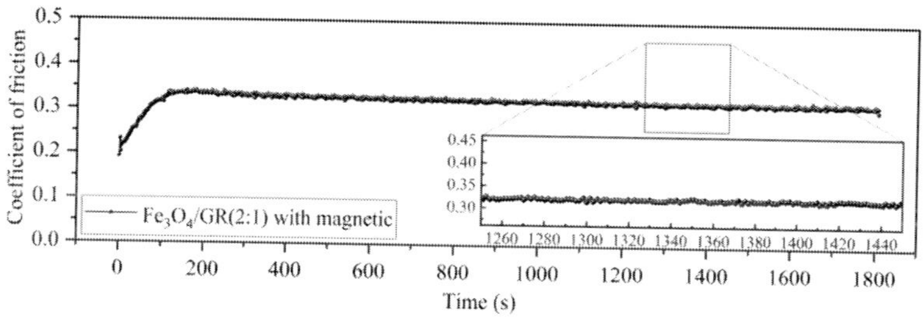

(d)

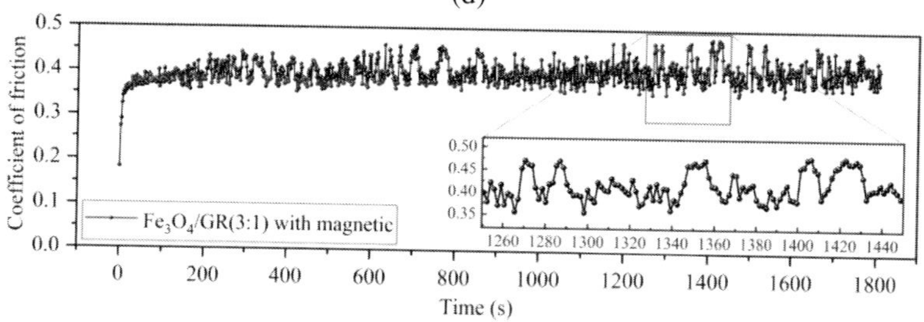

(e)

Figure 12. CoF curves of nanolubricant with different mixing ratios. (a) Fe_3O_4/GR(1:3) with magnetic, (b) Fe_3O_4/GR(1:2) with magnetic, (c) Fe_3O_4/GR(1:1) with magnetic, (d) Fe_3O_4/GR(2:1) with magnetic, (e) Fe_3O_4/GR(3:1) with magnetic.

When mixing ratios are 1:2, 1:1 and 2:1, the fluctuation of CoF curves is improved. However, when Fe_3O_4 content is lower (1:2) or equal to (1:1) GR content, there's some mild fluctuations at some time. When GR content is slightly higher than Fe_3O_4 content (2:1), the CoF curve is the most stable, without obvious fluctuation. It can conclude from CoF curves that the lubrication performances of three mixing ratios are relatively remarkable. Furthermore, mean CoF of stable stages on curves under different lubrication conditions was calculated. The box plot of CoF with different mixing ratios is gained (Figure 13).

It can be seen from Figure 13 that with the increase of Fe_3O_4 volume fraction, the mean CoF decreases firstly and then increases. The box height and dispersion range show the same variation laws. The minimum CoF (0.32) is achieved under Fe_3O_4/GR(2:1), which is 18.6% and 19.8% lower than those under Fe_3O_4/GR(1:3) and Fe_3O_4/GR(3:1), respectively. Given the fixed magnetic field strength, viscosity of magnetofluid increases with the increase of volume fraction of magnetic particles. This is because magnetic nanoparticles may collide and rub with other nanoparticles under the action of magnetic fields, thus increasing energy consumption of magnetic nanolubricant and increasing the viscosity [16]. Hence, there's a high volume content of magnetic Fe_3O_4 nanoparticles when the mixing ratio is 2:1, which increases viscosity of lubricant, improves antifriction and antiwear performances, and decreases the CoF.

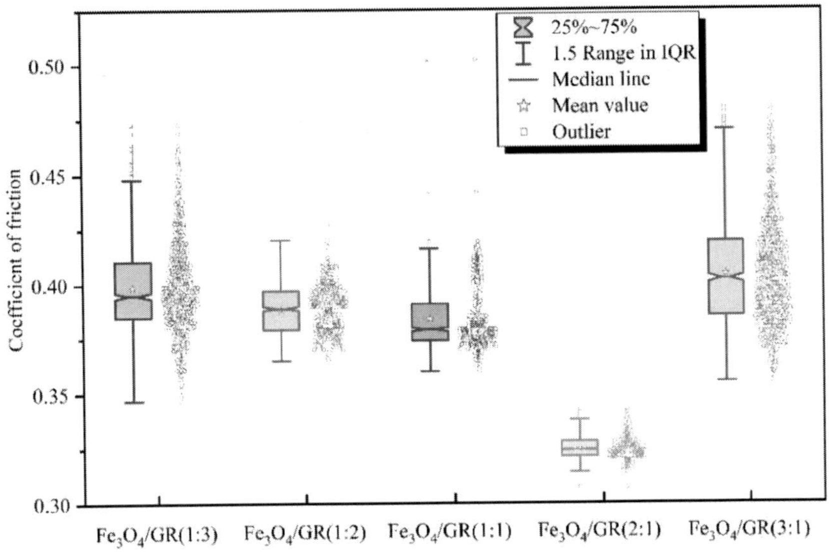

Figure 13. Box diagram of CoF in nanolubricant with different mixing ratios.

7.4.2. Wear Contour Curve and Wear Area

The scratch profile curves and wearing areas when using Fe_3O_4/GR nanolubricant with different mixing ratios under the application of magnetic field are shown in Figure 14. The wearing area of scratch also decreases firstly and then increases with the increase of Fe_3O_4. It achieves the minimum (6,146 μm^2) at Fe_3O_4/GR(2:1), which decreases by 63.4% and 51.5% than those of Fe_3O_4/GR(1:3) and Fe_3O_4/GR(3:1). This agrees with the variation laws of the CoF. Besides, it can be seen from the scratch curve that the curve valley is relatively deep at the most and lowest Fe_3O_4 content. Differently, the wearing area curve of Fe_3O_4/GR(2:1) is relatively stable. The highest valley depth is only 17.46 μm, which is 52.7% lower than the maximum depth (36.88 μm) of Fe_3O_4/GR(3:1).

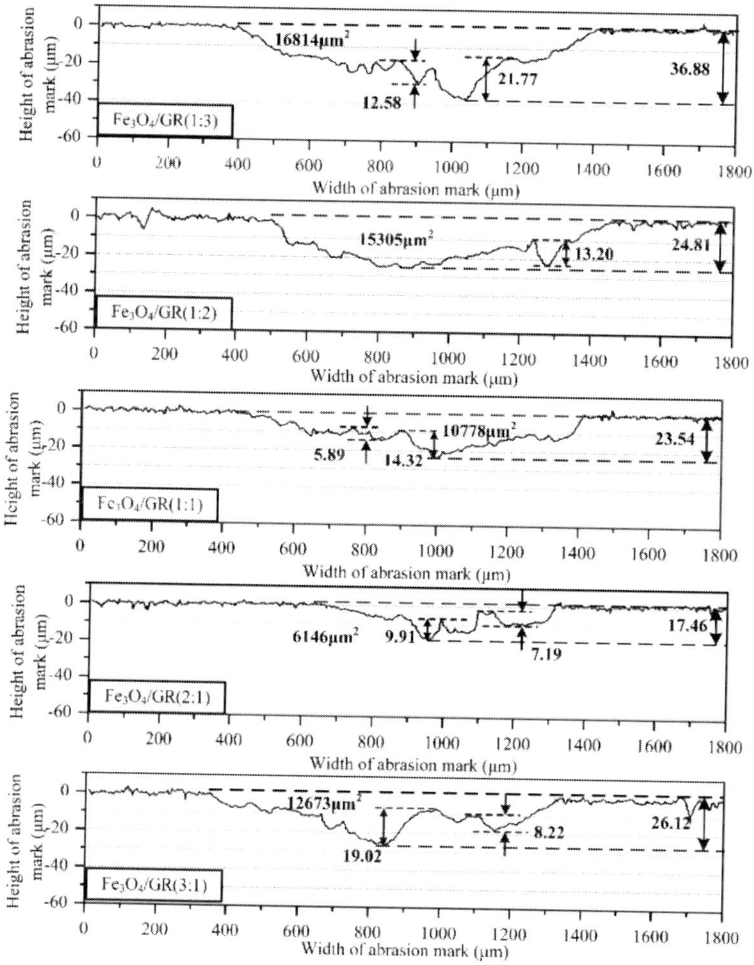

Figure 14. Wear curve under different nanolubricant with different mixing ratios under magnetic field. (a) Fe_3O_4/GR(1:3), (b) Fe_3O_4/GR(1:2), (c) Fe_3O_4/GR(1:1), (d) Fe_3O_4/GR(2:1), (e) Fe_3O_4/GR(3:1).

7.4.3. The Surface Morphology of Wear Marks

Moreover, the disc surface was observed by a laser microscope in order to recognize scratch surface morphology and surface defect. The scratch surface morphologies under the use of Fe_3O_4/GR nanolubricant with different mixing ratios were observed and surface roughness was calculated (Figure 15). In view of scratch width, it achieves the minimum (665.12 μm) at Fe_3O_4/GR(1:1), and the scratch width at Fe_3O_4/GR(2:1) is 687.63 μm, showing a small difference. In view of surface morphology, there are the least surface defects under Fe_3O_4/GR(2:1). The surface under the use of Fe_3O_4/GR(1:1) also has periodic ripple scratches.

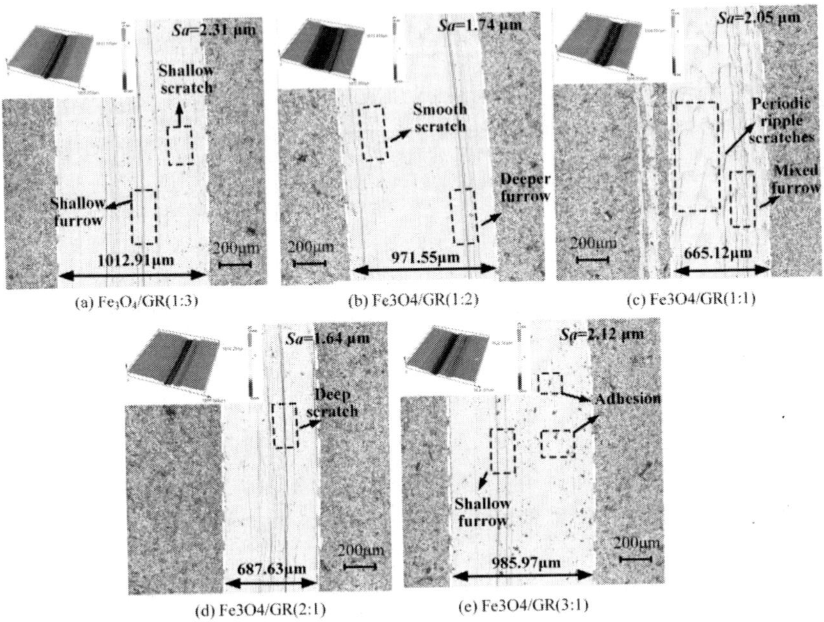

Figure 15. Scratch topography and surface roughness *Sa* under magnetic field. (a) Fe_3O_4/GR(1:3), (b) Fe_3O_4/GR(1:2), (c) Fe_3O_4/GR(1:1), (d) Fe_3O_4/GR(2:1), (e) Fe_3O_4/GR(3:1).

Based on the above analyses, surface morphology and surface roughness, it can be concluded that the best antifriction and antiwear effects are achieved under Fe_3O_4/GR(2:1) conditions, showing the optimal lubrication effect.

Conclusion

In this chapter, an experimental study on tribological properties of different nanolubricants was carried out and some major conclusions could be drawn:

i. According to comparison of Fe_3O_4 nanolubricant with and without magnetic fields, the CoF curve under the application of magnetic fields fluctuates significantly less than that without magnetic fields. The fluctuation range decreases to about 0.4~0.5 and the CoF decreases. The wearing area of scratch declines significantly after the application of magnetic fields compared to that without magnetic fields. Moreover, the furrows and scratches on surface are relatively ordered and straight.

ii. The lubrication performances of GR nanolubricant and Fe_3O_4 nanolubricant under the application of magnetic fields are compared. It found that the fluctuation range of CoF curve of the Fe_3O_4 nanolubricant is smaller compared to that of GR nanolubricant. For scratch surface morphology, magnetic fields improve the original surface damage of Fe_3O_4 nanolubricant and they can achieve a similar effect with GR nanolubricant. However, Sa=3.19 μm by using Fe_3O_4 nanolubricant, which is still increased by 62.8% than that by using GR nanolubricant.

iii. A comparative study on antifriction and antiwear performances of Fe_3O_4@GR prepared by chemical modification method and Fe_3O_4/GR (mix ratio=1:1, 5vol%) prepared by physical mixing method was carried out. It is found that Sa of Fe_3O_4@GR and Fe_3O_4/GR values 1.71 μm and 2.05 μm, respectively. The CoF curve of Fe_3O_4/GR is significantly stable. Scratch width decreases significantly, and surface adhesion is improved. However, periodic ripple scratches are developed on surface under the use of Fe_3O_4/GR compared to that under Fe_3O_4@GR. This is because Fe_3O_4 content in Fe_3O_4/GR nanolubricant is higher than that in Fe_3O_4@GR nanolubricant. Moreover, the relatively active Fe_3O_4 nanoparticles are in regular arrangement in Fe_3O_4/GR nanolubricant under the action of magnetic field, and GR particles can form regular distribution and infiltration among regular spherical nanoparticles, making them easier to enter the grain-workpiece interface and thereby improving the infiltration capacity.

iv. In the experiment of mixing ratio optimization for Fe_3O_4/GR nanolubricant, it is found that with the increase of Fe_3O_4, the mean CoF decreases firstly and then increases. The minimum CoF (0.32) is achieved at Fe_3O_4/GR(2:1), and the distribution range is the smallest. It is 18.6% and 19.8% lower than those at Fe_3O_4/GR(1:3) and Fe_3O_4/GR(3:1), respectively. Similarly, wearing area and surface roughness at Fe_3O_4/GR(2:1) are also the lowest, valuing 6,146 μm^2 and 1.64 μm, respectively. The surface defects are improved obviously.

References

[1] Singh T, Dureja JS, Dogra M, et al. (2018). Environment Friendly Machining of Inconel 625 under Nano-Fluid Minimum Quantity Lubrication (NMQL). *Int. J. Precis. Eng. Manuf.*, *19*(11), 1689-1697. 10.1007/s12541-018-0196-7.

[2] Mao C, Zhang J, Huang Y, et al. (2013). Investigation on the Effect of Nanofluid Parameters on MQL Grinding. *Mater. Manuf. Process.*, *28*(4), 436-442. 10.1080/10426914.2013.763970.

[3] Makhesana MA, Patel KM, Krolczyk GM, et al. (2023). Influence of MoS_2 and graphite-reinforced nanofluid-MQL on surface roughness, tool wear, cutting temperature and microhardness in machining of Inconel 625. *CIRP J. Manuf. Sci. Technol.*, *41*, 225-238. 10.1016/j.cirpj.2022.12.015.

[4] Kim WY, Senguttuvan S, Kim SH, et al. (2021). Numerical study of flow and thermal characteristics in titanium alloy milling with hybrid nanofluid minimum quantity lubrication and cryogenic nitrogen cooling. *Int. J. Heat Mass Transf.*, *170*, 14. 10.1016/j.ijheatmasstransfer.2021.121005.

[5] Shahrivar K, Ortiz AL, de Vicente J. (2014). A comparative study of the tribological performance of ferrofluids and magnetorheological fluids within steel-steel point contacts. *Tribol. Int.*, *78*, 125-133. 10.1016/j.triboint.2014.05.008.

[6] Hu ZD, Yan H, Qiu HZ, et al. (2012). Friction and wear of magnetorheological fluid under magnetic field. *Wear*, *278*, 48-52. 10.1016/j.wear.2012.01.006.

[7] Emami M, Sadeghi MH, Sarhan AAD, et al. (2014). Investigating the Minimum Quantity Lubrication in grinding of Al_2O_3 engineering ceramic. *J. Clean Prod.*, *66*, 632-643. 10.1016/j.jclepro.2013.11.018.

[8] Silva LR, Corrêa ECS, Brandao JR, et al. (2020). Environmentally friendly manufacturing: Behavior analysis of minimum quantity of lubricant - MQL in grinding process. *J. Clean Prod.*, *256*, 1-9. 10.1016/j.jclepro.2013.01.033.

[9] Liu W, Deng ZH, Huang YC. (2017). Experimental on surface integrity of titanium alloy TC4 by high speed cylindrical grinding. *Aerospace Materials & Technology*, *47*(4), 66-70.

[10] Xu J, Lu WZ, Wang HZ, et al. (2014). Characteristics and wear properties of grinding surface of titanium alloy TC4-DT. *Acta Aeronautica et Astronautica Sinica*, *35*(2), 567-573.

[11] Mao C, Zhou FJ, Hu YL, et al. (2019). Tribological behavior of cBN-WC-10Co composites for dry reciprocating sliding wear. *Ceram. Int.*, *45*(5), 6447-6458. 10.1016/j.ceramint.2018.12.132.

[12] Mao C, Ren YH, Gan HY, et al. (2015). Microstructure and mechanical properties of cBN-WC-Co composites used for cutting tools. *Int. J. Adv. Manuf. Technol.*, *76*(9-12), 2043-2049. 10.1007/s00170-014-6410-6.

[13] Zhang YB, Li CH, Ji HJ, et al. (2017). Analysis of grinding mechanics and improved predictive force model based on material-removal and plastic-stacking mechanisms. *Int. J. Mach. Tools Manuf.*, *122*, 81-97. 10.1016/j.ijmachtools.2017.06.002.
[14] Peng Z, Kwang-Hee L, Chul-Hee L. (2016). Friction Behavior of Magnetorheological Fluids with Different Material Types and Magnetic Field Strength. *Chin. J. Mech. Eng.*, *29*(1), 84-90. 10.3901/cjmc.2015.1126.139.
[15] Zhang P, Dong YZ, Choi HJ, et al. (2018). Tribological and rheological tests of core-shell typed carbonyl iron/polystyrene particle-based magnetorheological fluid. *Journal of Industrial Engineering Chemistry*, *68*, 342-349.
[16] Li Q, Xuan YM, Wang J. (2005). Experimental investigation on viscosity of magnetic fluids. *Journal of Engineering Thermophysics*, *26*(5), 859-861.

Chapter 8

Modelling Infiltration Behavior and Grinding Force in Magnetic Traction Nanolubricant Grinding

Abstract

The MTN method was proposed for improving infiltration performance of nanolubricant in the grinding zone. However, the micro-interface transport mechanics of magnetic nanolubricant under the magnetic field is not clear, and the grinding force model with the influence of magnetic field has not been established. Firstly, the mechanical law of magnetic traction lubricant transport in the grinding zone was revealed, and the model of lubricant infiltration velocity and flow rate with the influence of the magnetic field was established. Secondly, a grinding wheel model is established based on the truncated hexahedron abrasive, which revealed the interference mechanical behavior of materials for a single abrasive and established the mechanical model. Finally, a grinding force model of magnetic traction nanolubricant was established for titanium alloy grinding and verified by experiments. The results show that the introduction of magnetic field can significantly increase the wetting speed and flow rate of lubricant. When the magnetic field intensity is 5×105 A/m, the normal and tangential grinding forces decrease by 31.8% and 74.3%, respectively, compared with no magnetic field. And the minimum mean deviation of tangential and normal grinding force is 9.2% and 5.7%, respectively. A theoretical basis for magnetic traction nanolubricant grinding and technical support for high surface integrity requirements of large contact length grinding of difficult materials were provided.

Keywords: grinding, magnetic nanolubricant, magnetic traction, dynamics behavior, mechanical model

8.1. Introduction

On the micro-interface with complicated geometric shapes and high relative shear speed, the transportation and infiltration performances of lubricating medium are important factors that decide development of the cooling and lubrication performances [1-3]. For traditional flood grinding technique, the high-pressure continuous supply of lubricating media brings strong infiltration kinetic energy. In MQL, lubricant is jet into the grinding zone by the high-pressure air after it is atomized. Due to the kinetic energy attenuation characteristics, the infiltration capacity is relatively poor. Tang [4] describes the lubricant transport behaviors in the grinding zone by the capillary infiltration mechanism and gained the mathematical formula for infiltration depth. However, he only considered the influences of surface tension and initial kinetic energy on infiltration depth but ignored the rotating speed of the grinding wheel. As a result, a mathematical model of lubricant transport speed and flow rate under the influences of linear speed of grinding wheel and magnetic field strength was built in this study, in which boundary slippage was taken into account. Furthermore, effects of magnetic field were introduced into the grinding force model through the quantitative relation between flow rate and coefficient of friction (CoF) in the grinding zone.

CBN and diamond grinding wheels are irreplaceable in deep grinding and high-speed grinding of difficult-to-machine materials because of their high hardness and strength. Theoretical modeling of material processing based on CBN and diamond grinding wheels depends on precision of grain modeling. Existing studies usually employ sphere [5-7], cones [8] and others [9] as geometric models of grains. However, they cannot reflect the stress state on the grain-workpiece interface accurately when they are applied to grinding force modeling. On this basis, a statistical analysis on form, and size of CBN grains was carried out. The geometric modeling technology of truncated hexahedronal grains is proposed and a material removal mechanical model considering geometric and pose characteristics of grains is built. Furthermore, a new model of magnetic field assisted grinding force was constructed by combining with the CoF model and experimental verification was carried out.

8.2. Modelling Nanolubricant Infiltration Behavior in the Grinding Zone's Magnetic Field

8.2.1. Physical Mechanism

The grinding zone is composed of the grinding wheel surface and workpiece surface. Thickness of the grinding zone is at the magnitudes of 10 μm and length (grinding arc) is at the magnitudes of 1~10 mm. Due to the great scale difference, the arc characteristics of the grinding zone can be ignored, and it is defined as the microscale space formed by two plates. Furthermore, the linear speed of the grinding wheel is generally 10~30 m/s and the workpiece feeding speed is 0.1~1 m/min in ordinary grinding. It can be defined as the microscale space where the upper plate moves, and the lower plate is static (Figure 1).

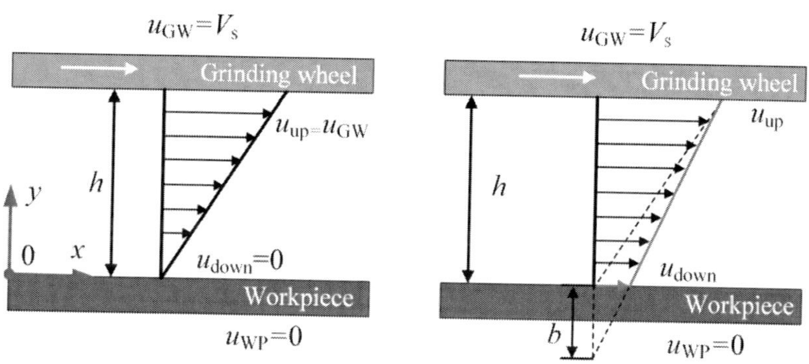

Figure 1. Physical modeling of lubricant transport. (a) No slip, (b) Schematic diagram of slip coefficient.

$$\frac{d^2u}{dy^2} = \frac{1}{\eta}\frac{dp}{dx} \tag{1}$$

In Figure 1, u is the flow rate of fluid (m/s) and it is a function of coordinate y. η represents the dynamic viscosity of fluid (N·s/m^2). p denotes the pressure against the fluid along the flowing direction of lubricant (N). Lubricant transport belongs to incompressible flowing:

$$\frac{dp}{dx} = -\frac{(p_2 - p_1)}{L} = -\frac{\Delta p}{L} \tag{2}$$

Through integral of the governing Eq.(1), the expression of u can be gained:

$$u = \frac{1}{2\mu}\frac{dp}{dx}y^2 + C_1 y + C_2 \tag{3}$$

With further considerations to boundary slippage, the boundary slippage coefficient (b) that characterizes boundary slippage degree was introduced. It is defined as follows: distance from the extended line of lubricant speed curve profile and points with equal speed to the wall surface. The value of b is related to contact state and pressure of the solid-liquid interface. Slippage coefficients of grinding wheel surface and workpiece surface are different due to the significant differences in materials and morphologies. In previous studies, scholars only set the same slippage coefficient to characterize upper and lower solid-liquid interface, which disagrees with practical engineering conditions in this study. Hence, the slippage coefficients of grinding wheel-lubricant interface and workpiece-lubricant interface were set b_1 and b_2, respectively. Their physical significance is shown in Figure 2.

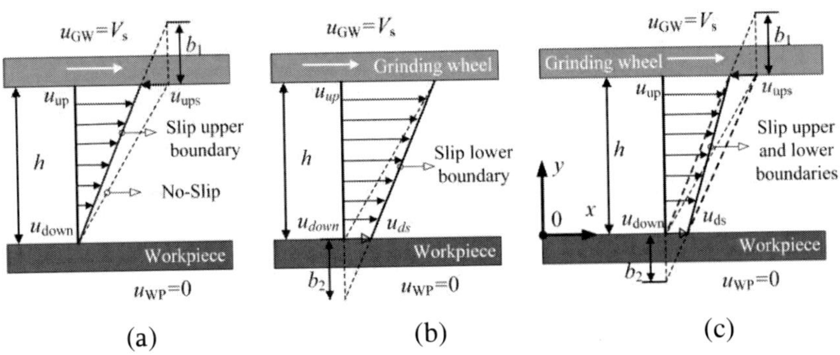

Figure 2. Physical model of lubricant transport considering different slip coefficients. (a) Consider the grinding wheel interface slip (b) consider the workpiece interface slip (c) the grinding wheel and the workpiece interface slip simultaneously.

In this base, the boundary conditions of lubricant flow rate u_L related to b_1 and b_2 can be expressed as:

When

$$y=0,\ u_L = u_{ds} = \frac{V_s \cdot b_2}{h+b_2} \tag{4}$$

When

$$y{=}h,\ u_L = V_s - u_{ups} = V_s - \frac{V_S \cdot b_1}{h+b_1} \tag{5}$$

Specifically, u_L is the flow rate of lubricant (m/s). b_1 and b_2 are slippage coefficients of grinding wheel-lubricant interface and workpiece-lubricant interface, respectively. u_{up} and u_{down} are slip speeds of lubricant-grinding wheel interface and lubricant-workpiece interface, respectively. u_{ds} refers to the variation of slip speed of lubricant on the lower wall surface (workpiece surface) (m/s). u_{ups} is the variation of slip speed of lubricant on the upper wall surface (grinding wheel surface) (m/s). V_s expresses the speed of grinding wheel (m/s). h shows the distance between the grinding wheel and workpiece (m). The boundary conditions Eq.(4) and Eq.(5) were brought into Eq.(3), getting the distribution formula of lubricant flow rate:

$$u_L = \frac{V_S}{h} \cdot \left(\frac{h}{h+b_1} - \frac{b_2}{h+b_2}\right) y + \frac{V_S \cdot b_2}{h+b_2} \tag{6}$$

If the width of the grinding zone is b, the lubricant flow (Q) in the grinding zone can be deduced through integral:

$$Q = \frac{V_S \cdot h \cdot b}{2} \cdot \left(\frac{h}{b_1+h} - \frac{b_2}{b_2+h}\right) + \frac{V_S \cdot h \cdot b_2 \cdot b}{b_2+h} \tag{7}$$

8.2.2. Nanolubricant Force Analysis and Velocity Distribution

For fluid layer with speed gradient, it can be known from the constitutive model of Newtonian fluid that the shear stress (τ) between fluid layers can be expressed by the following formula:

$$\tau = \eta \frac{du}{dy} \tag{8}$$

During grinding, since fluid develops shear flowing under the dragging force of grinding wheel, the shear stress among liquid layers reaches the critical shear stress value τ_0 (the positive pressure between lubricant element and solid interface is 0), accompanied with slippage. Under the critical shear stress, the speed gradient can be calculated according to Eq. (8):

$$\frac{du}{dy} = \frac{\tau_0}{\eta} \tag{9}$$

τ_0 is the initial critical shear yield stress of lubricant and it is a function of viscosity (η) and speed gradient. In this study, the initial speed of lubricant from entrance of grinding zone into the micro-interface is 2~5 m/s, while the linear speed of grinding wheel can be adjusted within the range of 10~30 m/s (for ordinary grinding). It can be calculated from Eq. (8) that τ_0 ranges within 13000~65000 N/m^2.

Calculate derivate of the Eq. (6) along the y direction, it found that the speed gradient is a function of b_1 and b_2. Further combining with Eq. (9), it can get:

$$\frac{\tau_0}{\eta} = \frac{V_s}{h}\left(\frac{h}{h+b_1} - \frac{b_2}{h+b_2}\right) \tag{10}$$

In the following text, the slippage coefficients of grinding wheel-lubricant interface and lubricant-workpiece interface are calculated, respectively. It is important to note that these slippage coefficients are solved independently in the system. This also agrees with the definition of slippage coefficient. Hence, let b_2=0 for the grinding wheel-lubricant interface and it calculates that:

$$b_1 = \frac{\eta V_s}{\tau_0} - h \tag{11}$$

Similarly, the slippage coefficient of the lubricant/workpiece interface can be calculated:

$$b_2 = \frac{\eta V_s}{\tau_0} - h \tag{12}$$

The magnetic fields perpendicular to the grinding wheel interface were further introduced into the grinding zone. The positive pressure of lubricant element and solid interface will be affected by magnetic field force. It was reported that when the positive pressure of lubricant element and solid interface is p, the critical shear stress of solid-liquid interface can be expressed as [10, 11]:

$$\tau_c = \tau_0 + kp \tag{13}$$

Where k is the CoF of solid-liquid interface, and it is a constant related with solid-liquid intermolecular forces.

It can be seen from Eq. (13) that the application of magnetic fields will increase or decrease pressure of the solid-liquid interface, thus changing the critical shear stress and speed gradient. As a result, the slip speed changes accordingly. The positive pressure that lubricant element bears is the resultant force of magnetic field volume force (F_m) and gravity (G). Since the order of magnitude of magnetic field force and viscous force is far higher than that of gravity, the gravity is ignored in this chapter. Hence, $p=F_m$.

For the grinding wheel-lubricant interface, the interface pressure increases by applying magnetic fields and the critical shear stress (τ_{c1}) is:

$$\tau_{c1} = \tau_0 + k_1 F_m \tag{14}$$

For the lubricant/workpiece interface, the interface pressure decreases by applying magnetic fields and the critical shear stress (τ_{c2}) is:

$$\tau_{c2} = \tau_0 - k_2 F_m \tag{15}$$

Where k_1 and k_2 are CoFs of the grinding wheel-lubricant interface and lubricant-workpiece interface. They are constants related to adsorption forces of interfaces. Given the same liquid, the value of k is positively related to surface roughness. Since the grinding wheel surface roughness is higher than workpiece surface, there's $k_1>k_2$.

Combining Eqs. (11), (12), (14) and (15), the slippage coefficients of grinding wheel-lubricant interface and lubricant-workpiece interface after the application of magnetic fields can be calculated.

$$b_1 = \frac{\eta V_s}{\tau_0 + k_1 F_m} - h \tag{16}$$

$$b_2 = \frac{\eta V_s}{\tau_0 - k_2 F_m} - h \tag{17}$$

Lubricant is used as magnetofluid. The magnetic field force against unit volume of lubricant can be calculated by the following equation:

$$F_m = \mu_0 (\boldsymbol{M} \cdot \nabla) \boldsymbol{H} \tag{18}$$

Where $\boldsymbol{M}$ is the magnetization intensity of lubricant (A/m). $\boldsymbol{H}$ expresses the magnetic field strength of grinding zone (A/m). μ_0 shows the vacuum permeability ($4\pi\times10^{-7}$ N/A^2). On the grinding zone interface, $\boldsymbol{M}$ and $\boldsymbol{H}$ are parallel. In other words, there's:

$$\boldsymbol{M} = \frac{M\boldsymbol{H}}{|\boldsymbol{H}|} \tag{19}$$

Furthermore, according to relation of vectors:

$$\nabla(\boldsymbol{H}\cdot\boldsymbol{H}) = 2(\boldsymbol{H}\cdot\nabla)\boldsymbol{H} \tag{20}$$

The Eq.(18) can be transformed into a scalar expression:

$$F_m = \mu_0 M \nabla H \tag{21}$$

Where, $M=|\boldsymbol{M}|$, $H=|\boldsymbol{H}|$。

Eqs. (16), (17) and (21) are brought into Eqs. (6) and (7), and calculation formulas of lubricant speed distribution and flow under the influences of magnetic fields can be gained:

$$u_L = \frac{y}{\eta h}[2h\tau_0 - \eta V_s + h\mu_0 M\nabla H(k_1 - k_2)] + \left[V_s - \frac{h}{\eta}(\tau_0 - k_2\mu_0 M\nabla H)\right] \tag{22}$$

$$Q = \frac{yb}{2\eta}[2h\tau_0 - \eta V_s + h\mu_0 M\nabla H(k_1 - k_2)] + hb\left[V_s - \frac{h}{\eta}(\tau_0 - k_2\mu_0 M\nabla H)\right] \tag{23}$$

Where the lubricant viscosity (η) is a function of magnetic fields and it can be brought into Eq. (1).

8.2.3. The Numerical Solution of Infiltration Velocity and Flow

The lubricant infiltration speed curves under different magnetic field strengths which are gained according to Eq. (22) are shown in Figure 3. When the magnetic field strength is 0, the lubrication rate of the upper boundary is 11.96 m/s and the lubrication rate of the lower boundary is 8.04 m/s. Compared to the linear speed of grinding wheel (20 m/s), lubricant has too low infiltration speed to cover the whole grinding zone. After magnetic fields are applied, the lubrication rate curves generally move rightward. For the upper boundary, the growth rate of lubrication rate decreases. The lubrication rate reaches 16.27 m/s when the magnetic field strength is 110,000 A/m. Due to influences of magnetic field forces, the lubricant adsorption force onto the grinding wheel interface is strengthened, thus decreasing the slipping trend. It is predicted that the lubrication rate can reach 20 m/s when the magnetic field strength increases to enough high. However, the maximum magnetic field strength in this study can only reach 110,000 A/m. For the lower boundary, the growth rate of lubrication rate is kept a constant and it reaches 12.4 m/s when the magnetic field strength is 110,000 A/m. This is because the involvement of magnetic field forces offsets the adsorption forces and frictional forces of lubricant-workpiece interface, thus increasing the slipping trend. As a result, applying magnetic fields perpendicular to the movement direction of lubricant on the grinding wheel-workpiece interface can strengthen infiltration capacity of lubricant.

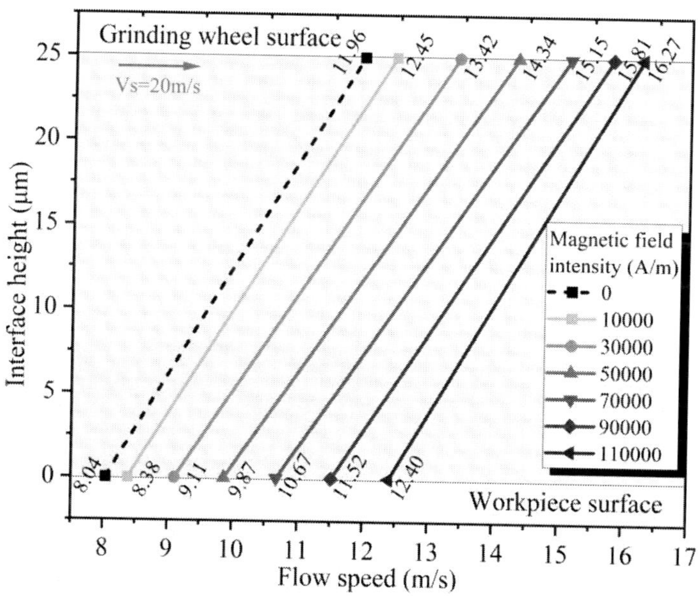

Figure 3. Curve of lubricant infiltration velocity under different magnetic field intensity.

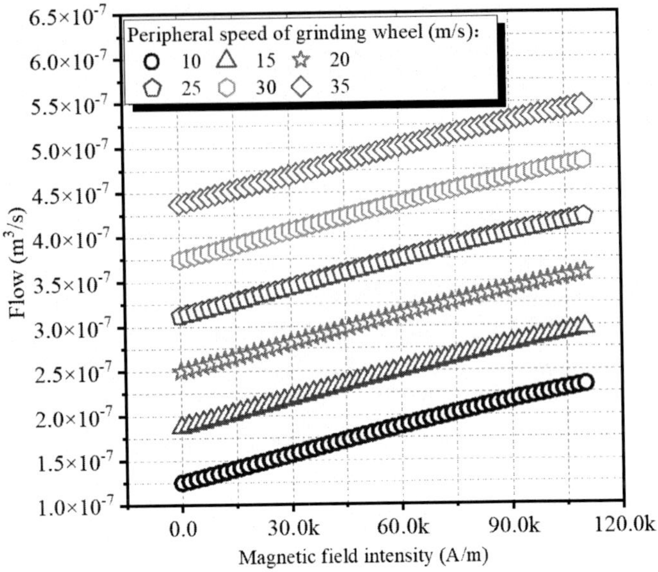

(a)

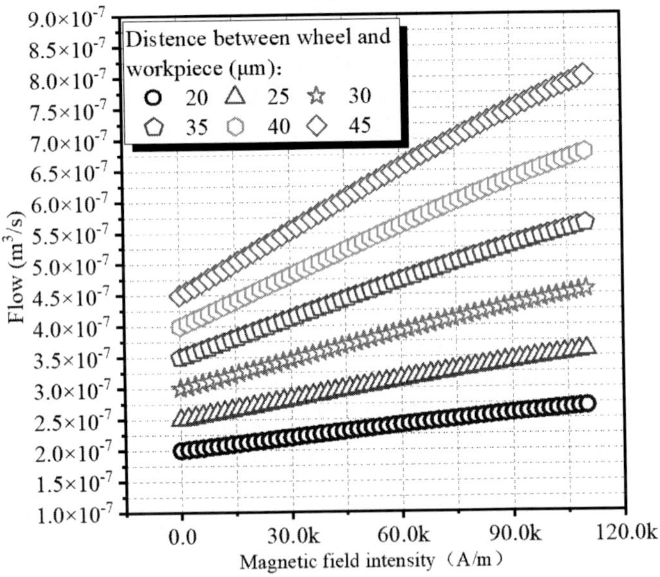

(b)

Figure 4. Lubricant infiltration flow under different conditions. (a) Consider grinding wheel linear speed and magnetic field, (b) Consider interface height and magnetic field.

Changes of grinding parameters are finally manifested by changes of lubrication rate and height of the upper interface in the grinding zone. Therefore, the linear speed of grinding wheel (10 m/s~35 m/s) and height of interface (20 μm~45 μm) are chosen as independent variables. The variation laws of lubricant flow in unit time under different magnetic field strengths can be calculated from Eqs. (22) and (23).

With the increase of magnetic field strength, the infiltration flow of lubricant in unit time presents an approximately new increasing trend. This is the same with variation laws of the speed curve. As shown in Figure 4(a), the lubricant flow presents a linear growth trend with the increase of linear speed of the grinding wheel, because increasing linear speed of the grinding wheel brings stronger inertia force of lubricant. It can be seen from Figure 4(a) that the lubricant flow increases with the increase in height of interface. When the magnetic field strength is 0, the increased lubricant flow is attributed to the expansion of infiltration space. When the magnetic field strength is 110,000 A/m, the increased lubricant is caused by the expanded infiltration space and magnetic field together. This can be proved by different slopes of flow curves.

8.3. Geometric Modelling of the Grinding Wheel Based on a Truncated Hexahedron

8.3.1. Abrasive Morphology of the CBN Grinding Wheel

Although CBN grains are less hard (microhardness: HV7200~9800) than diamond, they have strong thermostability and chemical stability, and they won't react with iron and iron alloy. Besides, thermal conductivity of CBN grains is dozens and even hundreds of times that of corundum grains. They apply to grinding of difficult-to-machine materials (especially Ti alloy) with low thermal conductivity. In the grinding zone, form and pose of CBN grains decide geometric characteristics of the micro-contact zone. For accurate characterization of tribological characteristics of grain-workpiece interface, a statistical model of CBN grain form was constructed in this study (Figure 5).

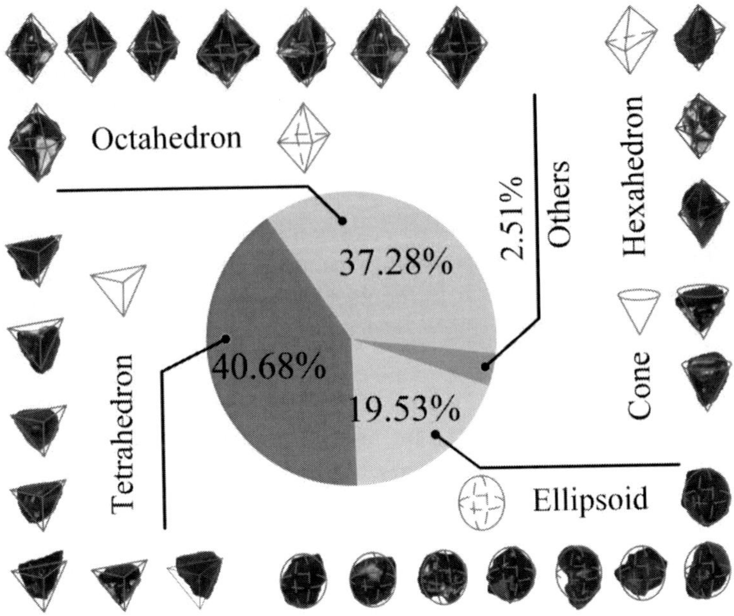

Figure 5. Shape statistics of CBN abrasives.

Shapes of CBN grains are random. According to statistics on shapes of 500~600 grains (Figure 5), there are tetrahedron, hexahedron, octahedron, spheroid and cone shapes. The Tetrahedron/Octahedron/Spheriod grains account for 97.49% of total grains. CBN grinding wheel was prepared with grains and binding agent through high-temperature sintering technology. The center point of grains that can cut materials is in the binding agent plane. If the center point of grains is out of the binding agent plane, it will fall off during the finishing or cutting process of the grinding wheel. Hence, grains out of the binding agent can grind materials. Tt can be found by introducing tetrahedron/ octahedron/spheriod grains into the grinding conditions that all grains have three surfaces participating in material removal (interfering surfaces with materials) (Figure 6), which are two planes (rake face) and one arc surface (flank surface). Therefore, modeling of CBN grains shall focus on these three surfaces. Truncated hexahedron model, a new grain model, is proposed in this study to describe shapes of grains. It is determined that the model has 3 important parameters according to practical grinding conditions of grain-workpiece interface: included angles of two rake faces (vertex angle of grains, α), flank surface area (S) and grain size (D) (Figure 6).

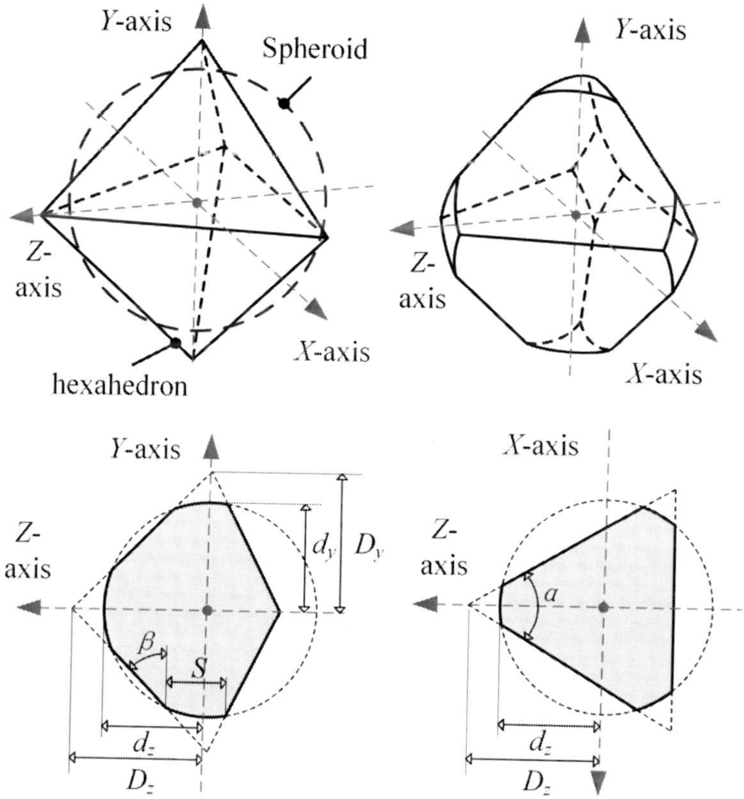

Figure 6. Geometric model of truncated hexahedron abrasives.

(i) Vertex angle of grains (α). Although grains with different forms (T/S/O) participate in material removal during grinding, they all can be simulated by using the truncated hexahedron model. α can represent different grain forms and ranges within 60°~90°. It is found in statistics that 60° grains (corresponding to Tetrahedron) account for 37% and 90°grains (corresponding to Octahedron) account for 40%, which are approximately equal. Hence, α conforms to standard uniform distribution. In other words, there's α~U (60, 90). The cutting states of two rake faces can represent the practical interference conditions between different-shaped grains and materials.

(ii) Grain size (D). Grain size observes normal distribution. Obviously, $D=2D_z$. According to internal standard ISO 6106-1979, the mesh number of grinding wheel used in this study is 240# and grain size ranges between 53μm

(D_{min}) and 63μm (D_{max}). Hence, the probability density distribution function of grain size can be expressed as:

$$\phi(D) = \frac{1}{\sqrt{2\pi}\cdot\sigma} e^{\left[\frac{(D-D_{mean})^2}{2\sigma^2}\right]} \tag{24}$$

Where σ=(D_{max}-D_{min})/6, D_{mean}=(D_{max}+D_{min})/2. D_z is the grain size along the Z direction (mm). D_{max}, D_{min} and D_{mean} are maximum, minimum and mean of grain size along the Z direction (mm), respectively.

(iii) Area of flank surface (*S*). The arc flank surface is caused by wearing of the grain angle during processing. In the beginning, grains have sharp angles. However, grains will experience three stages during grinding, including rapid wearing, stable wearing and material peeling. Among them, the stable wearing stage is the normal working stage. Under this state, the flank surface of grains is an arc surface. The value of *S* is related to sphere diameter ($2d_z$). d_z can be gained from statistics of grains: $d_z \approx$ (0.82~0.9) D_z and it observes the uniform distribution. In other words, d_z ~U(0.82D_z, 0.9D_z).

Furthermore, *S* can be calculated from the truncated hexahedron model (approximately equal to spherical crown area) and it can be expressed as:

$$S \approx 2\pi RH = 2\pi d_y \left[d_y - d_y \cdot cos\left(\frac{\pi}{4} - arccos\frac{D_y}{\sqrt{2}d_y}\right)\right] \tag{25}$$

To build grain poses in the program, a coordinate matrix (G_v) of 6 vertexes of the grain hexahedron can be built:

$$G_v = \begin{pmatrix} 0 & D_y & 0 \\ -\frac{\sqrt{3}}{2}D_z & (1-\frac{\sqrt{3}}{2\,tan(\frac{2}{\alpha})})D_z & 0 \\ \frac{\sqrt{3}}{2}D_z & (1-\frac{\sqrt{3}}{2\,tan(\frac{2}{\alpha})})D_z & 0 \\ 0 & 0 & D_z \\ 0 & 0 & -D_z \end{pmatrix} \tag{26}$$

The truncated hexahedron model was further solved through spheres. The grained calculated using MATLAB software are shown in Figure 7.

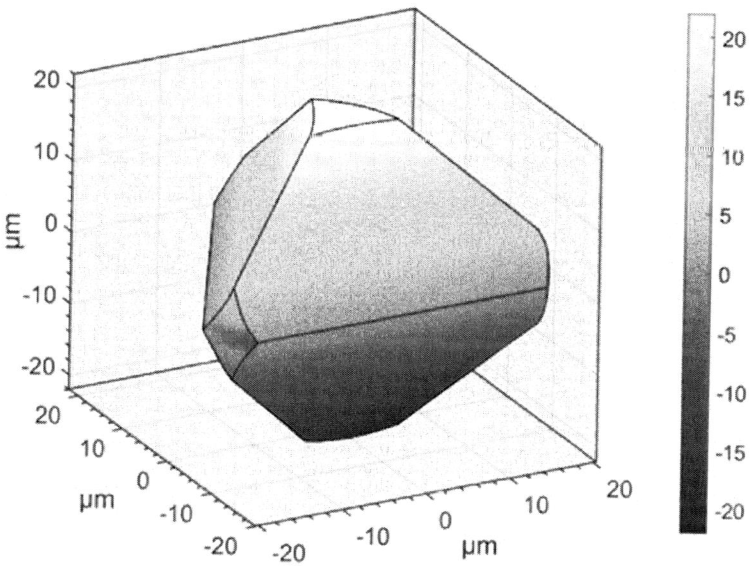

Figure 7. Truncated hexahedron abrasives model generated by Matlab.

8.3.2. Abrasive Distribution and the Attitude Model

During grinding, grains and workpiece interact to remove workpiece materials. There are two factors that influence geometric model of the grain-workpiece contact zone: poses of grains and protrusion height of grains over the binding agent. The distribution and poses of grains on the grinding wheel surface are shown in Figure 8.

(i) Poses of grains. A Cartesian coordinate system of grains was established by using center point of grains (O) as the origin. Since the grinding wheel was prepared with full mixture of CBN grains and binding agent through high-temperature sintering, the pose state of grains in the binding agent is completely random. For a group of grains, they rotate around the X-axis by γ_x, rotate around the Y-axis by γ_y, and rotate around the Z-axis by γ_z, all of which shall conform to the uniform distribution from 0° to 360°. Moreover, if poses of grains in Figure 8 are set $\gamma_x=\gamma_y=\gamma_z=0°$, it can be known by combining shape characteristics of truncated hexahedron that it can represent all conditions of grains when γ_y and γ_z range within [$-\alpha$, α]. When γ_x varies within [$-\alpha$, β], it can represent all situations of grains. Specifically, β is the dip angle of rake faces. Hence, it can determine that γ_x~U((α-π)/2, π/2-β), γ_y~U((α-π)/2, (π-α)/2) and γ_z~U((α-π)/2, (π-α)/2).

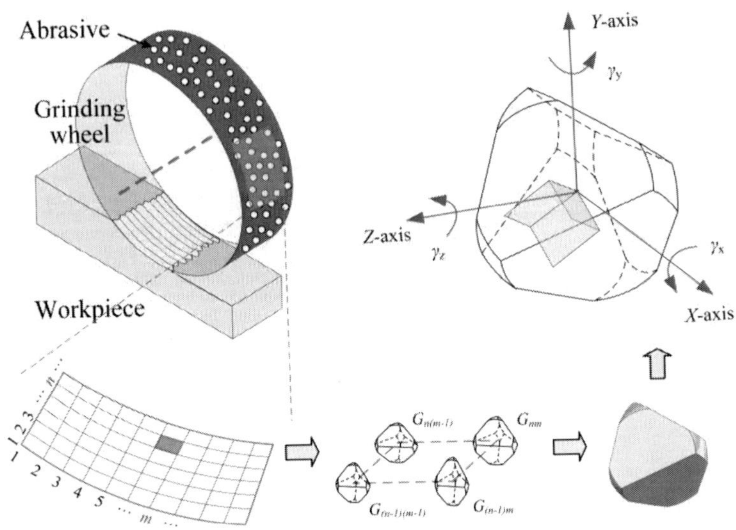

Figure 8. Distribution and attitude of abrasives on grinding wheel surface.

Based on above analysis, the rotation matrixes of six fixed vertexes around X-axis, Y-axis and Z-axis ($R_x(\gamma_x)$, $R_y(\gamma_y)$ and $R_z(\gamma_z)$) are:

$$R_x(\gamma_x) = \begin{bmatrix} 1 & 0 & 0 \\ 0 & \cos\gamma_x & \sin\gamma_x \\ 0 & -\sin\gamma_x & \cos\gamma_x \end{bmatrix} \tag{27}$$

$$R_y(\gamma_y) = \begin{bmatrix} \cos\gamma_y & 0 & -\sin\gamma_y \\ 0 & 1 & 0 \\ \sin\gamma_y & 0 & \cos\gamma_y \end{bmatrix} \tag{28}$$

$$R_z(\gamma_z) = \begin{bmatrix} \cos\gamma_z & \sin\gamma_z & 0 \\ -\sin\gamma_z & \cos\gamma_z & 0 \\ 0 & 0 & 1 \end{bmatrix} \tag{29}$$

The coordinates of six vertexes after rotation of grains can be described by $G_{v\gamma}$:

$$G_{v\gamma} = G_v R_x R_y R_z \tag{30}$$

(ii) Protrusion height of grains over the binding height (Z). With reference to the paper of Wang Sheng, the protrusion height of grains obeys to uniform distribution and it ranges within [$-L_r$, L_r], where L_r is the interval of grains, and it can be calculated by the following formula [12]. Meanwhile, the positions of grains were simulated by "grain oscillation method," thus getting the theoretical space among grains:

$$L_r = D_{mean}\left(\sqrt{\frac{\pi}{8(31-S_0)}} - 1\right) \tag{31}$$

Where S_0 is the structure number of the grinding wheel.

Based on above modeling, geometric information of a single grain in the grinding zone, including shape, position and pose, can be determined by 7 parameters. Values of these 7 parameters have been calculated in the above text. Therefore, if there's a grain (G_{nm}) in the grinding zone, it can be expressed by arrays:

$$G_{nm} = [\alpha, D, d_z, \gamma_x, \gamma_y, \gamma_z, Z] \tag{32}$$

8.3.3. The Abrasive Matrix on the Grinding Wheel Surface

The grain number along circumferential direction and radial direction of the grinding wheel in the grinding zone were set as N_x and N_y, respectively. These two parameters are related with grain space on the grinding wheel surface, contact arc length (L_g) of the grinding zone and grinding wheel width (b). Their expressions are:

$$N_x = \frac{L_g}{D_{mean}+L_r} \tag{33}$$

$$N_y = \frac{b}{D_{mean}+L_r} \tag{34}$$

Furthermore, the total grain number in the grinding zone is:

$$N = N_x \cdot N_y \tag{35}$$

A $N_x \times N_y$ matrix $G(D)$ of the grain group can be produced by using the normal distribution function of grain size and total grain number in the grinding zone (MATLAB software):

$$G(D) = \begin{bmatrix} D_{11} & D_{12} & \cdots & D_{1N_y} \\ D_{21} & D_{22} & \cdots & D_{2N_y} \\ \vdots & \vdots & \ddots & \vdots \\ D_{N_x 1} & D_{N_x 2} & \cdots & D_{N_x N_y} \end{bmatrix} \tag{36}$$

Due to random distribution characteristics of grains, the protrusion height of each grain on the grinding wheel surface can be acquired through a random function. According to previous research conclusions, the protrusion height observes to the uniform distribution. To further eliminate excessive standard deviation of grain section area brought by random distribution, the protrusion height of grains was further processed by "grain vibration iteration method." In this study, the 240# CBN grinding wheel was applied. The standard deviation of grain section area tends to be stable when the oscillation times are higher than 2,000. The iterative equation is:

$$z_{gjn} = z_{gj0} + \delta \cdot z_{gj0} + \delta \cdot z_{gj1} + \cdots + \delta \cdot z_{gjn} \tag{37}$$

Where j is the grain number and n is the times of oscillation which are designed positively. z_{gj0} and z_{gjn} are coordinates of the initial and final Z-axis of grains. δz_{gj1} is the random number that meets uniform distribution of $[-L_r, L_r]$.

The axial position matrix of finally produced grains is:

$$G(z) = \frac{G(D)}{2} + G(z_g) = \begin{bmatrix} z_{11} & z_{12} & \cdots & z_{1N_y} \\ z_{21} & z_{22} & \cdots & z_{2N_y} \\ \vdots & \vdots & \ddots & \vdots \\ z_{N_x 1} & z_{N_x 2} & \cdots & z_{N_x N_y} \end{bmatrix} \tag{38}$$

8.3.4. Dynamic Abrasive and Cutting Depths

The protrusion height of grains decides whether they contact with the workpiece materials. According to different cut-in depths of grains, the contact state between grain and workpiece can be divided into cutting, ploughing, sliding and non-contact. Firstly, the grain-workpiece contact or noncontact state can be known from grinding depth of grinding wheel. In the experiment, the relative zero point for cutter searching program of grinding wheel is determined when the grinding wheel just touches the workpiece surface. Given the grinding depth (a_p), the grain with the maximum protrusion height shall be used as the zero point. On this basis, protrusion heights of grains which interact with workpiece material shall be within [z_{max}-a_p, z_{max}]. According to this mathematical relationship, the matrix of static effective grains ($G(c)$) and cutting depth matrix ($G(a_p)$) of each grain can be calculated:

$$G(a_p) = G(c) - z_{max} + a_p \quad (39)$$

Since movement of grains is not considered in the solving process, the solved effective grains in the grinding zone are called "static effective grains." If movement of grains are considered, the workpiece surface that the next grain faces with has been cut by one or even several previous grains. The space and height difference between grains are mathematical variables that determine whether the next grain cuts the workpiece surface. Additionally, different grinding parameters also may influence cutting depth of grains. It can be seen from Figure 9 that the static effective grain 2 doesn't contact with the workpiece after it is cut by the previous grain.

With considerations to movement of grains and shielding relation of previous and next grains during movement, Zhang et al. proposed the solving model of "dynamic effective grains" [13]. In the grinding theory, the cutting depth of a single grain is expressed as:

$$a_{gmax} = 2\lambda \frac{V_w}{V_s} \sqrt{\frac{a_p}{D}} \quad (40)$$

Where a_{gmax} is the cutting depth of grains, or known as thickness of the maximum undeformed debris (μm). Λ is the space between two successive cutting grains (μm). V_w is the workpiece speed (m/s). D_s is the diameter of grinding wheel (mm).

For engineering grinding wheel (e.g., brazed CBN grinding wheel), the grain space and protrusion height are parameters that can be set and are equal. Under this circumstance, all grains can contact with workpiece materials: λ=L_r+D. For ordinary grinding wheel with random arrangement of grains, λ refers to the space of effective cutting grains. With considerations to differences of protrusion height among dynamic effective grains, it can further deduce that cutting depth of grains is:

$$a_{g\,max(n)} = 2 \cdot \lambda_{(n\sim n-1)} \cdot \frac{V_w}{V_s} \cdot \sqrt{\frac{a_p}{D}} + \left(a_{p(n)} - a_{p(n-1)}\right) \tag{41}$$

Where $a_{gmax(n)}$ expresses the cutting depth of the nth dynamic effective grain. $\lambda_{(n\sim n-1)}$ is the space between the n-1th and n^{th} dynamic effective grains. $a_{p(n)}$ and $a_{p(n-1)}$ are protrusion heights of the nth and n-1th dynamic effective grains.

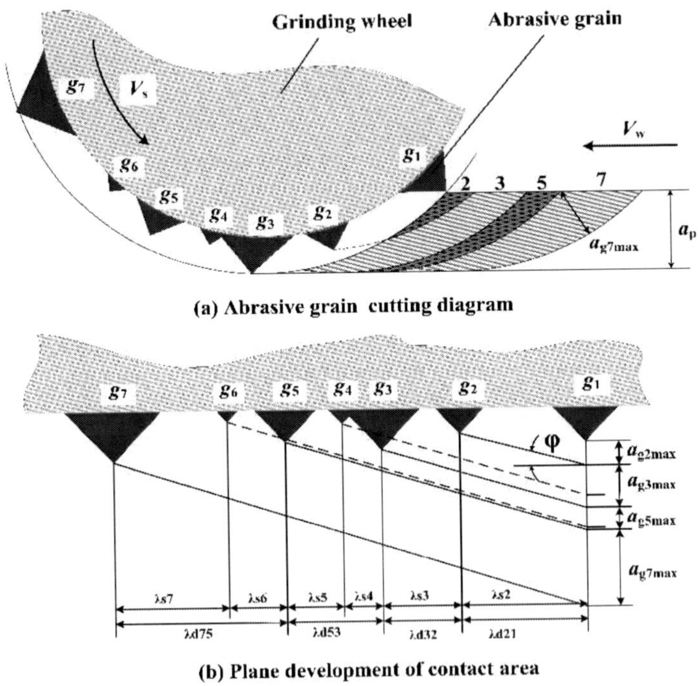

Figure 9. Maximum undeformed chip thickness of dynamic effective abrasives.

8.4. Force Modelling a Single Abrasive in the Micro-Contact Region

8.4.1. Geometric Modelling of an Abrasive/Workpiece Interference

Given the cutting depth of each grain (a_g), there are at most 3 contact surfaces between the grain and workpiece by analyzing their rotation angles around X, Y and Z (Figure 10), which are rake face I, rake face II and the flank surface III.

The interaction areas of rake faces I and II with the workpiece (S_{RI} and S_{RII}) are equal:

$$S_{RI} = S_{RII} \approx \left(a_g \cdot \tan\beta + (D_z - d_z) \cdot \tan\frac{\alpha}{2}\right) \cdot \frac{a_g}{d_z} \cdot D_z \cdot \sin\frac{\alpha}{2} \quad (42)$$

Based on the coordinate system, a cuboid was built by using the normal line of surface as the side. Three angles (θ_x, θ_y and θ_z) were defined (Figure 10).

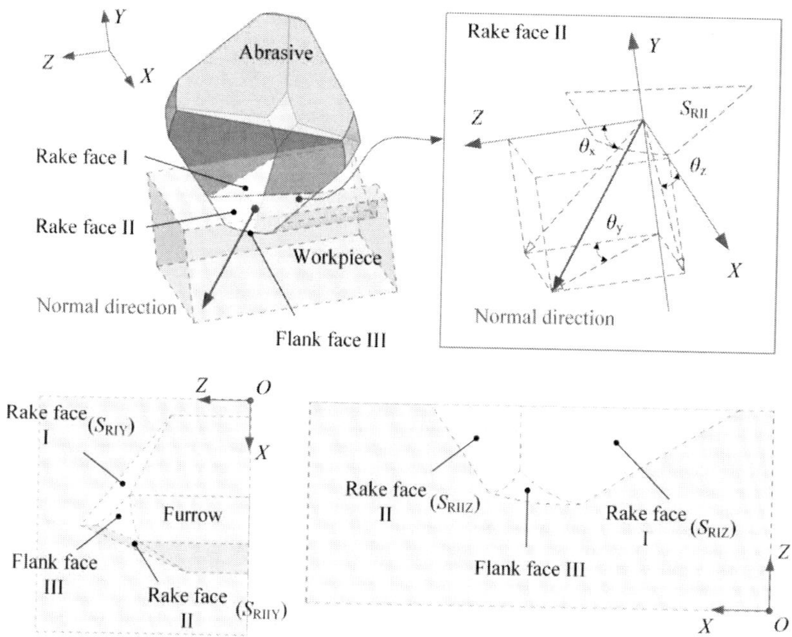

Figure 10. Interference surface and geometric relationship of abrasive/workpiece.

Three angles of rake faces I and II can be solved by shape parameter and pose parameter of grains:

$$\begin{cases} \theta_{xI} = \gamma_x + \beta \\ \theta_{yI} = \frac{\pi-\alpha}{2} - \gamma_y \\ \theta_{zI} = \frac{\alpha}{2} - \gamma_z \end{cases} \quad (43)$$

$$\begin{cases}\theta_{xII} = \gamma_x + \beta \\ \theta_{yII} = \gamma_y + \frac{\pi-\alpha}{2} \\ \theta_{zII} = \gamma_z + \frac{\alpha}{2}\end{cases} \tag{44}$$

Planes passing through O can be built by using X, Y and Z axes as the normal vectors. Then, the projection of rake faces I and II on three planes are:

$$\begin{cases}S_{RIX} = S_{RI} \cdot sin\,\theta_{\mathrm{yI}} \cdot cos\,\theta_{zI} \\ S_{RIY} = S_{RI} \cdot sin\,\theta_{xI} \cdot sin\,\theta_{zI} \\ S_{RIZ} = S_{RI} \cdot cos\,\theta_{xI} \cdot cos\,\theta_{yI}\end{cases} \tag{45}$$

$$\begin{cases}S_{RIIX} = S_{RII} \cdot sin\,\theta_{\mathrm{yII}} \cdot cos\,\theta_{zII} \\ S_{RIIY} = S_{RII} \cdot sin\,\theta_{xII} \cdot sin\,\theta_{zII} \\ S_{RIIZ} = S_{RII} \cdot cos\,\theta_{xII} \cdot cos\,\theta_{yII}\end{cases} \tag{46}$$

Where the projection areas of rake faces are positive, while the interaction area of the flank surface III and workpiece can be solved by the formula of S_{III}. This is not introduced in this section.

8.4.2. Mechanical Behavior and Stress Distribution of an Abrasive/Workpiece Interface

Due to interaction with grains, workpiece materials may develop fracture removal, plastic deformation and elastic sliding. Meanwhile, the workpiece material flow on grain surface and produce frictional force. The frictional force can be solved by stress and CoF of the grain surface. Hence, this section discusses the grain surface stresses under three different conditions (Figure 11).

(i) Plastic deformation of materials

Influenced by grains, materials develop plastic deformation along normal direction of the rake faces. Hence, the normal stress (σ_n) of rake faces of grains against the material is the yield limit (σ_s). According to geometric characteristics of cuboid, the following formula can be gotten:

$$\begin{cases} \tan\theta_x = \frac{\sigma_y}{\sigma_z} \\ \tan\theta_y = \frac{\sigma_x}{\sigma_z} \\ \tan\theta_z = \frac{\sigma_y}{\sigma_x} \end{cases} \quad (47)$$

$$\sigma_x{}^2 + \sigma_y{}^2 + \sigma_z{}^2 = \sigma_n{}^2 \quad (48)$$

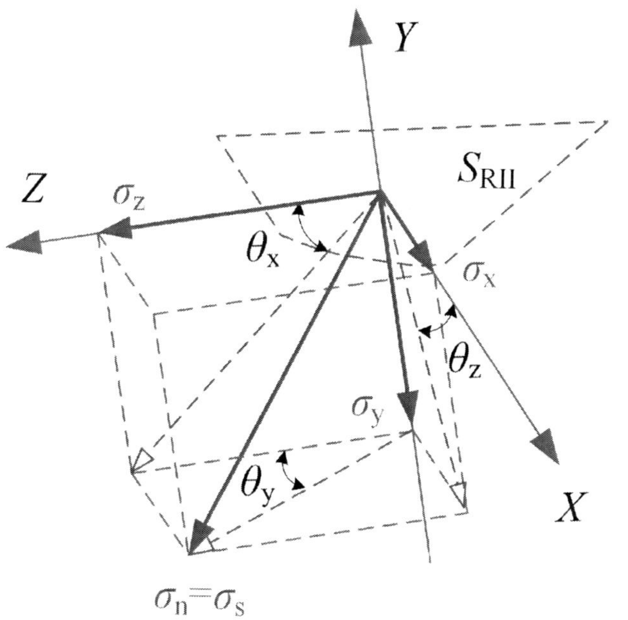

(a)

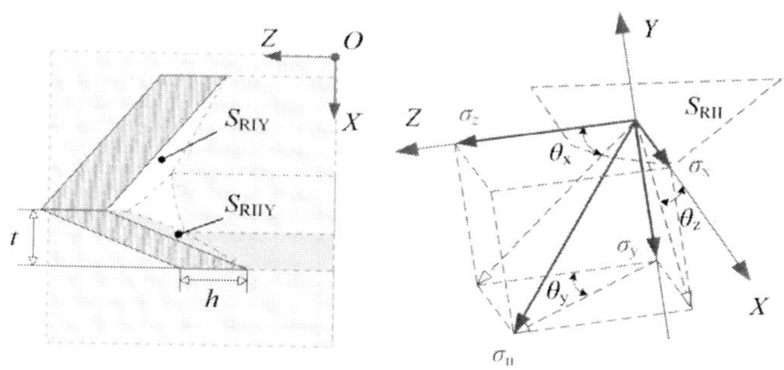

(b)

Figure 11. Surface stress of abrasives and projection in coordinate direction. (a) material plastic deformation, (b) material fracture removal.

Furthermore, the stress components (σ_x, σ_y and σ_z) of rake faces I and II along three directions can be deduced:

$$\begin{cases} \sigma_{xI} = \frac{\sigma_n}{\sqrt{1+(1+\tan\theta_{xI})^2(\tan\theta_{xI}\tan\theta_{zI})^2}} \\ \sigma_{yI} = \sigma_{xI}\tan\theta_{zI} \\ \sigma_{zI} = \frac{\sigma_{xI}\tan\theta_{zI}}{\tan\theta_{xI}} \end{cases} \tag{49}$$

$$\begin{cases} \sigma_{xII} = \frac{\sigma_n}{\sqrt{1+(1+\tan\theta_{xII})^2(\tan\theta_{xII}\tan\theta_{zII})^2}} \\ \sigma_{yII} = \sigma_{xII}\tan\theta_{zII} \\ \sigma_{zII} = \frac{\sigma_{xII}\tan\theta_{zII}}{\tan\theta_{xII}} \end{cases} \tag{50}$$

(ii) Fracture removal of materials

Influenced by grains, workpiece materials break along the feeding direction of grains. Energy for fracture removal of materials is provided by rake faces of grains. If the movement distance of grains toward Z-axis in unit time is h, the projection of debris on the X-Z plane is shown in Figure 11(b). The following formula can be gained according to energy for material breakage and stress on grain surface:

$$\tau_b \cdot h \cdot b = \tau_z \cdot S_{RIIZ} \tag{51}$$

Where h and b are height and length of rake face II on the projection plane along the Y direction. τ_b is the shear break limit and its relationship with break limit (σ_b) is:

$$\tau_z = \tau_b = (0.5 \sim 0.58) \cdot \sigma_b \tag{52}$$

Furthermore, stresses on grain surface along other directions can be gained according to geometric relations of the cuboid:

$$\begin{cases} \tau_{xII} = \frac{\tau_z \cdot \tan\theta_{xII}}{\tan\theta_{zII}} \\ \tau_{yII} = \tau_{xII}\tan\theta_{zII} \\ \tau_{nII} = \sqrt{\tau_{xII}^2 + \tau_{yII}^2 + \tau_z^2} \end{cases} \tag{53}$$

Similarly, the stress relations of rake face I can be acquired:

$$\begin{cases} \tau_{xI} = \frac{\tau_z \cdot \tan\theta_{xI}}{\tan\theta_z} \\ \tau_{yI} = \tau_{xI} \tan\theta_{zI} \\ \tau_{nI} = \sqrt{\tau_{xI}^2 + \tau_{yI}^2 + \tau_z^2} \end{cases} \tag{54}$$

(iii) Elastic sliding of materials

The elastic sliding of materials only occurs on the flank surface of grains. Modeling of forces is discussed in the next section.

8.4.3. The Single Abrasive Interference Model

Due to changes in forms and poses of grains, workpiece material has two stages under the action of rake faces: ploughing and cutting. Stress distributions on micro-contact surfaces under ploughing and cutting states are analyzed, respectively. Meanwhile, a mechanical model is built (Figure 12). The rake face II is chosen for calculation.

(i) Ploughing

When the rake face is at the ploughing state, the deformation degree of workpiece material in contact with the rake face is too small and it only develops plastic deformation, without material removal. The stresses on the rake face are $\sigma_{nI}=\sigma_{nII}=\sigma_s$. Hence, forces caused by yield limit of materials are:

$$F_{pnII} = S_{RII} \cdot \sigma_S = \left[a_g \tan\beta + (D_z - d_z)\tan\frac{\alpha}{2}\right]\frac{a_g}{d_z} D_z \cdot \sin\frac{\alpha}{2} \cdot \sigma_S \tag{55}$$

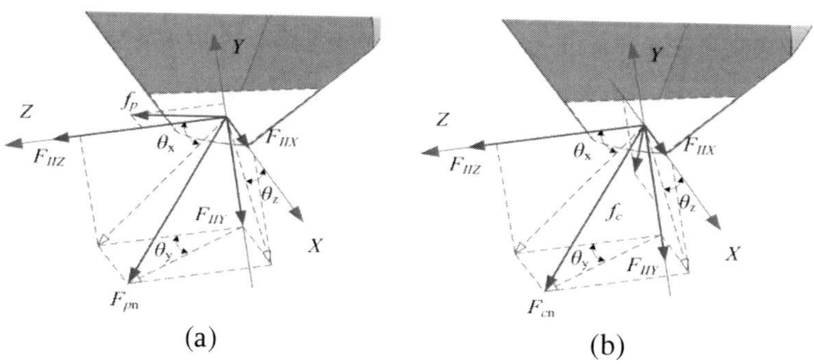

Figure 12. Projection of abrasive surface forces and coordinate directions. (a) ploughing, (b) cutting.

Forces of material plastic deformation along X/Y/Z directions are:

$$\begin{cases} F_{pXII} = -S_{RIIX} \cdot \sigma_{xII} \\ F_{pYII} = -S_{RIIY} \cdot \sigma_{yII} \\ F_{pZII} = -S_{RIIZ} \cdot \sigma_{zII} \end{cases} \tag{56}$$

Similarly, the plastic deformation force of rake face I is:

$$\begin{cases} F_{pXI} = -S_{RIX} \cdot \sigma_{xI} \\ F_{pYI} = -S_{RIY} \cdot \sigma_{yI} \\ F_{pZI} = S_{RIZ} \cdot \sigma_{zI} \end{cases} \tag{57}$$

Besides, workpiece material flows opposite with the rake face during plastic deformation since grains move toward the Z direction, thus causing frictional force. The direction of frictional force is shown in Figure 12(a). The frictional force (f_{pII}) can be calculated according to following equation:

$$f_p = F_{pn} \cdot \mu \tag{58}$$

Where μ is CoF.

Furthermore, components of frictional force of rake faces II and I along X/Y/Z directions are:

$$\begin{cases} f_{pXII} = -f_{pII} \cdot cos\,\theta_{yII} \\ f_{pZII} = f_{pII} \cdot sin\,\theta_{yII} \\ f_{pYII} = 0 \end{cases} \tag{59}$$

$$\begin{cases} f_{pXI} = f_{pI} \cdot cos\,\theta_{yI} \\ f_{pZI} = f_{pI} \cdot sin\,\theta_{yI} \\ f_{pYI} = 0 \end{cases} \tag{60}$$

(ii) Cutting

When rake faces are in the cutting stage, materials which interact with rake faces develop fracture removal along Z-axis and the rake faces have to provide stress (τ_n) for fracture removal of materials. The removed material also will develop plastic deformation. Hence, deformed debris can be observed. The stress that rake faces have to provide shall be the same as the previous part (σ_s). Therefore, stresses of rake faces are σ_n=τ_n+σ_s. The model of stresses can be expressed as:

$$F_{cnI} = \left[a_g \tan\beta + (D_z - d_z)\cdot \tan\frac{\alpha}{2}\right]\frac{a_g}{d_z}D_z \cdot \sin\frac{\alpha}{2}\cdot(\tau_{nI}+\sigma_s) \tag{61}$$

$$F_{cnII} = \left[a_g \tan\beta + (D_z - d_z)\cdot \tan\frac{\alpha}{2}\right]\frac{a_g}{d_z}D_z \cdot \sin\frac{\alpha}{2}\cdot(\tau_{nII}+\sigma_s) \tag{62}$$

Forces of material plastic deformation along X/Y/Z directions are:

$$\begin{cases} F_{cXI} = -S_{RIX}\cdot(\tau_{xI}+\sigma_s) \\ F_{cYI} = -S_{RIY}\cdot(\tau_{yI}+\sigma_s) \\ F_{cZI} = S_{RIZ}\cdot(\tau_z+\sigma_s) \end{cases} \tag{63}$$

$$\begin{cases} F_{cXII} = -S_{RIIX}\cdot(\tau_{xII}+\sigma_s) \\ F_{cYII} = -S_{RIIY}\cdot(\tau_{yII}+\sigma_s) \\ F_{cZII} = -S_{RIIZ}\cdot(\tau_z+\sigma_s) \end{cases} \tag{64}$$

Since debris turns upward along the rake faces after being removed, grains bear downward frictional forces along the rake faces (Figure 12(b)). Frictional force can be calculated according to the following equation.

$$f_c = F_{cn}\cdot\mu \tag{65}$$

Furthermore, components of the frictional force along X/Y/Z directions are:

$$\begin{cases} f_{cXI} = f_{cI}\cdot sin\,\theta_{zI} \\ f_{cYI} = -f_{cI}\cdot cos\,\theta_{zI} \\ f_{cZI} = 0 \end{cases} \tag{66}$$

$$\begin{cases} f_{cXII} = -f_{cII}\cdot sin\,\theta_{zII} \\ f_{cYII} = -f_{cII}\cdot cos\,\theta_{zII} \\ f_{cZII} = 0 \end{cases} \tag{67}$$

(iii) Sliding

The elastic sliding exists on the flank surface-workpiece interface. The arc flank surface and the workpiece material only have elastic sliding behaviors, providing frictional forces. Durgumahanti et al. [14] deduced the calculation formula of frictional force of the flank surface, which was positively related with flank surface area (S) and the average contact stress (P). Moreover, Malkin and Guo [15] deduced the expression of P. On this basis, the normal frictional force and tangential frictional force can be expressed as follows:

$$f_n = S \cdot P = \frac{4 \cdot P_0 \cdot S \cdot V_w}{V_S \cdot D_S} \tag{68}$$

$$f_t = \mu_1 \cdot f_n = \mu_1 \cdot S \cdot P \tag{69}$$

Where D_s is the diameter of grinding wheel and P_0 is a constant related with S_0.

8.4.4. The Identification of Plastic Deformation and Material Removal

The ploughing and cutting states between grains and workpiece can be judged according to following two aspects:

(i) When the cut-in depth (a_g) of grains is smaller than the critical depth (a_{gc}), grains are in the ploughing state; otherwise, grains are in the cutting state. In this study, it determines that a_{gc}=1.18 μm with references to research conclusion of Zhang et al. [13].

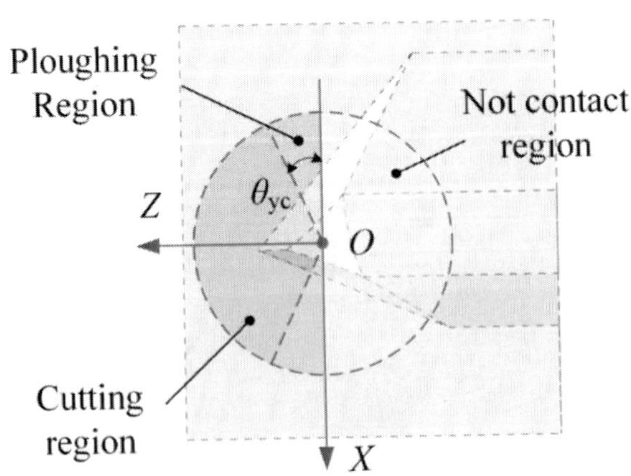

Figure 13. The Angle discriminant condition of abrasive interference state.

(ii) The rotation angle of grains along the Y axis changes the projection area of rake faces on the X-Y plane. When the projection area is smaller than the critical value, the deformation degree of workpiece materials in contact with rake faces is too low and only plastic deformation occurs, without material removal. The principle is the same as with the critical value of cutting depth of grains. As shown in Figure 13, the critical state can be determined by angle (θ, the included angle of projection of the rake face normal line on the X-Z plane and Z axis). It can calculates $\theta_{yc}\approx 63°$ according to the critical cutting depth of $a_{gc}=1.18\mu m$.

The interaction state between grains and workpiece can be judged as follows: ① Cutting: $\theta_y \leq \pi/2-\theta_{yc}$; ② ploughing: $\pi/2-\theta_{yc} < \theta_y < \pi/2$; ③ non-contact: others.

8.5. Modelling and Experimental Verification of the Grinding Force

8.5.1. The Grinding Force Model

As mentioned above, rake faces of dynamic effective grains in the grinding zone have two states of ploughing and cutting, and the grinding force is the sum of forces of all grains. The operation steps of grinding force prediction model are shown in Figure 14: ① build the geometric model and matrix of the grinding wheel according to its parameters; ② solve dynamic effective grains in the grinding zone and cutting depth matrix according to grinding parameters; ③ choose the calculation formula of a single grain according to judgment conditions of grain-workpiece interaction behaviors; ④ stresses of all grains are superposed to solve the general grinding force.

The grinding force model can be expressed as:

$$F_t = \Sigma_1^{N_d}\left[F_{pt(\mathrm{agn})} + F_{ct(\mathrm{agn})} + f_{pt(\mathrm{agn})} + f_{ct(\mathrm{agn})} + f_{t(\mathrm{agn})}\right] \quad (70)$$

$$F_n = \Sigma_1^{N_d}\left[F_{pn(\mathrm{agn})} + F_{cn(\mathrm{agn})} + f_{pn(\mathrm{agn})} + f_{cn(\mathrm{agn})} + f_{n(\mathrm{agn})}\right] \quad (71)$$

Where F_t is the normal grinding force of the grinding wheel. F_n is the normal grinding force of the grinding wheel. N_d is the number of dynamic effective grains in the grinding zone. a_{gn} is the cutting depth of the nth grain: $1 \leqq n \leqq N_d$.

Numerical values of parameters of a single grain are related with the grain-workpiece contact state and they value 0 under non-contact state. If the grain contacts with the workpiece, the contact state shall be further distinguished. If the grain is in the cutting state, calculate the cutting force and the ploughing force is 0. If the grain is in the ploughing state, calculate the ploughing force and the cutting force is 0.

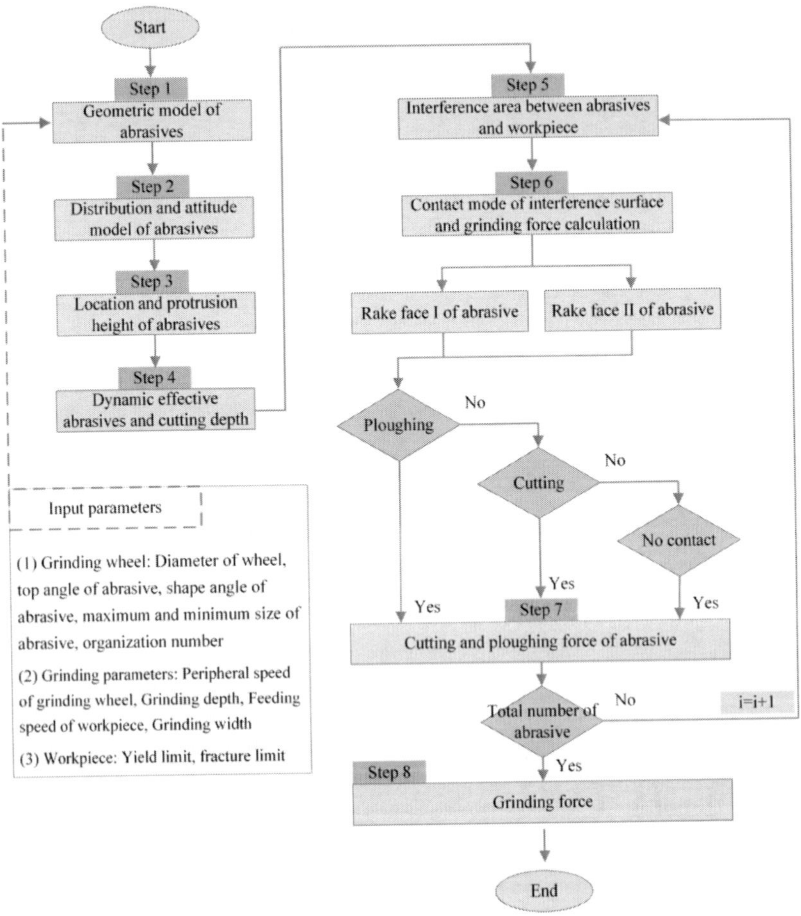

Figure 14. Flow chart of grinding force prediction.

Table 1. Input parameters

Parameter type	Parameter	Value
Grinding wheel parameter	Grinding wheel diameter D_s (mm)	300
	Abrasive tip Angle α (°)	60~90
	Grinding wheel tissue number	12
	Particle shape Angle β (°)	30~50
	Maximum particle size D_{max} (mm)	63
	Minimum particle size D_{min} (mm)	53
Grinding parameter	Grinding wheel line speed V_s (m/s)	10~30
	Feed speed V_w (m/s)	0.01~0.2
	Grinding depth a_p (μm)	10~40
	Grinding width b (mm)	20
Workpiece material parameter	Breaking limit (MPa)	1505
	Yield limit (MPa)	1120

In this simulation test, the specific parameters of the used 240# CBN grinding wheel (CBN240N100B1A1) include: dimension=300 mm × 20 mm × 76.2 mm. The workpiece uses Ti-6Al-4V material. At present, there's only one unknown parameter (P_0) in the model and it can be solved by grinding force measured by a group of experiments. For the grinding condition with a grinding depth of 25 μm, workpiece feeding rate of 0.1 m/s and linear speed of 20 m/s, P_0 values 2×10^{11}. The rest of the parameters are listed in Table 1. The experimental apparatus for grinding process is shown in Chapter 9.

8.5.2. Results for Different Grinding Depths

Given the fixed feeding speed of workpiece and linear speed of grinding wheel, the tangential force and normal force under different grinding depths in MQL grinding were calculated. The prediction results and experimental results are shown in Figure 15.

With the increase of grinding depth, grinding force presents a rising trend, but the growth rate declines gradually. This is because increasing grinding depth expands the grinding zone area significantly, which increases the number of dynamic effective grains and cutting depth to some extent, thus increasing the grinding force. The predicted results of grinding force agree well with experimental results. The average deviation of prediction is 9.2% for tangential force, and 5.7% for normal force.

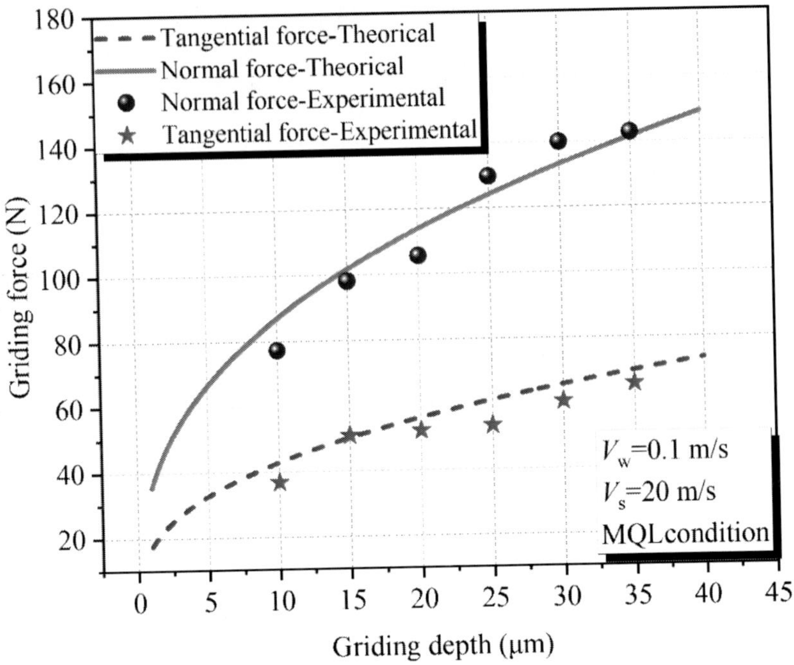

Figure 15. Grinding force under different grinding depths.

8.5.3. Results for Different Wheel Periphery Speeds

Given the fixed feeding speed and grinding depth of workpiece, the tangential force and normal force under different linear speeds of the grinding wheel in MQL grinding were calculated. Prediction results and experimental results are shown in Figure 16. With the increase of grinding depth, the grinding force presents a descending trend, and the reduction rate decreases gradually. It won't change the area of grinding zone by changing linear speed of the grinding wheel alone and the number of dynamic effective grains at the fixed moments may not change. The number of involving grains in unit time increases with the increase of linear speed of the grinding wheel, while the cutting depth of grains decreases, thus showing a descending trend of the grinding force. On this basis, the high-speed/ ultra high-speed grinding has better performances than ordinary grinding. The predicted results of grinding force agree well with experimental results. The average deviation of prediction is 9.7% for tangential force, and 8.8% for normal force.

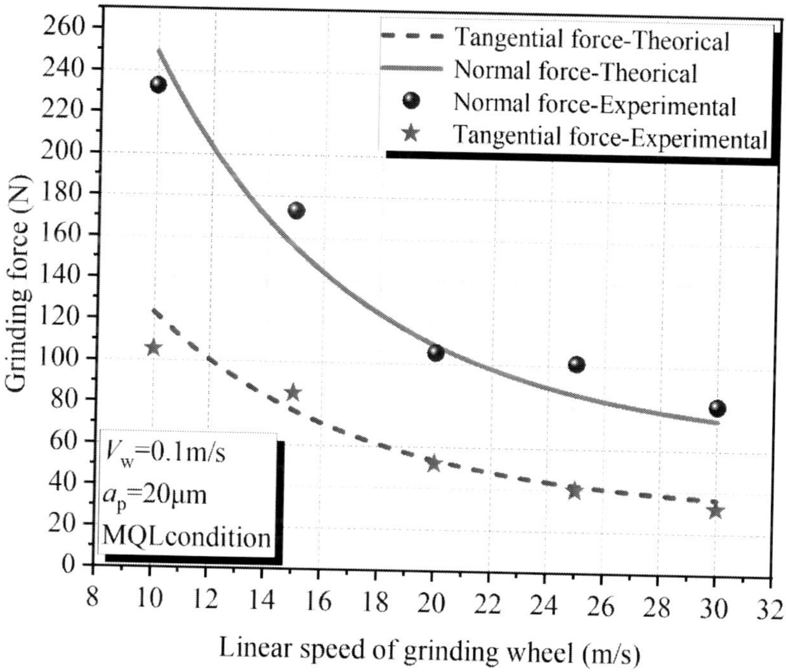

Figure 16. Grinding force under different linear speed of grinding wheel.

8.5.4. Results for Different Feed Speeds

Given the fixed linear speed and grinding depth of grinding wheel, the tangential force and normal force under different feeding speeds of the workpiece in MQL grinding were calculated. Prediction results and experimental results are shown in Figure 17.

As feeding speed of the workpiece increases, the grinding force presents a rising trend and the growth rate increases gradually. The area of grinding zone will not change by changing the feeding speed of the workpiece alone. Hence, the number of dynamic effective grains at fixed moments will be kept the same. However, cutting depth of grains increases with the increase of feeding speed of the workpiece, thus bringing a rising trend of grinding force. due to grain shapes, increasing cutting depth of grains may cause an exponential growth of their contact area with the workpiece. This explains the increased growth rate. Although increasing feeding speed of the workpiece can improve grinding efficiency significantly, wearing of grains is intensified upon the increase of grinding force. The predicted results of grinding force agree well with experimental results. The average deviation of prediction is 9.4% for tangential force, and 9.1% for normal force.

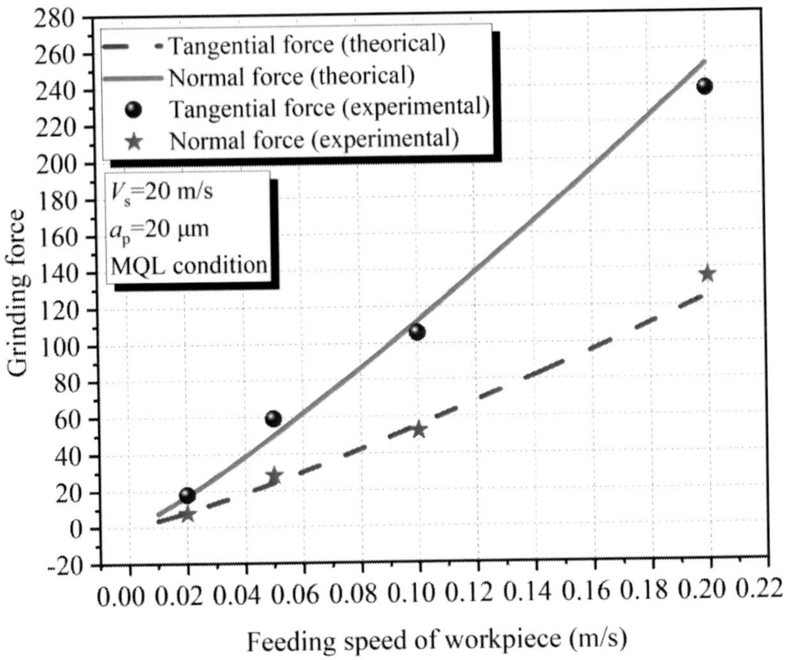

Figure 17. Grinding force under different feed speed of workpiece.

8.5.5. Results for Different Magnetic Field Strengths

According to fitting process of predicted results and experimental results of grinding force, it can be concluded that the CoF in MQL grinding reaches 0.51. Accordingly, the infiltration rate of grinding liquid before application of magnetic fields is 10m/s, which is 0.5 times of linear speed of the grinding wheel (movement speed of grains, 20 m/s). In other words, only half the number of grains in the grinding zone participate in cutting behaviors under lubrication conditions, while the other half of the grains are in dry cutting state. The CoF (0.51) gained from the grinding force model is the numerical sum of these two regions. To establish the relationship between numerical values of magnetic field strength and grinding force, a dry grinding experiment was carried out in this section. The normal force and tangential force were brought into the force model, thus inferring that the CoF is 0.67. Hence, the CoF under complete infiltration condition of lubricant is 0.34. Furthermore, the variation of infiltration flow with magnetic field strength is linear. When the magnetic field strength is 110,000 A/m, the infiltration rate of lubricant is 14.34 m/s. It was applied as the combined boundary condition with the infiltration rate of lubricant with a magnetic field strength of 0. The interfacial CoF can be expressed as:

$$\mu = -7 \times 10^{-7} \cdot \boldsymbol{H} + 0.507 \tag{72}$$

The magnetic fields can be brought into the grinding force model through Eq.(72), thus disclosing influencing laws of magnetic field strength on grinding force (Figure 18). It can be known from the results that grinding force presents a descending trend with the increase of magnetic field strength. This agrees with the theoretical analysis.

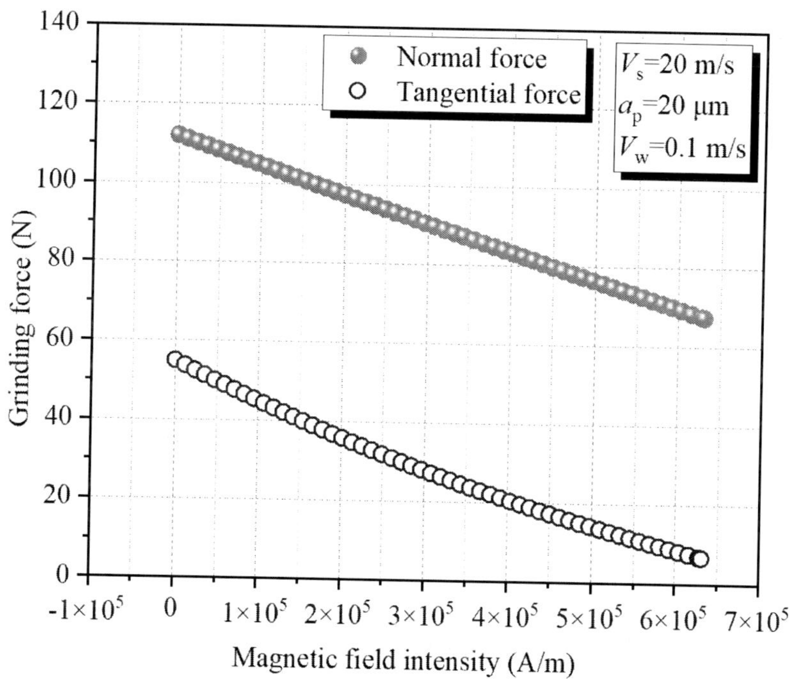

Figure 18. Grinding force under different magnetic field intensity.

Conclusion

This chapter pays key attention to influencing laws of magnetic field strength on infiltration rate and flow of lubricant. For grinding of difficult-to-machine materials, a new grinding force model based on truncated hexahedron grain model was proposed. The research details are introduced as follows:

i. It discloses the shear transport kinetics mechanism of lubricant in the grinding zone and strengthened infiltration mechanism by magnetic field adsorption, and builds a mathematical model about shear infiltration speed and flow of lubricant. The magnetic fields are further introduced. The influencing laws of magnetic field strength on slippage coefficient of interfaces are established by analyzing stresses of lubricants. As a result, the magnetic field strength is introduced into the mathematical model of lubricant infiltration speed and flow. It can be known from the numerical solutions that the lubricant infiltration speed and flow present a rising trend with the increase of magnetic field strength.
ii. According to forms and size distribution laws of grains on the CBN grinding wheel, a method to construct the truncated hexahedron grain model is proposed. Furthermore, the stress distribution model on the grain-workpiece interface based on truncated hexahedron model and mechanical model of a single grain are constructed. The critical criteria for interfering angle of rake faces between ploughing and cutting behaviors are established. Combined with the dynamic effective grain model, a new grinding force model is built.
iii. A model of quantitative relations among magnetic field strength, lubricant infiltration rate and interfacial CoF of grinding zone is built. Hence, the magnetic field strength is introduced into the grinding force model. According to calculation results of grinding force, increasing magnetic field strength can lower grinding force significantly.
iv. The grinding force model is verified by different grinding parameters. According to comparison, predicted results and experimental results agree well. The average deviation of prediction is 9.7% for tangential grinding force, and 9.1% for normal grinding force.

References

[1] Godlevski VA, Volkov AV, Latyshev VN, et al. (1998). A description of the lubricating action of the tribo-active components of cutting fluids. *Lubrication Science*, *11*(1), 51-62.

[2] Williams JA. (1977). The action of lubricants in metal cutting. *Journal of Mechanical Engineering Science*, *19*(5), 202-212.

[3] Smith T, Naerheim Y, Lan MS. (1988). Theoretical analysis of cutting fluid interaction in machining. *Tribology international*, *21*(5), 239-247.

[4] Tang LZ, Geometric model of microchannel in grinding area and microdroplet infiltration mechanism and experimental verification, Qingdao University of Technology, Qing Dao, 2022.

[5] Koshy P, Jain VK, Lal GK. (1997). Stochastic simulation approach to modelling diamond wheel topography. *International Journal of Machine Tools Manufacture*, *37*(6), 751-761.

[6] Koshy P, Iwasald A, Elbestawl MA. (2003). Surface generation with engineered diamond grinding wheels: insights from simulation. *CIRP Annals*, *52*(1), 271-274.

[7] Koshy P, Jain VK, Lal GK. (1993). A model for the topography of diamond grinding wheels. *Wear*, *169*(2), 237-242.

[8] Cooper WL, Lavine AS. (2000). Grinding process size effect and kinematics numerical analysis. *Journal of Manufacturing Science and Engineering*, *122*(1), 59-69.

[9] Williams JA, Xie Y. (1992). The generation of wear surfaces by the interaction of parallel grooves. *Wear*, *155*(2), 363-379.

[10] Bair S, Winer WO. (1990). The high shear stress rheology of liquid lubricants at pressures of 2 to 200 MPa.

[11] Winer WO, Bair S. (1979). Shear strength measurements of lubricants at high pressure. *Journal of Lubrication Technology*, *101*(3), 251-257.

[12] Wang S, Li CH, Zhang DK, et al. (2014). Modeling the operation of a common grinding wheel with nanoparticle jet flow minimal quantity lubrication. *Int. J. Adv. Manuf. Technol.*, *74*(5-8), 835-850. 10.1007/s00170-014-6032-z.

[13] Zhang YB, Li CH, Ji HJ, et al. (2017). Analysis of grinding mechanics and improved predictive force model based on material-removal and plastic-stacking mechanisms. *Int. J. Mach. Tools Manuf.*, *122*, 81-97. 10.1016/j.ijmachtools.2017.06.002.

[14] Durgumahanti USP, Singh V, Rao PV. (2010). A New Model for Grinding Force Prediction and Analysis. *Int. J. Mach. Tools Manuf.*, *50*(3), 231-240. 10.1016/j.ijmachtools.2009.12.004.

[15] Malkin S, Guo C. (1989). Theory and applications of machining with abrasives. *Grinding technology*, 107-245.

Chapter 9

Enhanced Grindability and Mechanism in the Magnetic Traction Lubricant Grinding of Ti-6Al-4V

Abstract

In this chapter, the magnetic-assisted device of MTN is developed to generate traction energy for increasing the infiltration capacity of magnetic nanolubricant in grinding. The effectiveness and mechanism of MTN grinding have also been studied. Firstly, results showed that in general grinding, the Fe_3O_4/graphene nanolubricant showed better cooling and lubrication performance than palm oil, Fe_3O_4 nanolubricant, and graphene nanolubricant. In MTN grinding, grinding force and grinding temperature had reduced by 35.8% and 66.4%, respectively, with the arithmetical mean height Sa decreasing by 27.5%. Plastic uplift and debris adhesion on the workpiece surface were also eliminated. Secondly, the improvement in infiltration performance under magnetic field traction was further verified by the large contact length grinding (LCLG) experiment. Finally, the unique infiltration and antifriction mechanism of MTN were analyzed. The MTN technology solves the technical bottleneck of poor surface integrity and provides a solution for the LCLG of materials with high hardness and toughness and low thermal conductivity.

Keywords: magnetic traction nano-lubrication, grindability, antifriction mechanism, titanium alloy

9.1. Introduction

Titanium alloy occupies a large proportion in the application of difficult-to-machine materials in aerospace, especially in the parts of fan blades, impels, blades and casings in aero-engines, where the weight ratio of titanium alloy is as high as 33%. Due to the high requirements of parts processing accuracy and surface quality, grinding is indispensable in the processing process. However, due to the significant characteristics of titanium alloy-low thermal conductivity, results in a large amount of grinding heat output in the material removal process, resulting in a sharp rise in the temperature of the grinding zone. Serious residual stress, harsh surface roughness and even workpiece burn problems will arise, especially in large contact length grinding (LCLG) [1-3].

In the previous study, the feasibility of magnetic field on the collision and adsorption behavior of magnetofluid microdroplet jets on the grinding wheel surface was verified, and the magnetic nanolubricant was prepared and its excellent tribological properties were verified. Therefore, according to the above research results, further research on magnetic traction nanolubricant (MTN) grinding is carried out.

Therefore, this chapter takes titanium alloy (Ti-6Al-4V) as the workpiece material for grinding experimental research, and analyzes the grinding performance parameters in the process of grinding Ti-6Al-4V with magnetic traction nanolubricant. Grinding force, grinding temperature, surface roughness and workpiece surface quality were investigated to explore the influence of different magnetic nanolubricants on the surface integrity of titanium alloy workpiece under the action of the magnetic field.

9.2. Experimental Setup of Ti-6Al-4V Grinding

9.2.1. Experimental Equipment

As a high-speed rotating tool, the wheel bears a large grinding force during material removal. Structural changes to the wheel should not be made. Otherwise, the wheel will break due to stress concentration under a large cutting force. Therefore, the grinding wheel is magnetized by using a noncontact magnetic field generator without changing its structure. Two permanent magnets are installed on the side of the wheel above the grinding zone. The distribution of magnetic field intensity could be changed by changing the permanent magnet parameters and relative position to the grinding wheel. The matrix material of the wheel is generally aluminum alloy, medium carbon steel (such as 45 [GB]/C45[ISO] steel), and low-carbon steel (such as Q235[GB]/E235 [IS] carbon steel). The direction of magnetic sensing line can be guided and changed by using the soft magnetic material as the matrix of the CBN wheel. Therefore, Q235(GB)/E235(ISO) carbon steel was designated as the matrix material of the wheel.

The Cartesian coordinate system is established as shown in Figure 1, and the origin was set at the lowest point and the center of the width of grinding wheel. Permanent magnets were fixed on the magnet fixing plate through a conical head bolt and nut and were located on the square slots, which were similar in size to the magnets, of the magnet fixing plate. Two pieces of the fixing plate were distributed on the front (+Z direction) and rear (− Z direction) sides of the wheel. The angle β (0–40°) between the magnet fixing plate and the connecting plate can be adjusted. The connecting plate was matched with the side plate through the chute structure, and the distance between the permanent magnet and the side of the wheel L (2–20 mm) can be adjusted. The U-shaped groove on the right side plate (location on +X direction) was fixed on the wheel guard by bolts. The distance between the bottom point of the grinding wheel and the permanent magnet H (10–30 mm) can be changed by adjusting the bolts. A similar mechanical structure was found on the left and right sides of the grinding wheel guard. The nozzle was connected to the MQL system for the microsupply of magnetic nanolubricants.

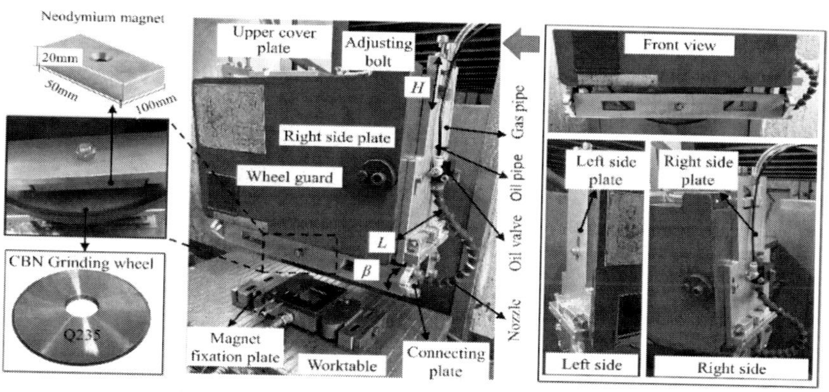

Figure 1. Mechanical structure of the magnetic-assisted device.

This research was undertaken by using a surface CNC precision grinding machine (K-P36) with a feed accuracy of 1 μm and equipped with a CBN wheel, as shown in Figure 1. Q235(GB)/E235(ISO) carbon steel with the relative permeability μ_r of 4000 was used as the substrate material of the wheel. The grinding wheel had dimensions of 300 (outside diameter) mm × 20 (thickness) mm × 76.2 (inside diameter) mm and grain size of 240#. The magnetic-assisted grinding device was designed and manufactured with aluminum alloy with a relative permeability μ_r of 1. A NdFeb magnet with a residual magnetization of 1.17–1.27 T and size of 100 mm × 50 mm × 20 mm was used in the experiment. Magnetic nanolubricants were supplied by an MQL device. A three-component dynamometer (YDM-III99) was used to measure grinding force, and a thermocouple (MX 200) was applied to measure grinding temperature.

9.2.2. Experimental Materials

The workpiece was Ti-6Al-4V with the size of 40×30×30mm. By using graphene, Fe_3O_4 and a mixture of nanoparticles as additives in lubricants (Palm oil). In the study, the size of graphene and Fe_3O_4 nanoparticles were 10 nm and 20 nm, and the volume ratio of mixture nanoparticles is 2:1. The volume fraction of nanolubricant is 5%.

9.2.3. Experimental Scheme

The lubrication conditions adopted in the experiment can be divided into the seven types presented in Table 1 in accordance with the presence or absence of an applied magnetic field.

Table 1. Experimental scheme

NO.	Nanolubricant	Magnetic field
1	Palm oil	No
2	Graphene (5 vol%)	No
3	Graphene (5 vol%)	Yes
4	Fe_3O_4 (5 vol%)	No
5	Fe_3O_4 (5 vol%)	Yes
6	Fe_3O_4/ Graphene (mix ratio=2:1, 5 vol%)	No
7	Fe_3O_4/ Graphene (mix ratio=2:1, 5 vol%)	Yes

The contact length (l_c) of grinding increases with grinding depth, which could be calculated as: $l_c = \sqrt{a_p \cdot D}$ [4]. In conventional speed grinding (general grinding), a_p ranges from 1μm to 50μm. In deep grinding, a_p ranges from 100μm to 30mm [5]. To distinguish general grinding from deep grinding by contact length, a_p = 100 μm was set as the critical value and $l_c \approx$ 5.5 mm with a grinding wheel diameter of 300 mm. In this study, LCLG was defined as l_c > 5.5mm.

Two groups of experiments were designed to evaluate the performance of MTN grinding: general grinding and LCLG. General grinding was conducted to evaluate the grindability of different nanolubricants under a magnetic field, and large contact length grinding was used to verify the infiltration and heat transfer properties of different nanolubricants under the influence of a magnetic field. In the general grinding experiment, a workpiece surface was formed by reciprocating the grinding mode with the grinding wheel speed V_s = 20 m/s, feed speed V_w = 0.1 m/s, and grinding depth a_p = 10 μm. In the LCLG experiment, the progressive grinding mode was used to form circular surfaces. Figure 2 shows that in the horizontal position, the center point of the wheel reciprocated between points A and C. In the vertical direction, when the grinding wheel moved from points A to C, the grinding wheel feed was 10 μm. In this way, a large contact length can be obtained for measuring grinding temperature and surface quality. In this experiment, the contact length of 25.17 mm was obtained. The grinding parameters used are shown in Table 2.

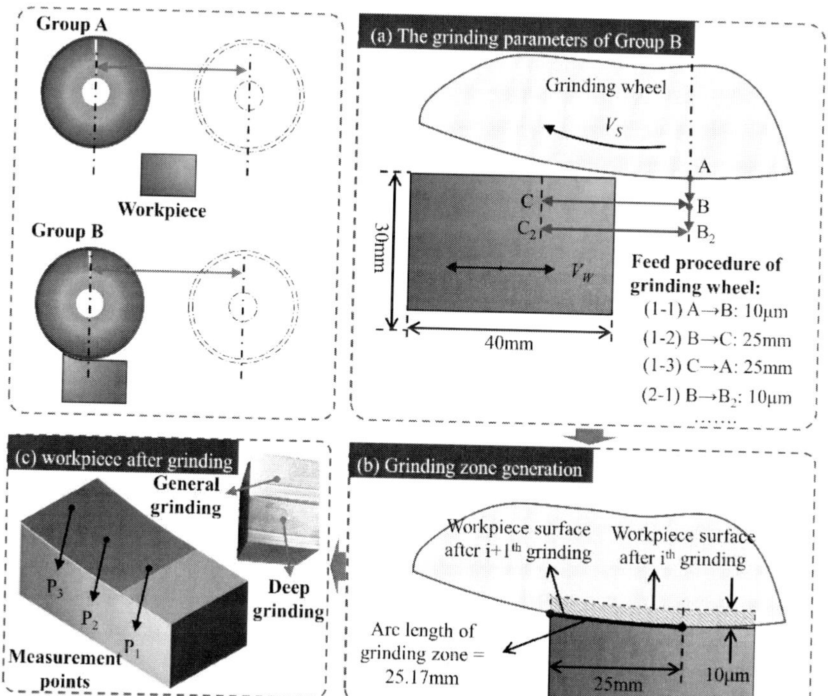

Figure 2. Feed strategy and grinding parameters of LCLG experiment. (a) Grinding parameter setting and grinding wheel feed steps, (b) Grinding zone generation, (c) Workpiece after grinding.

Table 2. Grinding process parameter

Parameter	Value
Grinding wheel speed V_s (m/s)	20
Feed speed V_w (m/s)	0.1
Grinding depth a_p (μm)	20
MQL flow rate (mL/h)	50
MQL nozzle distance (mm)	12
MQL nozzle angle (°)	15
MQL air pressure (Mpa)	0.6

9.3. Results of General Grinding

9.3.1. Grinding Force

The grinding force waveforms differ under different lubrication conditions, as shown in Figure 3. The black curve is the normal grinding force, and the red curve is the tangential grinding force. The axial force is ignored because it is close to 0. Grinding performance can be characterized by the amplitude and mean value of the grinding force curve. The mean value of grinding force reflects the lubrication performance of the lubricant. A low mean value indicates excellent lubrication performance. The amplitude represents the stability of the antifriction behavior of lubricants. When lubrication is unstable, the fluctuation in grinding force curve is large. The grinding wheel vibration under poor lubrication conditions is another cause of large amplitudes.

The use of palm oil results in the grinding force curve with the highest amplitude (Figure 3(a)), which indicates the poor lubrication performance of palm oil. In NMQL grinding, the amplitude of the grinding force curves with the application of graphene, Fe_3O_4, and mixed nanolubricants (Figure 3(b), (d), and (f)) is lower than that with palm oil. The grinding force curve of Fe_3O_4/graphene nanolubricants is more stable and uniform than the grinding force curves of other nanolubricants due to the unique infiltration mechanism of mixed nanoparticles, which is analyzed in detail in Section 6. In MTN grinding, the introduction of a magnetic field has further reduced the grinding force with Fe_3O_4 and mixed nanolubricants. The force curve obtained with Fe_3O_4 nanolubricants (Figure 3(e)) presents a regular arrangement, which indicates a stable lubrication state. This advantage is more obvious for the

Fe_3O_4/graphene nanolubricants than for other nanolubricants, and the best force curve is obtained in the presence of a magnetic field (Figure 3(g)).

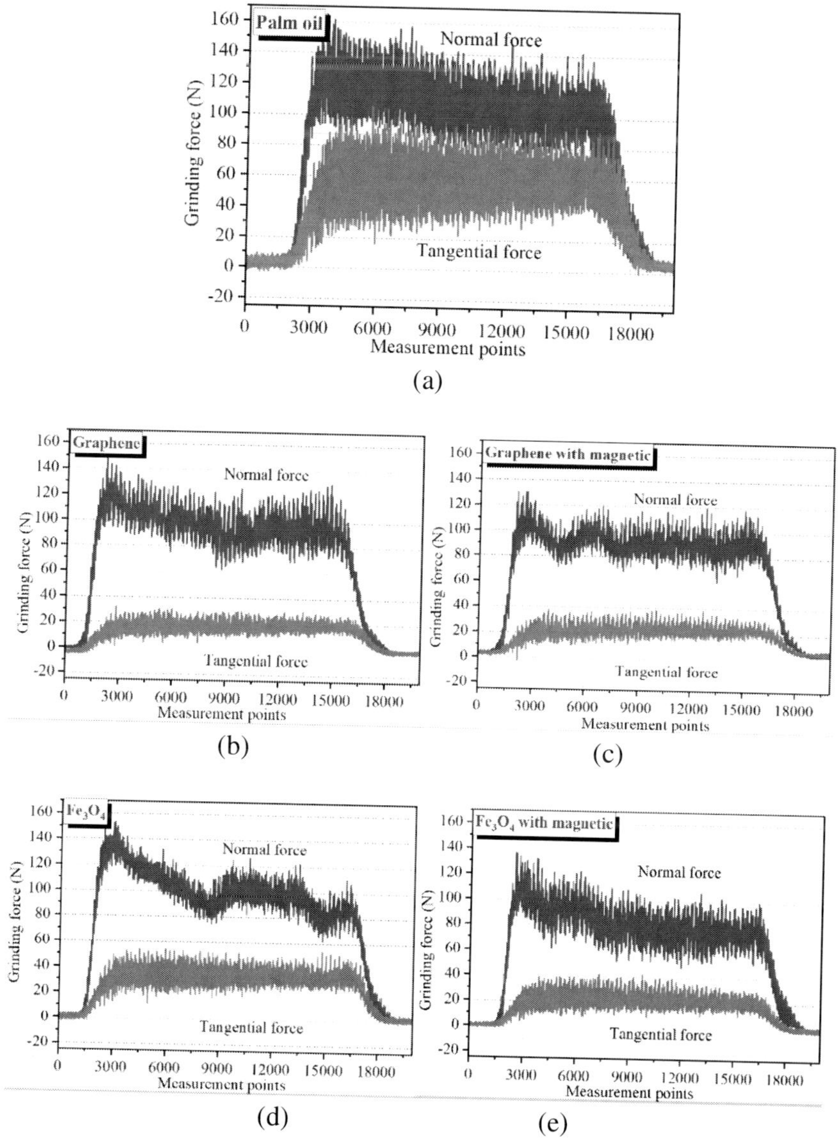

Figure 3. (Continued).

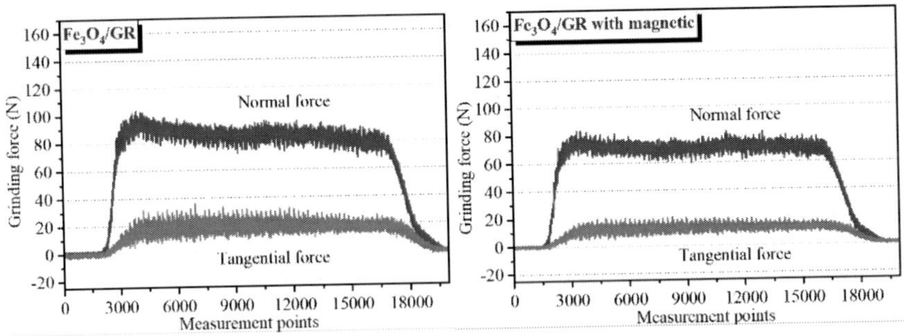

(f) (g)

Figure 3. Grinding force measurement results under different lubrication conditions. (a) Palm oil, (b) Graphene, (c) Graphene with magnetic, (d) Fe_3O_4, (e) Fe_3O_4 with magnetic, (f) Fe_3O_4/GR, (g) Fe_3O_4/GR with magnetic.

To acquire the mean value of the grinding force, grinding was carried out 20 times under each condition. The stable stage was taken from each force curve to calculate the mean value and standard deviation, as shown in Figure 4. Friction energy consumption accounts for more than 90% of the grinding energy, which is converted into grinding heat and has a significant effect on workpiece surface quality in grinding. Therefore, the mean value of the grinding force can be used to characterize the tribological properties of the grinding zone. The maximum normal and tangential grinding forces (109.9 and 52.58 N) are obtained with palm oil. In NMQL grinding, the grinding forces achieved by the three kinds of nanolubricants are lower than those acquired with palm oil, and the order of normal grinding forces from large to small is as follows: Fe_3O_4, graphene, Fe_3O_4/graphene nanolubricants. The lubrication performance of mixed nanolubricants is also proven to be better than that of a single nanolubricant, and the lubrication performance of layered graphene is better than that of spherical Fe_3O_4. The normal and tangential grinding forces found with Fe_3O_4/graphene nanolubricants have reduced by 22.4% and 63.9%, respectively, relative to those observed with palm oil.

In MTN grinding, the grinding force of two kinds of magnetic nanolubricants have decreased significantly under the action of the magnetic field. The normal and tangential grinding forces obtained with Fe_3O_4 nanolubricants have decreased by 17.8% and 33%, respectively, in the presence of the magnetic field relative to those in the absence of the magnetic field. The normal and tangential grinding forces attained with mixed nanolubricants under the magnetic field have reduced by 17.2% and 31.5%, respectively, relative to those acquired in the absence of the magnetic field.

Meanwhile, the grinding force obtained when graphene is used has changed negligibly. The grinding forces of the three kinds of nanolubricants in descending order are graphene, Fe_3O_4, Fe_3O_4/graphene nanolubricant. The normal grinding force of Fe_3O_4/graphene nanolubricants is 22.4% and 14.6% less than that of graphene and Fe_3O_4, respectively.

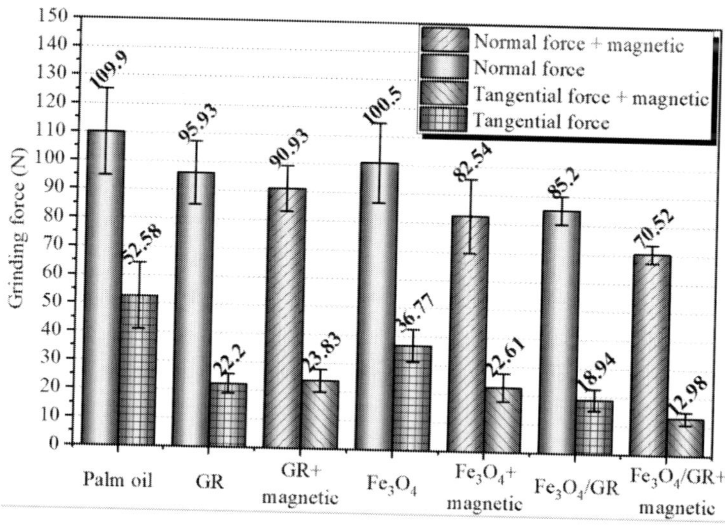

Figure 4. Mean value and variance of grinding force under different lubrication conditions.

The layered graphene nanomaterials present better antifriction performances due to their weak interlayer atomic bonding forces [6, 7]. When the upper and lower layers of layered nanoparticles move together, nanoparticles extend along the fracture layer of the low shear plane to form a film. When the interlayer bonding force is weak (that is, the interlayer bond energy is weak), the nanoparticles easily spread into a film. However, the film formation of layered graphene is relatively slow. Spherical Fe_3O_4 cannot easily form a stable lubricating oil film even if it is fully infiltrated in grinding. Rolling antifriction degrades the lubrication performance of spherical Fe_3O_4 relative to that of layered nanoparticles. In mixed nanolubricants, layered graphene infiltrates into the grinding zone with the help of Fe_3O_4 nanoparticles, which act as wheels. This mechanism can increase infiltration depth and efficiency. Therefore, the lubrication performance of mixed nanolubricants is better than that of Fe_3O_4 and graphene alone. Infiltration and antifriction performances further improve with the addition of a magnetic field.

9.3.2. Roughness of the Workpiece

In this work, the surface quality of the workpiece is characterized by using surface roughness *Sa*, developed interfacial area ratio (*Sdr*), and kurtosis (*Sku*). *Sa* is the extension of the 2D roughness *Ra* to 3D and is the arithmetic mean or geometric average of the distances between points within the contoured and center surfaces. *Sdr* indicates the ratio of the increase in surface area, and *Sku* is an extension of the line roughness parameter *Rku* to 3D for evaluating the sharpness of the height distribution ($Sku = 3$, normal distribution; $Sku > 3$, a sharp surface height distribution; $Sku < 3$, a flat surface height distribution). *Sa* can be used to measure workpiece surface quality comprehensively. *Sdr* and *Sku* can be utilized to measure the degree of plastic accumulation during material removal. As shown in Figure 5, surface roughness was subjected to statistical analysis.

Under the effect of the magnetic field, the maximum *Sa* (0.67 μm) is obtained with palm oil and the minimum *Sa* (0.49 μm) is acquired with Fe_3O_4/graphene nanolubricants and is 27.4% lower than the *Sa* observed when palm oil is used. *Sdr* decreases slightly when Fe_3O_4 nanolubricants are used in the absence of a magnetic field, indicating that the surface area of the workpiece has not reduced due to the unstable lubrication state and the material is not easily removed such that plowing becomes obvious. Moreover, poor lubrication performance leads to increased friction, which causes the surface area of the workpiece and *Sdr* to increase. Meanwhile, *Sdr* and surface area significantly decrease when graphene nanolubricants with good lubrication performance are used. Under a magnetic field, *Sdr* further reduces to 4.29% when mixed nanolubricants are used and is 59.1% less than that when palm oil is applied ($Sdr = 10.49\%$). At this time, the graphene infiltration performance of the Fe_3O_4/graphene nanolubricants is improved, and on the basis of the excellent heat transfer and lubrication performance of the Fe_3O_4/graphene nanolubricants, material removal is facilitated such that the surface of the workpiece smoothens. In terms of the *Sku* of the workpiece surface, the surface distribution is flat when $Sku < 3$. Under the magnetic field, the *Sku* acquired with mixed nanolubricants is the smallest ($Sku = 2.35$), which indicates that the surface of the workpiece is relatively flat.

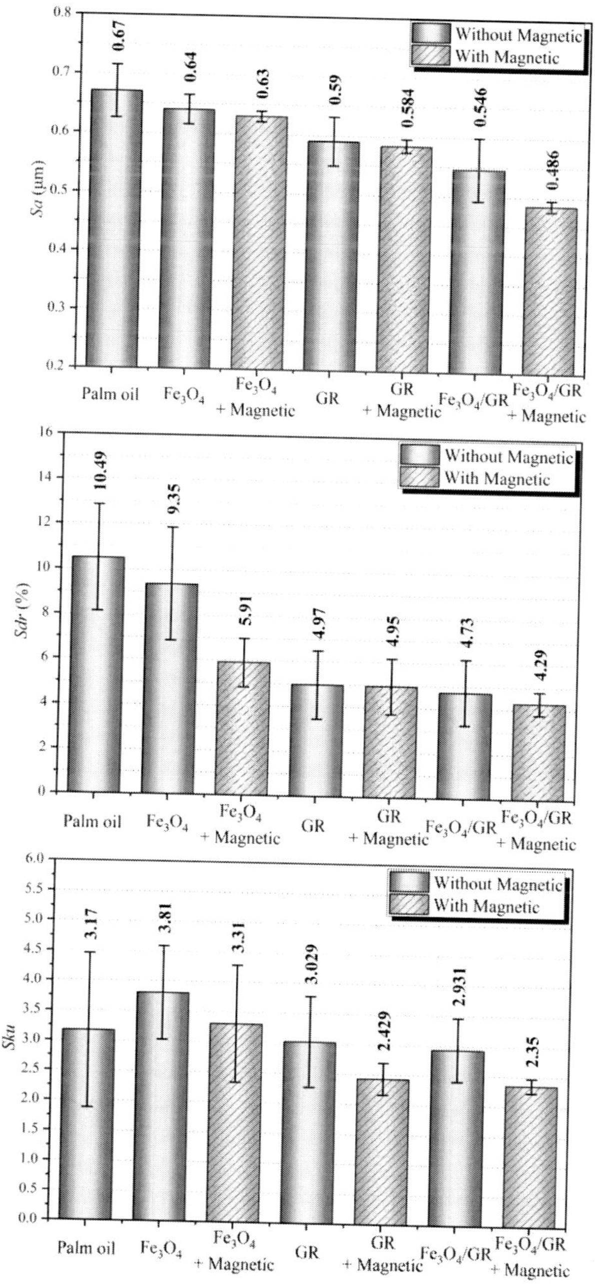

Figure 5. Surface roughness values under different lubrication conditions.

9.3.3. SEM and EDS Analysis

The SEM and EDS images of the workpiece under different lubrication conditions are provided in Figures 6 and 7. The surface morphology images show that the main defects are rolling adhesion, debris adhesion, material stripping removal, and micro pitting.

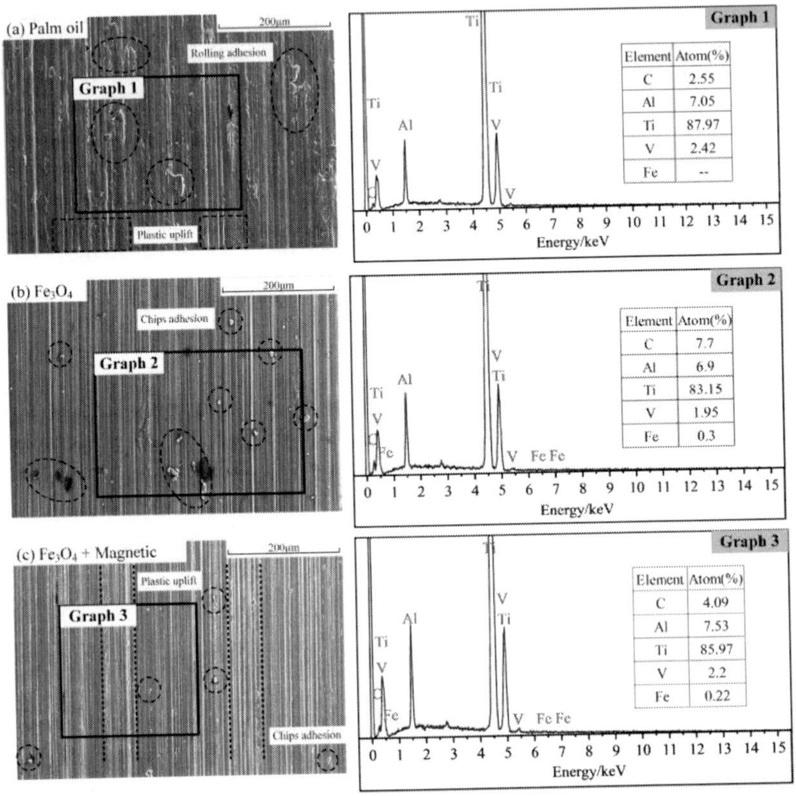

Figure 6. SEM and EDS analysis of the workpiece under (a) palm oil, (b) Fe_3O_4 and (c) Fe_3O_4+magnetic SEM.

Figure 6(a) illustrates that a large amount of plastic uplift and rolling adhesion have appeared on the surface with the usage of palm oil due to the instability of the lubricant film under poor lubrication conditions. Heavy mechanical and thermal loads lead to increased wheel grain wear and dulling, thus complicating workpiece material removal. Therefore, plastic uplift generates on both sides of the furrow. Furthermore, the plastic is crushed and adheres to the surface of the workpiece in the form of grains [8]. Figure 6(b) shows that debris adhesion on the surface increases when Fe_3O_4 nanolubricants are used due to insufficient heat exchange capacity. A large amount of grinding heat cannot be dissipated, causing the workpiece and the grinding debris to overheat and eventually adhere to the workpiece surface. As shown in Figure 6(c), in MTN grinding with Fe_3O_4 nanolubricants, the scratches on the workpiece are regular. However, the debris adhesion on the surface of the workpiece remains high, and burns even occur. This finding indicates that the low thermal conductivity and poor heat transfer capability of Fe_3O_4 nanoparticles have significantly improved.

Figure 7(a) and Figure 7(b) illustrate that debris adhesion has significantly reduced because of the high thermal conductivity of graphene. However, material stripping due to the insufficient infiltration performance of the graphene nanolubricant is found on the surface. The material removal behavior of grains occurs at a negative front angle, and the layered graphene nanoparticles cannot fill in the interface between the rear face and fresh surface of the workpiece in time. Figure 7(c) shows that the rolling adhesion of furrows and micropits persist even when Fe_3O_4/graphene is used. As presented in Figure 7(d), the surface defects of the workpiece have significantly reduced, and surface quality has significantly improved because the use of the magnetic field has changed the antifriction mechanism. In this case, graphene is carried and transported by the directionally and regularly arranged Fe_3O_4 at an accelerated rate. Therefore, the infiltration property in the grinding zone has improved significantly. In addition, graphene with high thermal conductivity can effectively transfer heat from the grinding zone, which effectively reduces the grinding temperature and alleviates burns on the workpiece surface, resulting in a smooth surface without debris adhesion.

Furthermore, the workpiece surface under different lubrication conditions was subjected to EDS analysis. The Fe element is observed when Fe_3O_4 or Fe_3O_4/graphene nanolubricants are used, indicating that Fe_3O_4 nanoparticles leave a small amount of residue on the workpiece surface and can form a certain lubricant film. The elemental Fe content on the surface is higher in traditional NMQL grinding than in MTN grinding. The adsorption effect of the magnetic field on Fe_3O_4 removes some nanoparticles, thus leaving a low amount of residue on the surface of the workpiece and also proves the effect of magnetic field from the side. The elemental C content is low when palm oil and Fe_3O_4 nanolubricants are used. By contrast, the elemental C content when the graphene nanolubricant is used increases to 13.45%, which is 81.04% greater than that when palm oil is utilized. The elemental C also reaches

12.63% in MTN grinding with the mixed nanolubricant, indicating that a stable lubricant film forms on the workpiece surface when graphene is involved in lubrication.

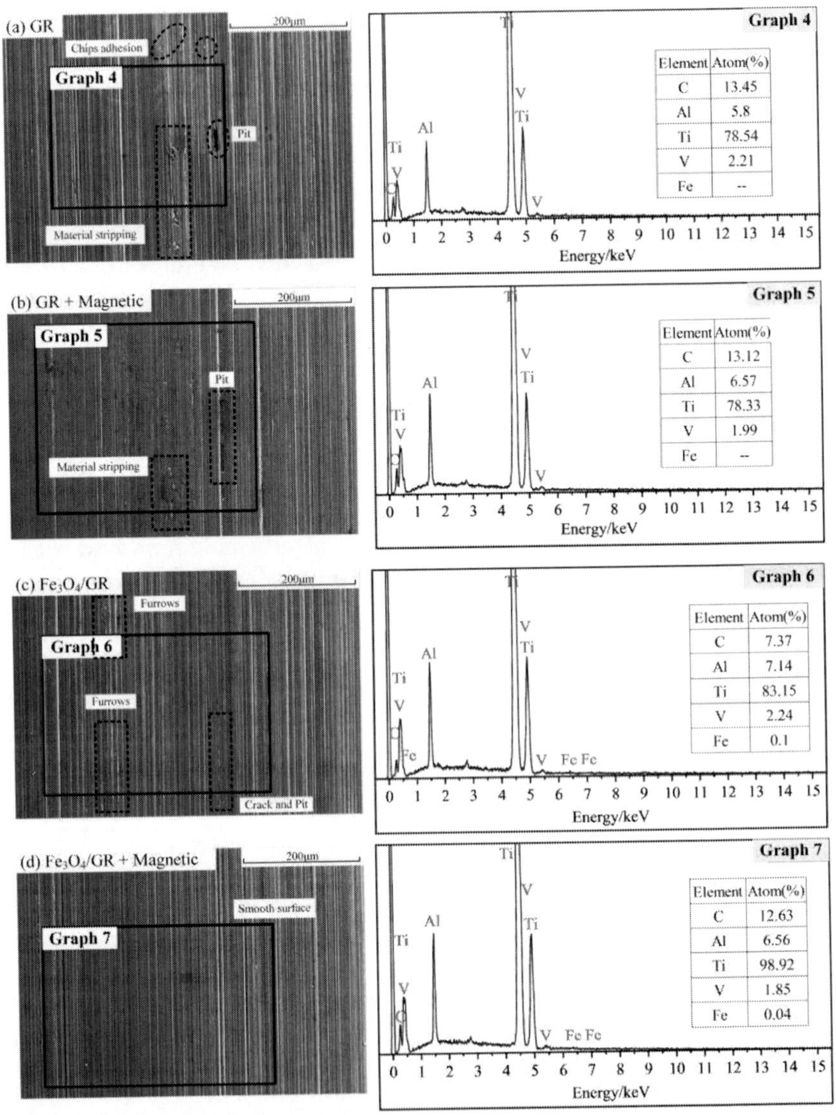

Figure 7. SEM and EDS analysis of the workpiece under (a) GR, (b) GR+magnetic, (c) Fe_3O_4/GR and (d) Fe_3O_4/GR+magnetic.

9.4. Experimental Results of Large-Arc Length Grinding

LCLG experimental studies were carried out to confirm the infiltration performance of MTN in LCLG. The grinding force and temperature measurement method is shown in Figure 8. In this study, the grinding temperatures at different points of the grinding zone (P1, P2, and P3) were measured separately to evaluate infiltration performance under different lubrication conditions. As shown in Figure 8(b), an array of through holes was manufactured in the workpiece, and three thermocouples (Channel 1 [P1], Channel 2 [P2], and Channel 3 [P3]) were arranged equally spaced in the grinding zone. The grinding temperatures at different heights in the grinding zone were recorded separately to characterize infiltration performance. When the infiltration performance is poor, the temperature differences among P1, P2, and P3 are large.

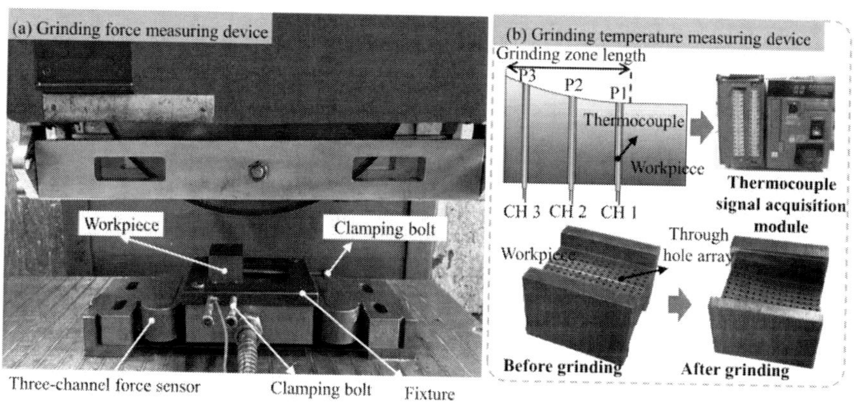

Figure 8. Grinding force and temperature measurement method for LCLG. (a) Grinding force measuring device, (b) Grinding temperature measuring device.

9.4.1. Grinding Force

Figure 9 shows that the trend of the normal grinding force curve differs from that of the general grinding force curve and presents a normal curve variation pattern determined by the grinding wheel feeding strategy. Figure 2 illustrates that during the ith feed, when the grinding wheel is fed from points A to C, the contact length rapidly reaches its maximum and the volume of material removed per unit time increases rapidly. Then, the force reaches its peak value.

The peak value of the grinding force obtained in LCLG is the same as that acquired in deep grinding.

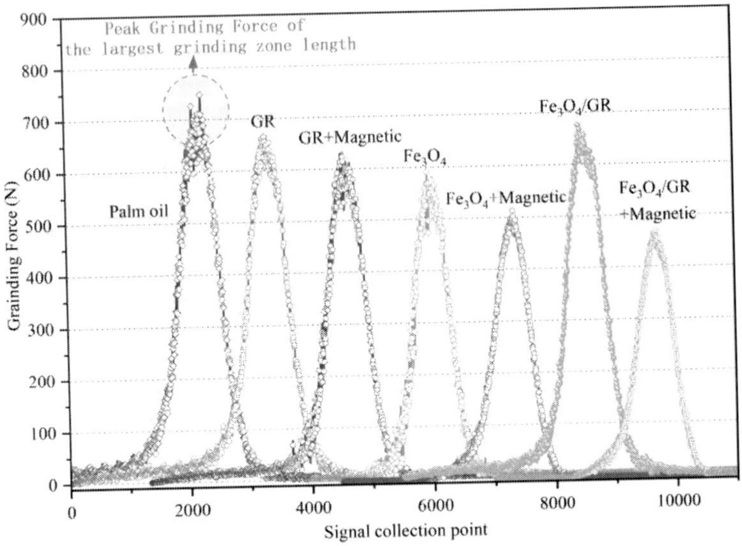

Figure 9. Normal grinding force curve under different lubrication conditions in LCLG.

When palm oil is used, the peak value of the grinding force reaches a maximum of 748.6 N, which is indicative of the highest friction consumption energy and worst lubrication performance. In NMQL, the peak value of the grinding force is reduced when graphene nanolubricants are used but is higher than that when Fe_3O_4 nanolubricants are utilized. Layered graphene cannot enter the highest point of the grinding zone quickly in a short period of time. By contrast, spherical Fe_3O_4 can easily roll into the highest point of the grinding zone and has a certain lubrication effect.

The peak value of the grinding force in MTN grinding with the Fe_3O_4 nanolubricant is 14.1% lower than that in NMQL grinding with the Fe_3O_4 nanolubricant because the traction action of the magnetic field improves the infiltration performance of Fe_3O_4. A similar situation is observed with the Fe_3O_4/graphene nanolubricant. In MTN, the peak value of the grinding force decreases to the minimum value of 469.4 N, which is 37.3%, 24.9%, and 8.36% less than the grinding force observed in grinding using palm oil, the graphene nanolubricant, and the Fe_3O_4 nanolubricant, respectively, because graphene is carried by Fe_3O_4 to the highest point of the grinding zone under a magnetic field.

Figure 10 shows that the variation in the peak value of the grinding force was recorded for 15 consecutive feeds to explore the friction reduction performance under continuous grinding. The change trend of the peak value of the grinding force differs under various lubrication conditions. In grinding, the wheel wears accelerate with the increase in growth rate.

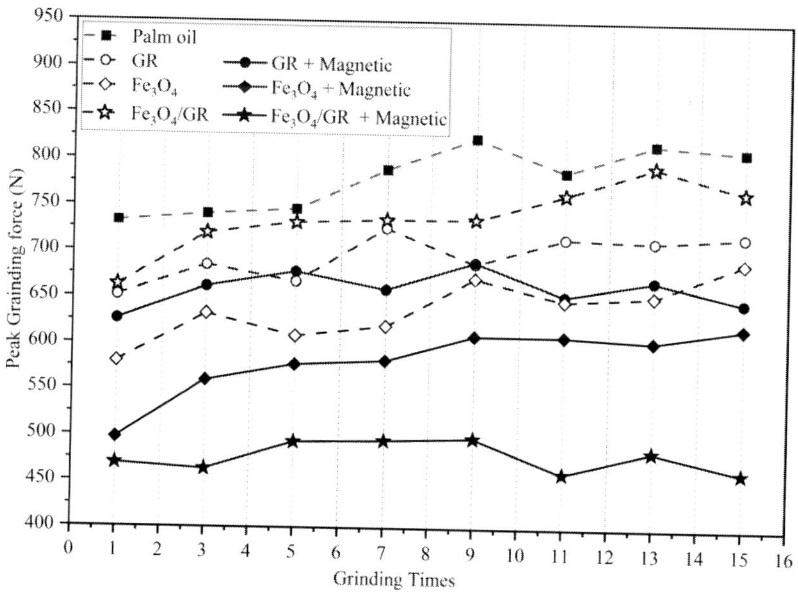

Figure 10. Variation in the peak value of normal grinding force with time in LCLG.

Conversely, if the lubricant has good antifriction properties, then the growth rate of the grinding force decreases when the lubrication effect is good. The grinding force shows the highest growth rate when palm oil is used, indicating that the wheel wears quickly and lubrication is ineffective. The growth rates when Fe_3O_4 and graphene nanolubricants are applied are similar and are better than those when palm oil is used. In MTN grinding with the Fe_3O_4 nanolubricant, the magnitude of the variation has significantly reduced, demonstrating that the magnetic field has improved the grinding force. A similar situation is also observed for the Fe_3O_4/graphene nanolubricant. The grinding force decreases significantly and the curve changes smoothly in the presence of a magnetic field. In this case, the lubricant presents good antifriction properties and protects the wheel from wear.

9.4.2. Grinding Temperature

Figure 11 shows the grinding temperature curves of the three channels detected during one grinding feed. The relationships among the heights of Channels 1, 2, and 3 are ranked from lowest to highest. The minimum grinding temperature is obtained at the lowest point of the grinding zone length (Channel 1), followed by that obtained in Channel 2, with the maximum value occurring at the highest point (Channel 3). The variation in grinding temperature is related to the flow rate of the lubricant. At the lowest point of the grinding zone (Channel 1), the lubricant has sufficient cooling performance and excellent lubrication performance, thus reducing the grinding temperature. As the height of the grinding zone increases, the flow rate of the lubricant decreases gradually, leading to the deterioration of cooling lubrication performance. Thus, the detected peak value of the grinding temperature increases. Therefore, the peak temperature of the highest point in the grinding zone (Channel 3) and the difference between Channels 3 and 1 can be used to characterize the infiltration performance [9, 10].

In addition, the temperature curves of the three channels have different trends. In Channels 1 and 3, the grinding temperature rapidly declines after it reaches its peak value. In Channel 1, the lubricant is sufficient for transferring out grinding heat quickly. In Channel 3, although the lubricant is insufficient, the high workpiece surface temperature also increases the grinding heat transfer efficiency. By contrast, in Channel 2, the temperature slowly peaks then decreases slowly, indicating that the heat transfer efficiency is low due to the insufficient lubricant.

Figure 12 shows the peak value of the grinding temperature and temperature difference (the difference between C1 and C3) under different conditions. The maximum grinding temperature (243.7 °C), which is indicative of the worst grinding performance, is obtained when palm oil is applied. When graphene and Fe_3O_4 nanolubricants are utilized, the peak values of the grinding temperature and temperature difference decrease significantly. Although the layered graphene presents better cooling performance in general grinding due to its high thermal conductivity, its infiltration performance is poor in LCLG and thus results in slightly higher grinding temperature compared with spherical Fe_3O_4 nanoparticles. In MTN grinding with Fe_3O_4 nanolubricants, the grinding temperature shows a significant decrease, confirming the adsorption and traction effect of the magnetic field on nanolubricants. In NMQL grinding, the grinding temperature is higher when using mixed nanolubricants than when using Fe_3O_4 or graphene alone, indicating that the mixing of the two nanoparticle structures restricts the performance of each structure. After magnetic application, the grinding temperature decreases obviously and reaches the minimum grinding temperature (81.9°C), which is 66.4% and 1.68% lower than the grinding temperatures achieved when palm oil and Fe_3O_4 nanolubricant are used, respectively.

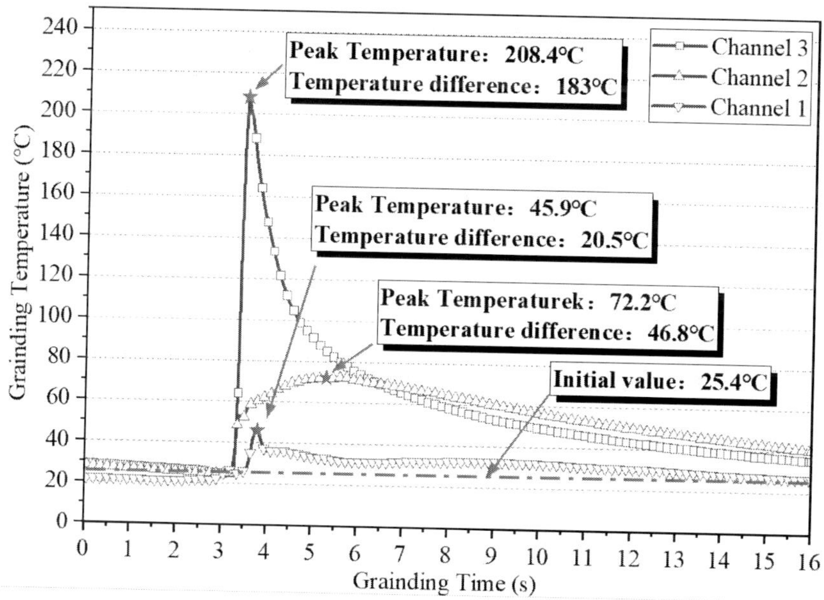

Figure 11. Typical grinding temperature curve in LCLG.

In addition, as can be inferred from the grinding temperature difference under different conditions (the grinding temperature difference of Channels 1 and 3), the maximum value (216.6°C) is obtained with palm oil. The temperature differences observed when Fe_3O_4 and graphene nanolubricants are used are similar and are 103.4°C and 119.1°C, respectively. In the presence of a magnetic field, the temperature difference of Fe_3O_4 and Fe_3O_4/graphene nanolubricants decreases significantly (68.5°C and 70.7°C, respectively). This result fully confirms that the magnetic field has a significant effect on lubricant infiltration.

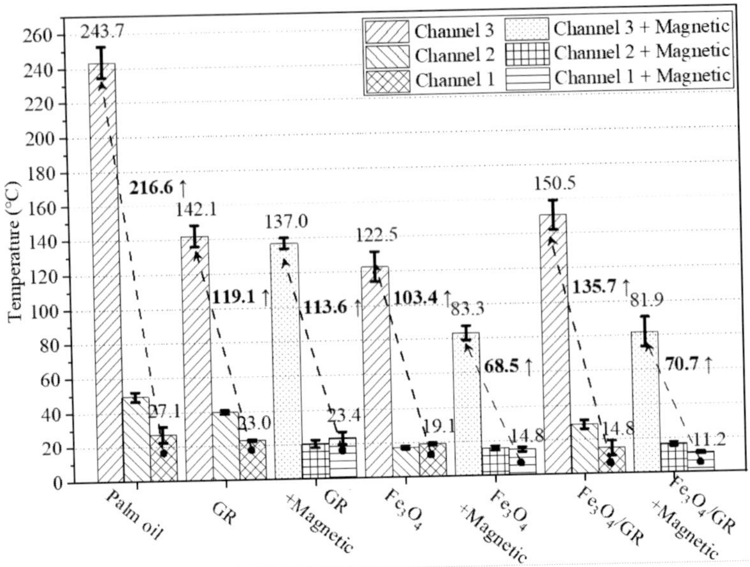

Figure 12. Variation in grinding temperature under different lubrication conditions.

9.4.3. Three-Dimensional Morphology and Surface Roughness

The analysis of grinding force and temperature has revealed that infiltration performance differs under various lubrication conditions. The material removal mechanism also changes under the action of thermodynamic coupling. Therefore, the 3D morphologies and roughness of the workpiece at P1, P2, and P3 were further analyzed, as shown in Figures 13 and 14.

When palm oil is used, plastic uplift occurs at all three points and becomes increasingly obvious from P1 to P3. Compared with the usage of palm oil, the plastic uplift with the usage of graphene nanolubricant reduces at P1 but becomes obvious again at P2 and P3. This phenomenon indicates that the insufficient infiltration of graphene results in poor lubrication. When using Fe_3O_4 and Fe_3O_4/graphene nanolubricants, the surface plastic accumulation at the different grinding zone lengths of the workpiece is not obvious because spherical Fe_3O_4 can also exhibit a certain infiltration ability in LCLG.

Furthermore, in MTN grinding with the Fe_3O_4 nanolubricant, the plastic uplift phenomenon at P1 and P2 is significantly reduced but continues at P3. Nonetheless, the result of MTN is even better than that of NMQL grinding. In MTN grinding using mixed nanolubricants, plastic accumulation has significantly reduced from P1 to P3, reflecting the infiltration and lubrication performance of MTN technology.

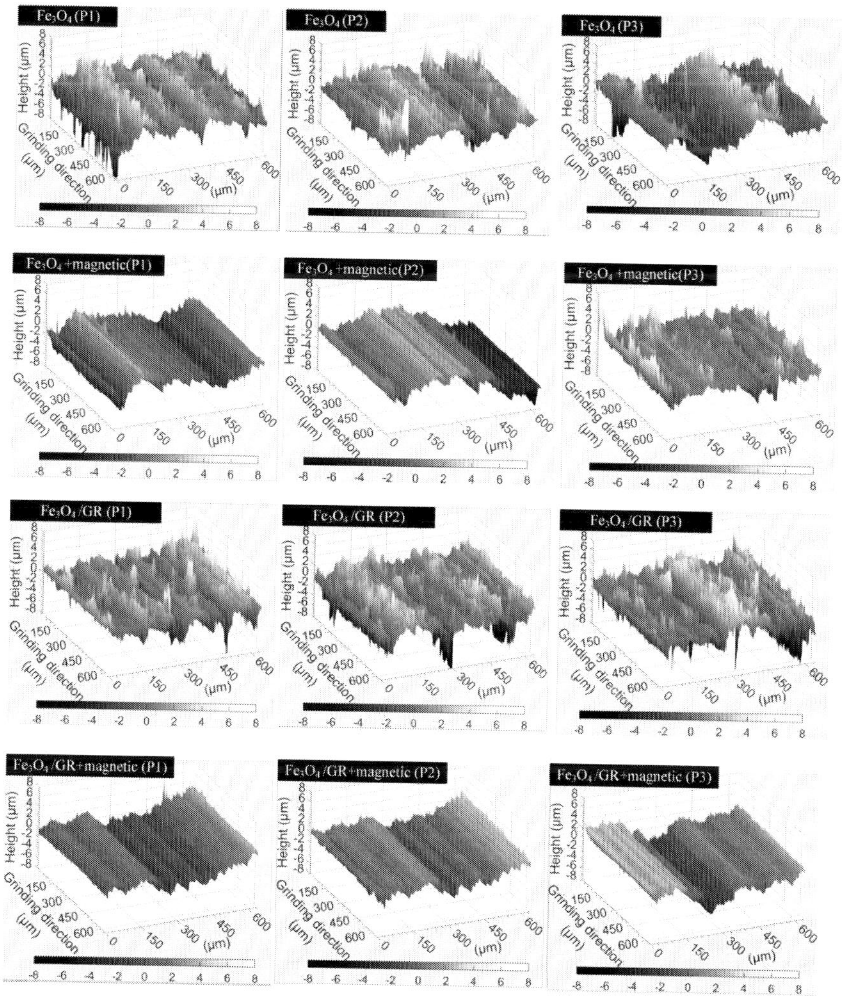

Figure 13. 3D morphology of the workpiece surface under the use of palm oil and graphene nanolubricant.

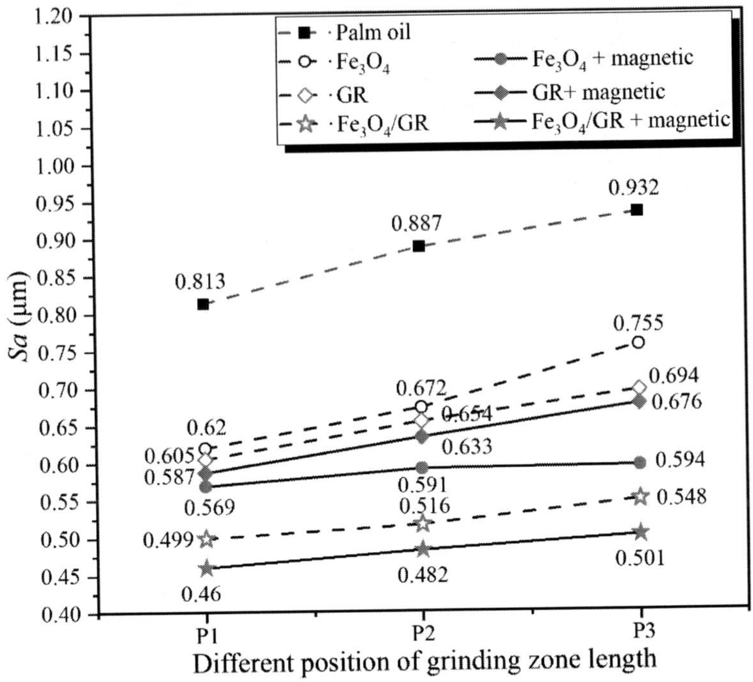

Figure 14. Surface roughness Sa under different lubrication conditions.

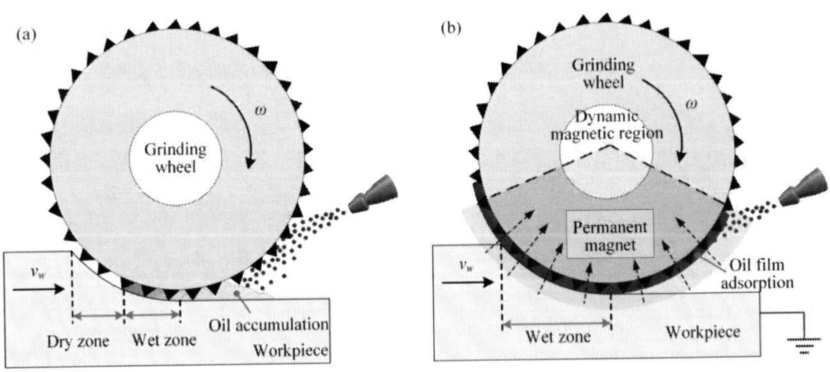

Figure 15. Infiltration mechanisms under (a) MQL condition, (b) MTN condition.

The values of surface roughness *Sa* from P1 to P3 were also recorded and are found to present similar trends as the 3D surface morphology, as shown in Figure 14. Under different conditions, *Sa* increases with the extension of the grinding zone length. In NMQL grinding, *Sa* increases more when palm oil and Fe_3O_4 nanolubricants are used than when the graphene nanolubricant is used. However, in MTN grinding with Fe_3O_4 nanolubricants, *Sa* slightly changes and decreases. Furthermore, in MTN, the best results are obtained when Fe_3O_4/graphene nanolubricants are applied. *Sa* shows the lowest value at each measuring point. *Sa* at P3 has decreased by 46.2% relative to the *Sa* acquired when palm oil is used.

As a result of the unique infiltration and antifriction mechanisms of MTN, MTN achieves better grindability than traditional NMQL. The infiltration mechanism was first analyzed as shown in Figure 15. In traditional NMQL technology, nanolubricants are sprayed at the entrance of the grinding zone after atomization, and oil accumulates. The infiltration of nanolubricants into the grinding zone is a typical shear flow. The driving force of nanolubricants depends solely on the viscous force between the nanolubricants and the wheel's surface. In this case, the lubricant has a relatively low infiltration depth and speed. Only one part of the grinding zone is under wet conditions, whereas the other part is under dry conditions (Figure 15(a)). When the magnetic field is introduced, it exerts an additional adsorption force, which is considerably higher than the viscous force and gravity, on the lubricant. The nanolubricants can be stored in the pores of the wheel and transported. The considerable increase in the infiltration flow rate and velocity of the magnetic nanolubricants significantly increases the effective flow rate (Figure 15(b)).

Conclusion

A novel MTN method was proposed to solve the problem of the insufficient infiltration performance of NMQL, and enhanced grindability was verified through experiments on the general grinding and LCLG of Ti–6Al–4V. The infiltration mechanism of nanolubricants was also revealed on the basis of the characteristics of grinding force, grinding temperature, and surface quality.

i. The grindability of graphene, Fe_3O_4, and Fe_3O_4/graphene nanolubricants in general grinding were studied. The minimum normal and tangential grinding forces of 70.52 and 12.98 N, respectively, were obtained for MTN technology with the Fe_3O_4/graphene nanolubricant. The minimum values of $Sa = 0.49$ μm and $Sdr = 4.29\%$ were also obtained, and surface defects were eliminated.

ii. The improvement in the infiltration performance of nanolubricants under magnetic field traction was further verified by the LCLG experiment. The lowest peak grinding force (469.4 N) and peak temperature (81.9 °C) were observed when the Fe_3O_4/graphene nanolubricant was used. Furthermore, the consistency of the 3D morphology and roughness values of the workpiece at different grinding zone lengths were enhanced.

iii. The magnetic nanolubricants can move with the grinding wheel under magnetic field traction, significantly improving infiltration performance. The regular arrangement of nanoparticles in the magnetic field was conducive to film formation in the grinding zone and therefore reduced grinding force and temperature. Roughness and plastic accumulation caused by strong thermal coupling were also eliminated.

References

[1] Jin T, Stephenson DJ. (2004). Three dimensional finite element simulation of transient heat transfer in high efficiency deep grinding. *CIRP Annals*, *53*(1), 259-262.

[2] Jin T, Stephenson DJ. (2003). Investigation of the heat partitioning in high efficiency deep grinding. *International Journal of Machine Tools Manufacture*, *43*(11), 1129-1134.

[3] Bell A, Jin T, Stephenson DJ. (2011). Burn threshold prediction for high efficiency deep grinding. *Int. J. Mach. Tools Manuf.*, *51*(6), 433-438. 10.1016/j.ijmachtools.2011.01.006.

[4] Darafon A, Warkentin A, Bauer R. (2013). 3D metal removal simulation to determine uncut chip thickness, contact length, and surface finish in grinding. *Int. J. Adv. Manuf. Technol.*, *66*(9-12), 1715-1724. 10.1007/s00170-012-4452-1.

[5] Ding WF, Zhang LC, Li Z, et al. (2017). Review on grinding-induced residual stresses in metallic materials. *Int. J. Adv. Manuf. Technol.*, *88*(9-12), 2939-2968. 10.1007/s00170-016-8998-1.

[6] Cheng JX, Xie F, Li B. (2016). Application of nano-carbon lubricating oil additives. *Contemp Chem Ind*, *45*(4), 843-846.

[7] Cui X, Li CH, Zhang YB, et al. (2019). Tribological properties under the grinding wheel and workpiece interface by using graphene nanofluid lubricant. *Int. J. Adv. Manuf. Technol.*, *104*(9-12), 3943-3958. 10.1007/s00170-019-04129-8.

[8] Liang XL, Liu ZQ, Wang B, et al. (2023). Friction behaviors in the metal cutting process: state of the art and future perspectives. *Int. J. Extreme Manuf.*, *5*(1), 37. 10.1088/2631-7990/ac9e27.

[9] Liu MZ, Li CH, Zhang YB, et al. (2021). Cryogenic minimum quantity lubrication machining: from mechanism to application. *Front. Mech. Eng.*, *16*(4), 649-697. 10.1007/s11465-021-0654-2.

[10] Cui X, Li CH, Zhang YB, et al. (2022). Grindability of titanium alloy using cryogenic nanolubricant minimum quantity lubrication. *J. Manuf. Process.*, *80*, 273-286. 10.1016/j.jmapro.2022.06.003.

Chapter 10

The Influence of Magnetic Field Parameters on Grinding Performance: An Orthogonal Optimization Experiment

Abstract

The magnetic field distribution has significant influence on grinding performance, which is related to the position relationship between the magnet and the grinding wheel. In chapter five, the theoretical research of magnetic field distribution was carried out. However, the comprehensive effects of magnetic field-assisted grinding device parameters on magnetic field distribution and antifriction characteristics of lubricant are unclear. In this chapter, a three-factor and four-level optimization experiment is carried out. The influence weights of horizontal distance, vertical distance, and angle between the permanent magnet and wheel on grinding force and surface roughness are obtained, and the relatively optimized process parameter combination is obtained. Furthermore, the surface power spectral density function and surface morphology are used as characterization parameters to obtain the optimization process parameters of titanium alloy grinding.

Keywords: magnetic-assisted device, orthogonal experiment, grinding performance, magnetic parameters, surface integrity

10.1. Introduction

The magnetic field strength of the grinding wheel is adjusted and controlled by changing parameters of the magnetic field assisted device, thus disclosing influencing laws of parameters on magnetic field strength and its distributions. Based on experimental verification of the previous chapter, the magnetic nanolubricant with good antifriction and antiwear performances has been gained. Parameters of the magnetic field assisted device influence lubrication performances of magnetic nanolubricant by affecting the magnetic field strength surrounding the grinding wheel, thus influencing the grinding performances. Therefore, L, H and β are important parameters to achieve antifriction in the grinding zone and acquire excellent workpiece surface. Nevertheless, although previous studies have proved that parameters of magnetic field assisted device influence the magnetic field strength surrounding the grinding wheel, the influencing laws of L, H and β on grinding performances still remain unknown. Hence, the optimal Fe_3O_4/GR nanolubricant (2:1) (volume fraction: 2%) gained in Chapter 6 was applied in the following experimental study.

Since L, H and β have many levels and they all can influence magnetic field distribution and strength, there are many experimental variables by using single-factor experimental method. Therefore, orthogonal experiment was employed in this chapter. On the one hand, influencing weights of L, H and β on grinding performances were analyzed through an experimental study. On the other hand, optimal magnetic field parameters were gained through experiments.

10.2. Design of Orthogonal Optimization Experiment

10.2.1. Experimental Equipment

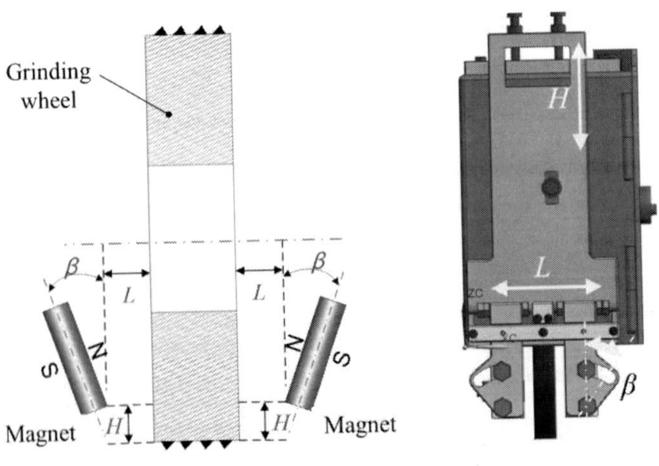

Figure 1. Experimental equipment and adjustment parameter settings.

The experimental apparatus and grinding parameters in this chapter are the same as those in the previous chapter. Differently, the positional parameters (L, H and β) of the magnetic field assisted device must be adjusted, thus realizing different distributions of magnetic fields. The three-way grinding forces were measured by the YDM—III99 three-way dynamometer. The workpiece surface morphology was observed with a laser scanning confocal microscope. The experimental apparatus and adjusting parameter settings are shown in Figure 1.

10.2.2. Experimental Scheme

Orthogonal experimental design is an efficient, fast and economical experimental design method [1]. In this chapter, the orthogonal experimental design was applied to explore magnetic field assisted NMQL grinding of Ti alloy. In the orthogonal experiment, the orthogonal test factors include *L*, *H* and *β* (Table 1). Specifically, *L* ranged within 6~16 mm and *β* ranged within 0°~30°. *H* was chosen 10~25 mm according to the experimental apparatus and technology. A three-factor and four-level $L_{16}(4^3)$ orthogonal experiment was carried out (Table 2). Each row in the matrix represents one experiment and all experiments were implemented randomly. The tangential grinding force (F_t) and normal grinding force (F_n) as well as surface roughness parameters (*Sa*, *Sdr* and *Sku*) in 16 experiments were measured, respectively.

Table 1. Horizontal table of grinding process parameter factors

Factor	Level			
	1	2	3	4
L (mm)	6	9	12	15
H (mm)	15	18	22	25
β (°)	0	10	20	30

Table 2. L_{16} orthogonal experiment table

NO.	*L* (mm)	*H* (mm)	*β* (°)
1-1	6	10	0
1-2	6	15	10
1-3	6	20	20
1-4	6	25	30
1-5	9	10	10
1-6	9	15	0
1-7	9	20	30
1-8	9	25	20
1-9	12	10	20

Table 2. (Continued)

NO.	L (mm)	H (mm)	β (°)
1-10	12	15	30
1-11	12	20	0
1-12	12	25	10
1-13	15	10	30
1-14	15	15	20
1-15	15	20	10
1-16	15	25	0

10.2.3. Experimental Materials

The workpiece material and size used in experiments are the same with those in Section 9.2.2 (Figure 2). The 2% Fe_3O_4/GR(2:1) was used as the nanolubricant. The nanofluids were prepared by a two step method [2].

Figure 2. Grinding surface under different groups.

10.3. Results of Orthogonal Optimization Experiment

10.3.1. SNR Analysis

In experiment, the geometric parameters of the magnetic field assisted device are optimized to get better grinding force and surface roughness. Therefore, the signal-to-noise ratio (S/N) was introduced in to evaluate grinding performances under different combination parameters [3], further getting the optimal parameter combination. The experiments in this chapter aim to get the minimum characterization parameters of grinding performances. The smaller-the-better characteristic evaluation experiment was employed. The calculation formula is [4]:

$$\frac{S}{N} = -10\, log \frac{1}{n}\left(\sum_{i=1}^{n} {y_i}^2\right) \tag{1}$$

Where y_i is the experimental data and n refers to amount of data. The highest S/R represents the optimal grinding performances, and it produces the lowest grinding force and surface roughness. The S/N responses of normal force, tangential force, and surface roughness (*Sa*, *Sdr* and *Sku*) are calculated from Eq. (1), which are listed in Table 3. The S/N values of corresponding factors are listed in Table 4.

Table 3. Corresponding S/N ratios for the observations (db)

NO.	F_n (N)	F_t (N)	*Sa* (μm)	*Sdr* (%)	*Sku*	F_n S/N	*Ft* S/N	*Sa* S/N	*Sdr* S/N	*Sku* S/N
1-1	43.47	15.89	0.43	9.857	3.118	-32.76	-24.02	7.33	-19.87	-9.88
1-2	42.32	15.39	0.477	15.15	3.748	-32.53	-23.74	6.43	-23.61	-11.48
1-3	55.03	16.10	0.766	20.857	3.888	-34.81	-24.13	2.32	-26.38	-11.79
1-4	57.13	16.06	0.739	18.728	3.389	-35.13	-24.11	2.63	-25.45	-10.60
1-5	66.58	16.75	0.64	15.626	3.292	-36.46	-24.48	3.87	-23.87	-10.35
1-6	69.58	18.24	0.681	17.059	3.428	-36.84	-25.22	3.34	-24.64	-10.70
1-7	64.36	16.78	0.808	21.984	3.631	-36.17	-24.49	1.85	-26.84	-11.20
1-8	67.34	17.45	0.763	20.858	3.468	-36.56	-24.83	2.34	-26.38	-10.80
1-9	47.51	16.30	0.513	16.678	4.69	-33.53	-24.24	5.79	-24.44	-13.42
1-10	60.49	17.79	0.586	15.638	3.426	-35.63	-25.00	4.64	-23.88	-10.69
1-11	57.29	17.53	0.401	11.775	3.433	-35.16	-24.87	7.94	-21.42	-10.71
1-12	57.07	16.84	0.532	14.261	3.247	-35.12	-24.52	5.48	-23.08	-10.23
1-13	56.38	16.36	0.695	21.109	3.976	-35.02	-24.27	3.16	-26.48	-11.98
1-14	53.46	17.56	0.484	17.586	5.928	-34.56	-24.89	6.30	-24.90	-15.45
1-15	47.77	16.64	0.588	19.121	3.718	-33.58	-24.42	4.61	-25.63	-11.40
1-16	59.62	17.56	0.551	15.518	3.826	-35.50	-24.89	5.17	-23.82	-11.65

(i) S/N analysis of grinding force.

The main effect of S/N of grinding force is shown in Figure 3. It can be seen from the main effect of S/N of normal grinding force (F_n) (Figure 3(a)) that F_n achieves the highest S/N when L=6 mm, H=10 mm and β=10°. Moreover, L is the primary influencing factor. With the increase of L, the magnetic fields below the grinding wheel distribute more evenly, which improves the adsorption performances of grinding wheel to magnetofluid. As a result, the magnetofluid distribution on the grinding wheel surface will be more uniform, thus developing the nanolubricant performances better.

Moreover, S/N of normal grinding force decreases continuously with the increase of H. In other words, the normal grinding force increases continuously, thus deteriorating the lubrication effect. The included angle (β) between permanent magnet and grinding wheel has insignificant influences.

Table 4. Signal-to-noise ratio response table for all factors

Factor	L (mm)	H (mm)	β (°)
Normal grinding force (F_n)			
1	-33.81	-34.45	-35.07
2	-36.51	-34.89	-34.43
3	-34.86	-34.93	-34.87
4	-34.67	-35.58	-35.49
Counter punch value	2.70	1.14	1.06
Rank	1	2	3
Tangential grinding force (F_t)			
1	-24.01	-24.26	-24.75
2	-24.76	-24.72	-24.29
3	-24.66	-24.48	-24.53
4	-24.62	-24.59	-24.47
Counter punch value	0.75	0.46	0.46
Rank	1	3	2
Sa			
1	4.676	5.041	5.945
2	2.854	5.178	5.100
3	5.965	4.179	4.191
4	4.813	3.909	3.070
Counter punch value	3.111	1.269	2.875
Rank	1	3	2
Sdr			
1	-23.83	-23.67	-22.44
2	-25.44	-24.26	-24.05
3	-23.21	-25.07	-25.53
4	-25.21	-24.68	-25.67
Counter punch value	2.23	1.40	3.23
Rank	2	3	1
Sku			
1	-10.94	-11.41	-10.74
2	-10.76	-12.08	-10.87
3	-11.27	-11.28	-12.87
4	-12.63	-10.82	-11.12
Counter punch value	1.86	1.26	2.13
Rank	2	3	1

According to analysis of the main effect of S/N of tangential grinding force (F_t), F_t achieves the highest S/N at L=6 mm, H=10 mm and β=10° (Figure 3(b)). Among them, L is the primary influencing factor, and its influences are consistent with those on normal grinding force. However, H exerts insignificant influences.

With comprehensive considerations to influences of parameters on normal and tangential grinding forces, a small grinding force can be gained when L=6 mm, H=10 mm and β=10°, indicating the good lubrication effect.

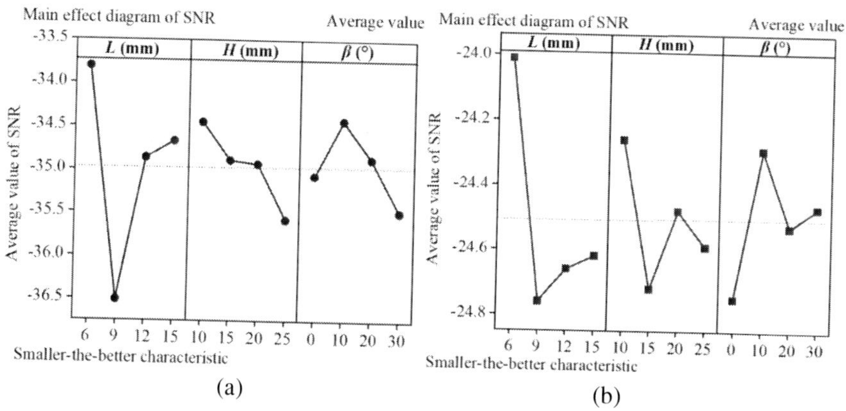

Figure 3. Main effects plot of S/N ratios for grinding force. (a) normal grinding force, (b) tangential grinding force.

(ii) S/N analysis of surface roughness (*Sa*, *Sdr* and *Sku*)

The surface roughness parameters measured in experiments are listed in Table 3. The main effects of S/N of *Sa*, *Sdr* and *Sku* are shown in Figure 4. It can be seen from main effect of S/N of *Sa* that *Sa* achieves the highest S/N when L=12 mm, H=15 mm and β=0° (Figure 4(a)). *Sa* is relatively sensitive to L and β. With the increase of β, S/N presents a linear reduction trend, indicating that a small, included angle can improve surface roughness (*Sa*). Whenβ=0°, the magnetic field strength surrounding the grinding wheel is relatively high and distributes uniformly, increasing adsorption of magnetofluid and improving infiltration performances of nanoparticles. The lubrication effect is good, thus improving the workpiece surface quality.

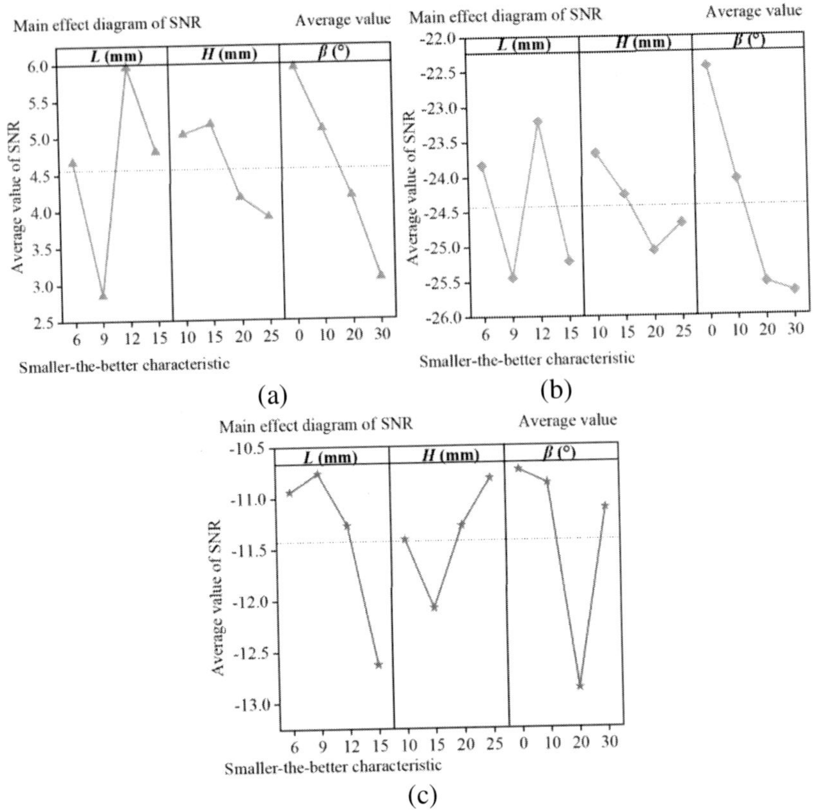

Figure 4. Main effects plot of S/N ratios for surface roughness. (a) *Sa*, (b) *Sdr*, (c) *Sku*.

According to main effect of S/N of *Sdr*, *Sdr* achieves the highest S/N when L=12 mm, H=10 mm and β=0° (Figure 4(b)). Same with the conclusion of *Sa*, *Sdr* achieves the highest S/N when the included angle is small, indicating that the workpiece surface area is increased slightly, and the surface is relatively smooth. In the main effect of S/N of *Sku*, *Sku* achieves the highest S/N when L=9 mm, H=25 mm and β=0° (Figure 4(c)).

Based on above surface roughness analysis, in the rank of S/N response table of parameters to surface roughness characterization parameters, L and β show the most obvious influences, with small differences. Hence, it shall consider setting appropriate L and small β to lower workpiece roughness. There are two parameter combinations to lower the surface roughness: L=12 mm, H=15 mm, β=0°, L=12 mm, H=10 mm, β=0° and L=9 mm, H=25 mm, β=0°.

10.3.2. Variance Analysis

To test the significance of positional parameters of permanent magnets on experimental results, the significant influences of three independent variables on dependent variables were calculated and analyzed through analysis of variance (ANOVA) [5].

(i) ANOVA of grinding forces

The three-factor ANOVA results of influences of three parameters (L, H and β) on normal and tangential grinding forces are shown in Table 5 and Table 6.

Table 5. ANOVA analysis results for F_n

Sources	Degree of freedom	Adj SS	P	Adj MS	F
L (mm)	3	657.12	0.021	219.04	7.09
H (mm)	3	94.22	0.448	31.41	1.02
B (°)	3	81.41	0.503	27.14	0.88
Error	6	185.32		30.89	
Total	15	1018.08			

Table 6. ANOVA analysis results for F_t

Sources	Degree of freedom	Adj SS	P	Adj MS	F
L (mm)	3	5.1277	0.006	1.7092	11.98
H (mm)	3	1.8118	0.063	0.6039	4.23
B (°)	3	1.6635	0.074	0.5545	3.89
Error	6	0.8558		0.1426	
Total	15	9.4587			

According to analysis results of normal grinding force, the significance levels (P value) of L, H and β are 0.021, 0.448 and 0.503, respectively. It can be seen from test results of three parameters that L is significant on the horizontal direction, and it can influence F_n significantly, showing the main effect. The analysis results of tangential grinding force show that P values of L, H and β are 0.006, 0.063 and 0.074, respectively. Same with normal grinding force, L is significant on the horizontal direction, and it influences F_n significantly, showing some main effect.

Grinding force reflects energy consumption and friction in the grinding process. According to analysis, L is significant on the grinding force level and shows the main effect. This is because L mainly influences distribution and uniformity of magnetic fields on the grinding wheel. Therefore, distribution uniformity of magnetic fields can affect grinding forces greatly. If nanolubricant distributes unevenly, it may increase frictions in local grinding zone due to the lack of lubrication and thereby increase the grinding force. H and β influence the magnetic field size obviously. Although changes in the magnetic field size may influence frictional force, it still can guarantee uniform lubricant distribution at each position of the grinding wheel. Hence, we adjust the value of L firstly from the perspective of antifriction and antiwear.

(ii) ANOVA of surface roughness (*Sa*, *Sdr* and *Sku*)

ANOVA results of surface roughness are shown in Table 7, Table 8 and Table 9.

Table 7. ANOVA analysis results for *Sa*

Sources	Degree of freedom	Adj SS	P	Adj MS	F
L (mm)	3	0.09591	0.047	0.031969	4.91
H (mm)	3	0.02613	0.347	0.008711	1.34
B (°)	3	0.08462	0.060	0.028206	4.33
Error	6	0.03908		0.006513	
Total	15	0.24573			

According to analysis results of *Sa*, *P* values of *L*, *H* and β are 0.047, 0.347 and 0.06, respectively. In view of test results of three parameters, *L* is significant on the horizontal direction, and it influences *Sa* greatly, showing the main effect. Moreover, *L* is a primary parameter that influences *Sa*, indicating that *Sa* is affected by antifriction of grinding force. This agrees with the influencing laws on grinding force. Hence, the value of *L* shall be adjusted firstly to get excellent *Sa*.

Table 8. ANOVA analysis results for *Sdr*

Sources	Degree of freedom	Adj SS	P	Adj MS	F
L (mm)	3	47.45	0.059	15.816	4.37
H (mm)	3	15.93	0.315	5.311	1.47
B (°)	3	89.52	0.015	29.840	8.25
Error	6	21.70		3.617	
Total	15	174.60			

According to analysis results of *Sdr*, *P* values of *L*, *H* and β are 0.059, 0.315 and 0.015, respectively. The test results of three parameters show that β is significant on the horizontal direction and it influences *Sdr* significantly, showing the main effect. According to previous studies, changes of β influence the magnetic field gradient greatly. The magnetic field gradient decreases with the increase of β and the adsorption force of magnetic fields to nanolubricant weakens. As a result, the adsorption capacity and infiltration depth of lubricant on the grinding wheel surface decreases. Therefore, there might be insufficient lubricant in the grinding zone, which further deteriorates antifriction and heat transfer performances, and makes it difficult to remove materials. Hence, some defects are developed on the workpiece surface, such as compressive adhesion and material stacking.

Table 9. ANOVA analysis results for *Sku*

Sources	Degree of freedom	Adj SS	P	Adj MS	F
L (mm)	3	2.0332	0.104	0.6794	3.22
H (mm)	3	0.8975	0.326	0.2992	1.42
B (°)	3	2.8966	0.054	0.9655	4.58
Error	6	1.2647		0.2108	
Total	15	7.0969			

According to analysis results of *Sku*, *P* values of *L*, *H* and β are 0.104, 0.326 and 0.054, respectively. The test results of three parameters show that none is significant on the horizontal direction, and they cannot affect *Sku* significantly. We have verified in Chapter 6 that adding magnetic fields can change kurtosis of surface and the workpiece surface is improved. This reflects that changes of *Sku* are attributed to the action of the magnetic field.

Based on above analysis, influences of *H* on grinding forces and surface roughness are not significant. This demonstrates that the magnetic field strength gained in the experiments under the chosen parameters has been high enough and it has reached the maximum magnetized saturation strength of magnetic nanolubricant. The effect of magnetic field on magnetic nanolubricant may not be changed even though increasing the magnetic field strength.

10.3.3. Optimization Results

According to above S/N analysis and ANOVA of normal grinding force, tangential grinding force and surface roughness (**Sa, *Sdr* and *Sku***), four groups of relatively optimal jetting parameters were gained: (1) L=12 mm, H=10 mm, β=0°; (2) L=12 mm, H=15 mm, β=0°; (3) L=9 mm, H=25 mm, β=0°; (4) L=6 mm, H=10 mm, β=10°.

10.4. Experimental Verification and Discussion

Based on experiments in the above section, four groups of relatively optimal magnetic field parameters have been gained. According to Table 3, both grinding forces and surface roughness achieve the highest comprehensive S/N when L=6 mm, H=15 mm and β=10°. They were used as the control group to verify the experiment. The experimental scheme is shown in Table 10. Based on the above studies, five groups of optimized experiment parameters were chosen. The grinding forces and three-dimensional parameters have no obvious differences under these five working conditions. Furthermore, the characterization parameters (including power spectral density function of surface profile and workpiece surface morphology) which haven't been applied in the previous section were used in order to recognize microscopic characteristics of workpiece surface and analyze grinding performances under five working conditions.

Table 10. Experimental design for verification

NO.	L (mm)	H (mm)	β (°)
2-1	12	10	0
2-2	12	15	0
2-3	9	25	0
2-4	6	10	10
1-2	6	15	10

10.4.1. Surface Profile Power Density Spectrum Analysis

To analyze surface profile under different parameter conditions, the time domain signals were analyzed by using the power spectral density function from the frequency domain. The low frequency in the power spectral density reflects the proportion of waviness on the profile, while the high frequency reflects the proportion of roughness on the profile [6-8]. The power spectral density function curves were gained from surface profile curves in five groups of experiments (Figure 5).

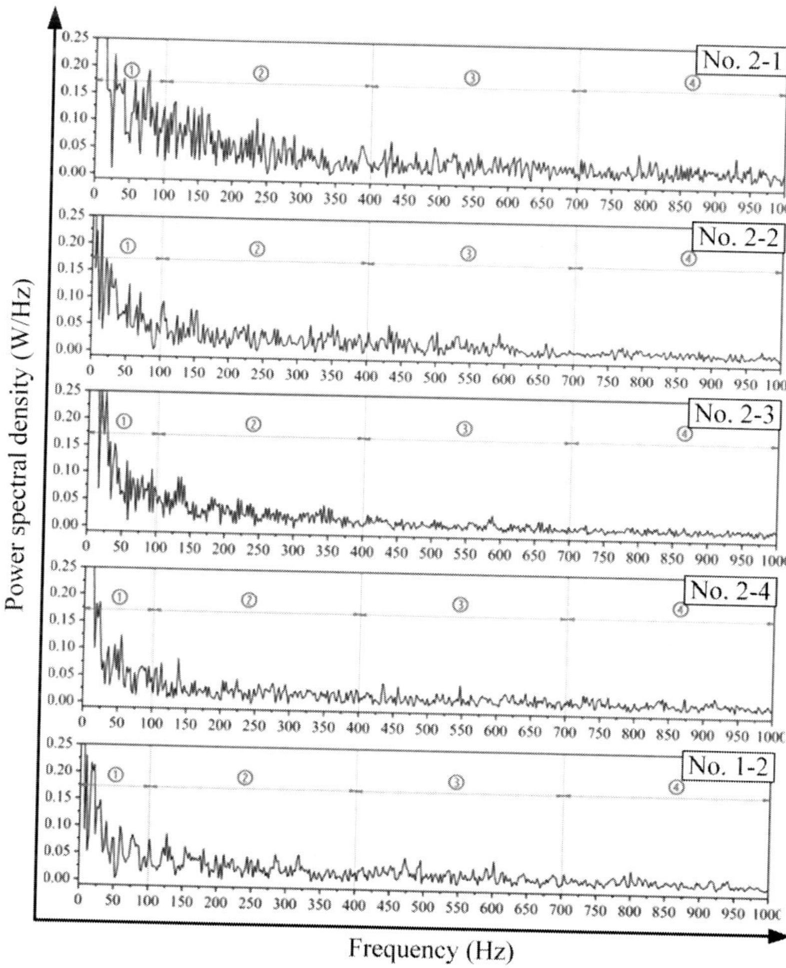

Figure 5. Power spectral density function curves under different conditions.

The power spectral density function curve was divided into four bands according to frequency: ① low frequency of 0-100 Hz; ② 101-400 Hz; ③ 401-700 Hz; and ④ 701-1000 Hz. The proportionality coefficients (*K*) four frequency bands were further calculated, which reflect the proportion of the frequency band in total curve. It is expressed as:

$$K_i = \frac{S_i}{S} = \frac{S_i}{S_1+S_2+S_3+S_4}, \text{ (i=1, 2, 3, 4)} \tag{2}$$

Where S_i is the area enclosed by the ith function curve and the X-axis. S is the area enclosed by the full-frequency density curve and the X-axis. i refers to a frequency band (①, ②, ③ or ④ in the curve). K of each frequency band can be calculated according to Eq. (7-1). Results are shown in Table 11.

According to analysis results of influencing laws of magnetic field assisted device parameters on magnetic field distribution in the grinding zone, the increasing L has small influences on magnetic field size surrounding the grinding wheel, but the magnetic field distribution becomes uniform gradually. In the verification experiment, magnetic field distribution at L=6 mm is the most uneven. As a result, the distribution of nanolubricant on the grinding wheel surface is the most uneven, thus resulting in poor lubrication conditions and poor waviness of the profile curve. The value of H influences the magnetic field distribution slightly, but the magnetic field size decreases with the increase of H. At this moment, the grinding wheel surface will adsorb more nanolubricant when H=10 mm and the viscosity is high. The grinding wheel-workpiece interface has good lubrication conditions, thus getting good workpiece surface quality and low surface roughness. β has no significant influences on distribution and size of magnetic fields. The magnetic field strengths at 0° and 10° differ slightly in experiments.

Table 11. The proportional coefficient K of power spectral density curves in different frequency segments

NO.	K①	K②	K③	K④
2-1	0.597	0.204	0.113	0.086
2-2	0.637	0.192	0.104	0.067
2-3	0.742	0.149	0.063	0.046
2-4	0.729	0.124	0.085	0.062
1-2	0.782	0.105	0.066	0.045

According to K values in the table, the proportionality coefficient (K①) gained by the Group 2-1 in the low-frequency band (①) is the lowest (0.597), and the proportionality coefficient (K④) gained in the high-frequency band (④) is the highest (0.086). This reveals that the power spectral density of

Group 2-1 in band ① is the lowest, and it is the highest in band ④. In other words, surface profile curve under this working condition is the profile feature dominated by roughness and the waviness is relatively low, getting the best workpiece surface. Different from Group 2-1, *H* increases and magnetic field strength decreases in Group 2-2, which will influence workpiece surface roughness obviously. It can be seen from Table 11 that $K_{①}$ is 0.637, which is next to that of Group 2-1, but $K_{④}$ is 0.067, which differs from that of Group 2-1.

Furthermore, the value of *L* is the lowest in Group 2-4 and the control group 1-2. The lubricant distribution is the most uneven and the lubrication effect is poor, accompanied by poor waviness of the profile curve. Therefore, $K_{①}$ is relatively high, valuing 0.729 and 0.782. This proves that profile curves of Group 2-4 and Group 1-2 have poor waviness, the average fluctuation space of workpiece surface profile is large, and the surface quality is not satisfying. The workpiece surface profile reflects the proportion of roughness of high-frequency band. In Group 2-4, $K_{④}$=0.062, which is higher than the control group ($K_{④}$=0.045). The value of *H* is different between Group 2-4 and the control group. Specifically, the value of *H* of Group 2-4 is lower than that of the control group. Hence, the workpiece surface quality of Group 2-4 is better than that of the control group. According to previous analysis, the value of *H* in Group 2-3 is the highest. Compared to other values of *H*, magnetic field strength of Group 2-3 is the lowest and the workpiece surface roughness is relatively high. It also can be seen from Table 11 that $K_{①}$ differs slightly with those under other working conditions and $K_{④}$ achieves the minimum (0.046).

10.4.2. Three-Dimensional Morphology of Workpiece Surface

The workpiece surface morphology under different working conditions is further observed (Figure 6). Compared to workpiece surface morphology in Chapter 9, relatively smooth workpiece surface has been gained under all five groups. There are very mild furrows in Group 2-1, while furrows in Group 2-2 are relatively more obvious and the overall workpiece surface is relatively smooth. According to power spectral density analysis results of surface profile, magnetic field strength is high and distributes uniformly at the maximum *L*, which can improve dispersion and adsorption of nanolubricant obviously. Compared to the other four groups, Group 2-3 develops some micropits and mild furrow compression, which is consistent with analysis results of power spectral density. In Group 2-4 and the control group, the workpiece surface is relatively smooth, only showing some mild adhesion and micropits. Based on the above analysis, we conclude that the best workpiece surface is achieved under experimental parameters of Group 2-1, which agrees with the analysis conclusion of profile power spectral density function. Hence, the optimal experimental parameters are L=12 mm, H=10 mm and β=0°.

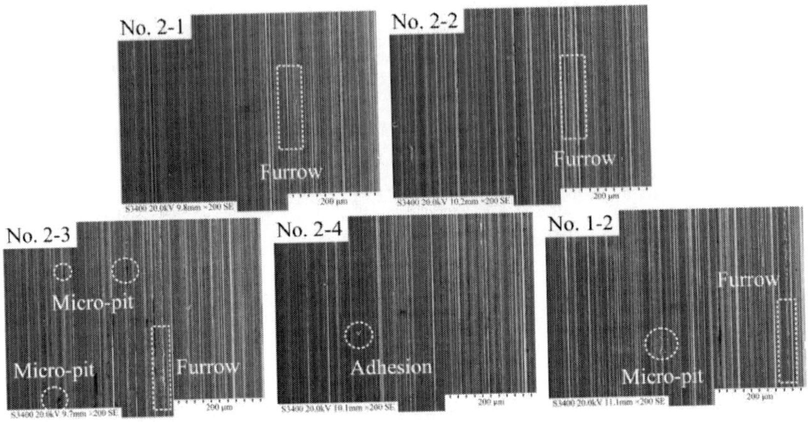

Figure 6. Surface morphology of workpiece under different conditions.

Conclusion

An experimental study of influences of geometric parameter combinations of magnetic field assisted device on grinding performances is carried out. Through S/N analysis and ANOVA of grinding forces and surface roughness (*Sa*, *Sdr* and *Sku*), the power spectral density of surface profile and workpiece surface morphology in verification experiments are further analyzed. Some major conclusions can be drawn:

i. When L=6 mm, H=10 mm and β=10°, both normal and tangential grinding forces (F_n and F_t) achieve the highest S/N. The P value of influences of L on F_n and F_t are 0.021 and 0.006. There's significance on the grinding force level, showing main effects. Hence, the distance from permanent magnet to two sides of the grinding wheel can influence improvement of magnetic field distribution in the grinding zone significantly, thus increasing adsorption capacity of lubricant and lowering grinding forces effectively.

ii. With respect to influences of L, H and β on surface roughness, it can be gained from the rank of their S/N responses to surface roughness, L and β have the most obvious influences and differ slightly. Moreover, β is 0° when surface roughness achieves the highest S/N. This proves that included angle plays an important role in lowering surface roughness.

iii. In verification experiments, $K_{①}$ in the low-frequency band (①) reaches the minimum (0.597) when L=6 mm, H=10 mm and β=10°, while $K_{④}$ in the high-frequency band (④) reaches the maximum (0.086). Compared to other four working conditions, the surface profile curve under L=6 mm, H=10 mm and β=10° is dominated by roughness, accompanied with small waviness and the best workpiece surface quality.

iv. According to observation of workpiece surface morphology in the verification experiment, it further proves that the workpiece surface quality under L=6 mm, H=10 mm and β=10° is better than other parameter combinations, which is attributed to the better cooling and lubrication performances.

References

[1] Liu RJ, Zhang YW, Wen CW, et al. (2010). Study on the design and analysis methods of orthogonal experiment. *Experimental Technology and Management*, *27*(9), 52-55.

[2] Li BK, Li CH, Wang YG, et al. (2015). Technological investigation about minimum quantity lubrication grinding metallic material with nanofluid. *Recent Patents on Materials Science*, *8*(3), 208-224.

[3] Fratila D, Caizar C. (2011). Application of Taguchi method to selection of optimal lubrication and cutting conditions in face milling of $AlMg_3$. *J. Clean Prod.*, *19*(6-7), 640-645. 10.1016/j.jclepro.2010.12.007.

[4] Pal RK, Garg H, Sarepaka RV, et al. (2016). Experimental Investigation of Material Removal and Surface Roughness during Optical Glass Polishing. *Mater. Manuf. Process.*, *31*(12), 1613-1620. 10.1080/10426914.2015.1103867.

[5] Cetin MH, Ozcelik B, Kuram E, et al. (2011). Evaluation of vegetable based cutting fluids with extreme pressure and cutting parameters in turning of AISI 304L by Taguchi method. *J. Clean Prod.*, *19*(17-18), 2049-2056. 10.1016/j.jclepro.2011.07.013.

[6] Wang QY, Liang ZQ, Bai SW, et al. (2021). Power spectrum density characterization of grinding wheel surface in ultrasonic vibration spiral grinding. *Diamond & Abrasives Engineering*, *41*(1), 58-64.

[7] Yu G, Li P, Zhao QL, et al. (2010). Characterization of ultra-precision machined surfaces with power spectral density. *Journal of Harbin Institute of Technology*, *42*(1), 29-32.

[8] Zhang XP, Li CH, Jia DZ, et al. (2019). Spraying parameter optimization and microtopography evaluation in nanofluid minimum quantity lubrication grinding. *Int. J. Adv. Manuf. Technol.*, *103*(5-8), 2523-2539. 10.1007/s00170-019-03642-0.

Index

A

abrasion, 56, 63, 85
abrasive, xviii, 24, 36, 78, 79, 80, 82, 83, 86, 88, 96, 100, 103, 104, 107, 108, 112, 113, 114, 115, 118, 122, 125, 130, 132, 133, 137, 147, 148, 149, 153, 155, 160, 198, 255, 265, 269, 271, 272, 274, 275, 276, 279, 282, 285
adhesion, 2, 21, 22, 43, 56, 58, 63, 64, 65, 71, 80, 85, 96, 104, 106, 108, 112, 114, 116, 128, 131, 227, 228, 235, 237, 243, 244, 251, 293, 304, 305, 329, 334
adsorption, xi, xii, xiii, 7, 12, 14, 18, 20, 26, 41, 43, 54, 95, 110, 119, 120, 131, 137, 138, 139, 169, 172, 175, 189, 194, 201, 202, 203, 207, 215, 221, 222, 223, 261, 263, 290, 305, 310, 315, 323, 325, 329, 333, 334
aeronautical, 46, 49
aerospace, xi, xii, xiii, 3, 5, 44, 49, 50, 56, 59, 84, 85, 88, 91, 92, 108, 119, 130, 137, 252, 293
aerospace materials, xii, xiii, 49, 92, 130, 137
airflow field, 139, 201, 202, 209, 212, 213, 215, 217, 221, 223
antifriction, xii, 1, 25, 27, 49, 50, 51, 54, 55, 60, 65, 66, 74, 75, 79, 80, 81, 82, 83, 84, 86, 91, 92, 95, 101, 108, 112, 119, 130, 176, 227, 243, 244, 245, 246, 248, 250, 251, 293, 298, 301, 305, 309, 315, 319, 328, 329
antifriction mechanism, 1, 79, 227, 293, 305, 315
antiwear, 17, 18, 25, 30, 32, 43, 45, 49, 50, 51, 54, 55, 60, 65, 66, 74, 75, 80, 81, 82, 83, 84, 86, 92, 95, 97, 110, 112, 119, 130, 132, 133, 227, 243, 244, 245, 246, 248, 250, 251, 319, 328
autooxidation, 29

B

bactericide, 7, 9, 15, 16, 45
base oil, 4, 8, 11, 12, 17, 18, 20, 27, 35, 43, 89, 91, 99, 110, 119, 132, 134, 135, 164, 166, 168, 173, 234
biological, 1, 9, 12, 16, 42, 44, 45, 86, 88, 92, 103, 133
bio-lubricant, xii, 1, 3

C

carbon, xii, xix, 2, 25, 26, 27, 29, 32, 43, 45, 46, 49, 50, 51, 52, 53, 54, 61, 65, 67, 74, 76, 81, 82, 83, 86, 87, 88, 94, 95, 97, 98, 101, 102, 131, 132, 133, 138, 164, 166, 172, 173, 179, 201, 294, 295, 316
carbon group nano-enhancers (CGNs), xii, xix, 49, 50, 51, 52, 59, 68, 75
carbon nanoparticles, 49, 166
chip curl, 57, 58, 61, 65, 74, 85
chip morphology, 55, 57, 61, 65, 69, 74, 79, 82
coefficient of friction (CoF), xiii, xix, 50, 53, 54, 56, 57, 58, 61, 63, 65, 78, 79, 82, 84, 85, 86, 100, 101, 102, 109, 110, 120, 131, 134, 227, 228, 232, 234, 235, 236, 237, 238, 239, 242, 243, 245, 247, 248, 251, 256, 261, 276, 280, 288, 290

collision and adsorption, xii, 201, 294
cooling mechanism, 1
cryogenic minimum quantity lubrication (CNMQL), xix, 59, 133, 317
cutting fluids, xii, 1, 3, 4, 5, 7, 8, 11, 12, 15, 18, 20, 21, 43, 44, 45, 46, 88, 91, 92, 100, 131, 133, 135, 225, 291, 335
cutting mechanism, 49, 54, 55

D

debris, 91, 96, 100, 103, 104, 105, 106, 108, 113, 115, 118, 119, 125, 126, 128, 129, 131, 230, 233, 273, 278, 280, 281, 293, 304, 305
deep grinding, xi, 137, 147, 149, 162, 256, 296, 308, 316
defects, 8, 27, 58, 106, 107, 108, 118, 129, 131, 227, 243, 250, 252, 304, 305, 315, 329
deformation, 57, 58, 64, 69, 70, 106, 118, 125, 127, 129, 148, 228, 276, 277, 279, 280, 281, 282, 283
degradability, 1
density, xiii, 24, 26, 27, 39, 53, 95, 98, 99, 103, 109, 120, 143, 144, 146, 164, 165, 167, 170, 171, 176, 177, 179, 180, 204, 205, 206, 233, 268, 319, 330, 331, 332, 333, 334, 335, 336
depths, 272, 285, 286
difficult-to-machine material, xi, 3, 16, 33, 49, 56, 58, 85, 91, 92, 132, 139, 227, 256, 265, 290, 293
distribution rule, 175
dynamics behavior, 255

E

eco-friendly, xii, 1, 11, 12, 14, 15, 17, 18, 20, 43, 47, 87, 89, 134, 135
equipment designs, 137
evolution, 175

F

failure, 1, 26, 33, 44, 49, 51, 55, 56, 70, 85, 95
fatty acid(s), 18, 19, 20, 21, 25, 26, 27, 29, 31, 32, 43, 46, 91, 94, 95, 98, 101, 110, 119, 131, 132, 167
Fe_3O_4/graphene nanolubricant, xiii, 227, 293, 298, 300, 301, 302, 305, 308, 309, 311, 312, 315, 316
finite element simulation, 175, 316
formation, xii, xiii, 25, 29, 39, 40, 46, 49, 50, 51, 56, 57, 58, 59, 66, 74, 75, 78, 81, 82, 84, 100, 102, 118, 119, 125, 127, 134, 175, 201, 210, 211, 215, 216, 218, 220, 224, 228, 236, 244, 245, 301, 316

G

geometric, 147, 176, 178, 196, 197, 206, 207, 256, 265, 267, 269, 271, 274, 275, 276, 278, 283, 291, 302, 322, 334
grindability, xii, xiii, 44, 45, 89, 91, 92, 132, 133, 293, 297, 315, 317
grinding performance, xiii, 86, 91, 92, 100, 101, 102, 110, 121, 122, 130, 131, 136, 294, 298, 310, 319, 320, 322, 323, 330, 334

H

high-temperature, 30, 31, 32, 34, 135, 266, 269

I

identification, 282
infiltration, xi, xii, xiii, 66, 74, 75, 81, 82, 84, 97, 137, 138, 139, 142, 167, 172, 175, 202, 227, 239, 251, 255, 256, 257, 262, 263, 264, 265, 288, 290, 291, 293, 297, 298, 301, 302, 305, 307, 308, 310, 311, 312, 313, 314, 315, 316, 325, 329
interference, 100, 118, 255, 267, 274, 275, 279, 282

J

jet parameters, 201

K

kinematics, 88, 201, 225, 291

L

large contact length grinding (LCLG), xiii, xix, 255, 293, 294, 296, 297, 307, 308, 309, 310, 311, 312, 315, 316
layered, 51, 53, 54, 60, 61, 63, 64, 66, 67, 69, 71, 74, 75, 77, 78, 79, 80, 81, 82, 84, 85, 86, 102, 112, 122, 165, 300, 301, 305, 308, 310
lubrication mechanism, xii, 1, 18, 49, 50, 61, 94, 225

M

machine materials, 50, 91, 92
magnet height (H), 25, 29, 30, 31, 44, 45, 46, 47, 86, 87, 88, 89, 94, 98, 99, 133, 134, 135, 136, 142, 146, 171, 177, 179, 188, 190, 191, 192, 196, 197, 199, 205, 225, 233, 252, 262, 295, 320, 321, 322, 323, 324, 325, 326, 327, 328, 329, 330, 332, 333, 334, 335
magnetic nanolubricant(s), xii, 137, 138, 139, 142, 143, 169, 170, 171, 172, 175, 185, 189, 201, 202, 203, 227, 234, 248, 255, 293, 294, 295, 296, 300, 315, 316, 319, 329
magnetic parameters, 319
magnetic traction, xi, xii, xiii, xix, 137, 138, 139, 140, 141, 142, 143, 175, 255, 293, 294
magnetic traction device, xii, 137, 140, 141
magnetic traction nanolubricant (MNT), xi, xii, xiii, xix, 137, 138, 139, 175, 255, 294
magnetic traction nano-lubrication, 293
magnetic-assisted device, 293, 295, 319
magnetorheological finishing (MRF), xix, 137, 138, 175
mechanical model, 255, 256, 279, 290
mechanism, xi, xii, xiii, 1, 9, 12, 16, 20, 30, 33, 35, 44, 46, 50, 51, 54, 55, 56, 57, 58, 59, 64, 65, 66, 67, 68, 74, 75, 76, 81, 82, 83, 84, 86, 92, 95, 96, 97, 98, 99, 104, 115, 122, 127, 130, 132, 133, 134, 147, 149, 152, 153, 155, 156, 157, 160, 172, 173, 175, 201, 205, 208, 214, 216, 223, 228, 243, 244, 256, 257, 290, 291, 293, 298, 301, 312, 315, 317
microdroplet(s), xiii, 96, 97, 114, 168, 201, 202, 206, 207, 208, 213, 214, 215, 216, 220, 221, 223, 291, 294
milling, xii, 20, 37, 44, 45, 47, 49, 50, 51, 68, 69, 70, 71, 73, 74, 75, 76, 78, 81, 82, 83, 84, 85, 88, 89, 133, 252, 335
minimum quantity lubrication (MQL), xii, xix, 1, 3, 18, 22, 24, 33, 44, 45, 46, 47, 49, 50, 52, 54, 55, 59, 61, 62, 63, 64, 68, 69, 70, 71, 75, 77, 78, 79, 80, 84, 85, 86, 87, 88, 89, 90, 92, 93, 100, 101, 103, 106, 109, 110, 111, 112, 114, 115, 116, 120, 122, 124, 125, 127, 128, 129, 130, 133, 134, 135, 136, 137, 173, 202, 225, 252, 256, 285, 286, 287, 288, 295, 296, 298, 314, 317, 335, 336
molecular, xii, 9, 10, 18, 19, 20, 25, 26, 30, 33, 34, 36, 42, 46, 49, 50, 51, 52, 54, 63, 95, 100, 101, 114, 238

N

nanolubricant minimum quantity lubrication (NMQL), xix, 3, 50, 51, 54, 55, 62, 64, 65, 67, 69, 70, 71, 73, 74, 75, 76, 77, 78, 79, 81, 83, 89, 103, 104, 107, 110, 113, 115, 116, 118, 120, 128, 130, 140, 166, 172, 202, 223, 227, 228, 231, 252, 298, 300, 305, 308, 310, 312, 314, 315, 317, 321
nanomaterial(s), 103, 110, 116, 164, 172, 227, 301

nickel-based alloy, xii, 45, 91, 92, 107, 108, 109, 112, 114, 118, 119, 129, 130, 131, 132, 173

O

optimization, xiii, 88, 101, 104, 134, 135, 201, 227, 234, 245, 251, 319, 320, 322, 330, 336
orthogonal, xiii, 47, 319, 320, 321, 322, 335
orthogonal experiment, 319, 320, 321, 335
oxidation, 26, 27, 28, 29, 30, 31, 32, 37, 43, 46, 54, 95, 108

P

PH value, 94, 97, 131
physical characteristic(s), 137
physical model, 175, 178, 206, 257, 258
physicochemical property(ies), xii, 27, 50, 52, 86, 89, 91, 92, 95, 114, 119, 130
plastic, 58, 64, 69, 71, 105, 106, 108, 118, 125, 127, 128, 131, 253, 276, 277, 279, 280, 281, 282, 283, 291, 293, 302, 304, 312, 316
pour point, 98, 131, 134
preparation, xii, 1, 3, 33, 37, 38, 39, 42, 45, 92, 97, 120, 137, 139, 164, 166, 167, 202, 244, 245
pressure, xii, 5, 6, 7, 8, 10, 16, 17, 20, 21, 32, 33, 38, 43, 44, 49, 50, 56, 66, 70, 74, 91, 103, 107, 119, 122, 125, 133, 138, 139, 159, 160, 164, 173, 202, 203, 204, 213, 214, 215, 216, 220, 221, 229, 234, 256, 257, 258, 259, 260, 261, 291, 298, 335

R

removal, 2, 49, 51, 54, 55, 56, 57, 58, 61, 65, 66, 68, 76, 79, 82, 84, 96, 100, 104, 105, 108, 114, 115, 118, 125, 128, 138, 228, 234, 235, 236, 237, 240, 253, 256, 266, 267, 276, 277, 278, 279, 280, 282, 283, 291, 294, 302, 304, 305, 312, 316, 335
roughness, xi, xiii, 20, 27, 35, 36, 44, 46, 47, 58, 64, 71, 80, 85, 86, 88, 89, 90, 93, 105, 107, 116, 128, 207, 209, 227, 228, 234, 242, 249, 250, 252, 261, 294, 302, 303, 312, 314, 316, 319, 321, 322, 323, 325, 326, 328, 329, 330, 331, 332, 333, 334, 335
rust inhibitors, 4, 5, 7, 12, 14, 20, 43

S

saturation magnetization, xii, 137, 146, 171
sawtooth, 57, 61, 65, 70, 74, 85
solution(s), 4, 11, 14, 15, 40, 49, 97, 107, 118, 119, 122, 129, 161, 164, 165, 262, 290, 293
spectrum, 222, 330, 335
speed(s), xiii, 4, 5, 7, 8, 22, 49, 50, 59, 66, 75, 84, 93, 96, 137, 138, 139, 147, 148, 161, 162, 167, 198, 202, 203, 204, 205, 206, 209, 210, 211, 212, 213, 214, 215, 216, 217, 218, 219, 220, 221, 222, 223, 224, 234, 252, 255, 256, 257, 258, 259, 260, 261,262, 264, 265, 273, 285, 286, 287, 288, 290, 294, 296, 297, 298, 315
spheroidal, 52, 165
steel, xii, 15, 35, 44, 47, 53, 87, 91, 92, 107, 119, 121, 124, 129, 130, 131, 132, 135, 136, 138, 172, 179, 198, 252, 294, 295
stress distribution, 276, 279, 290
structure(s), xii, 9, 18, 19, 20, 25, 27, 33, 34, 36, 43, 49, 50, 51, 52, 53, 54, 71, 84, 92, 93, 95, 100, 101, 112, 122, 130, 138, 140, 141, 153, 155, 160, 163, 164, 169, 229, 243, 245, 271, 294, 295, 311
suction, xii, 137, 139, 142, 215, 216
surface integrity, xi, xii, 35, 49, 58, 65, 74, 82, 84, 85, 88, 99, 106, 108, 129, 227, 228, 252, 255, 293, 294, 319
surface tension, 17, 20, 24, 30, 46, 94, 97, 104, 114, 131, 203, 204, 214, 215, 223, 256

T

temperature, xi, xii, 2, 3, 6, 7, 9, 16, 21, 22, 23, 24, 25, 31, 32, 34, 36, 42, 44, 46, 49, 50, 53, 54, 57, 58, 65, 87, 91, 92, 93, 95, 96, 98, 100, 103, 104, 106, 108, 109, 114, 116, 119, 122, 124, 126, 127, 130, 131, 134, 135, 161, 163, 165, 172, 203, 206, 234, 252, 293, 294, 296, 297, 305, 307, 310, 311, 312, 315, 316
thermal, xi, 2, 16, 22, 24, 26, 27, 30, 31, 34, 35, 36, 39, 45, 49, 50, 52, 53, 54, 56, 58, 86, 92, 95, 96, 97, 98, 99, 104, 105, 106, 108, 109, 112, 114, 118, 119, 120, 122, 127, 130, 131, 132, 164, 165, 233, 252, 265, 293, 294, 304, 305, 310, 316
thickness, 24, 52, 53, 57, 61, 65, 69, 70, 74, 85, 97, 106, 124, 163, 164, 178, 184, 188, 194, 197, 202, 204, 215, 230, 257, 273, 274, 295, 316
three-dimensional, 26, 87, 196, 312, 330, 333
titanium alloy, xi, xii, xiii, 44, 49, 58, 60, 63, 64, 69, 71, 80, 81, 85, 86, 88, 89, 91, 92, 99, 100, 102, 103, 106, 107, 108, 109, 118, 119, 120, 129, 130, 131, 132, 134, 252, 255, 293, 294, 317, 319
tool life, 2, 25, 34, 35, 46, 53, 54, 62, 63, 65, 70, 74, 79, 85, 95
tool wear (VB), xii, 6, 46, 47, 49, 50, 54, 56, 58, 63, 64, 65, 66, 67, 69, 70, 74, 84, 85, 86, 252
tribological test, xiii, 50, 227, 228
tubular, 51, 53, 61, 63, 64, 65, 66, 67, 70, 71, 74, 75, 77, 78, 79, 81, 82, 84, 85, 86, 112
turning, xii, 34, 35, 46, 47, 49, 50, 51, 59, 60, 61, 62, 63, 64, 65, 66, 67, 68, 71, 74, 76, 78, 81, 84, 85, 87, 88, 335

V

variance, 301, 327
verification, 89, 100, 134, 196, 256, 283, 291, 319, 330, 332, 334, 335
viscosity, xii, 20, 21, 22, 23, 24, 27, 30, 34, 36, 43, 44, 46, 91, 94, 96, 101, 102, 104, 110, 112, 114, 116, 119, 120, 122, 130, 131, 132, 163, 166, 167, 168, 169, 170, 171, 172, 203, 204, 209, 214, 215, 223, 247, 253, 257, 260, 262, 332

W

wear, xiii, 2, 6, 7, 16, 25, 28, 30, 34, 36, 43, 53, 56, 57, 62, 63, 64, 66, 67, 69, 70, 79, 80, 82, 87, 88, 91, 102, 104, 107, 112, 113, 116, 118, 122, 130, 131, 139, 161, 172, 224, 225, 227, 228, 229, 231, 232, 233, 234, 235, 236, 239, 240, 241, 242, 243, 248, 249, 252, 291, 304, 309
wetting film, 65, 66, 71, 74, 82
wheel wear, 44, 82, 91, 92, 102, 103, 112, 113, 118, 122, 123, 131, 137, 309